Yale Agrarian Studies

JAMES C. SCOTT, SERIES EDITOR

The Agrarian Studies Series at Yale University Press seeks to publish outstanding and original interdisciplinary work on agriculture and rural society—for any period, in any location. Works of daring that question existing paradigms and fill abstract categories with the lived-experience of rural people are especially encouraged.

JAMES C. SCOTT, *Series Editor*

For a complete list of titles in the Yale Agrarian Studies Series, visit www.yalebooks .com.

AMERICAN GEORGICS

Writings on Farming, Culture, and the Land

Edited by Edwin C. Hagenstein, Sara M. Gregg,
and Brian Donahue

Foreword by Wes Jackson

Yale UNIVERSITY PRESS   NEW HAVEN AND LONDON

Published with assistance from the Louis Stern Memorial Fund and the
Iowa State University Publication Endowment Fund.

Yale University Press books may be purchased in quantity for educational,
business, or promotional use. For information, please e-mail sales.press@yale.edu
(U.S. office) or sales@yaleup.co.uk (U.K. office).

Set in FontShop Scala type by Duke & Company, Devon, Pennsylvania.
Printed in the United States of America by Sheridan Books, Ann Arbor, Michigan.

Library of Congress Cataloging-in-Publication Data
American georgics : writings on farming, culture, and the land / edited by Edwin C.
Hagenstein, Sara M. Gregg, and Brian Donahue.
p. cm. — (Yale agrarian studies series)
Includes bibliographical references and index.
ISBN 978-0-300-13709-5 (hardcover : alk. paper) 1. Agriculture—United States—History.
I. Hagenstein, Edwin C. II. Gregg, Sara M. III. Donahue, Brian, 1955– IV. Title.
V. Series: Yale agrarian studies.

S441.A482 2001
630.973—dc22   2010045042

A catalogue record for this book is available from the British Library.

This paper meets the requirements of ANSI/NISO Z39.48–1992 (Permanence of Paper).

10 9 8 7 6 5 4 3 2 1

Brief Contents

Contents

This section features writing from the years in which the United States was founded and many of the nation's fundamental institutions and policies were put in place. Several of the writers we include took active roles in formulating, promoting, or attacking those policies. Many understood that their choices would determine the future course of American society. Alexander Hamilton's Report on Manufactures calls upon the central government to foster a strong manufacturing sector. The other writings express the agrarian vision that farming as a way of life was essential for building the kind of self-reliant character that is the mainstay of a democratic republic.

2. A Nation of Farmers: The Promise and Peril of American Agriculture, 1825–1860   57

*Here we draw on writings from the early to mid-nineteenth century, when
farming was growing rapidly, finding a new place in an increasingly specialized
and integrated national economy. The United States was still largely agricultural,
and westward expansion seemed to offer an unlimited future to the nation's farmers,
but in most regions farming and the agrarian vision that went with it had become
committed to market production. The agricultural press also grew, featuring articles
on improved farming techniques, better implements, and the importance of
maintaining soil fertility. Agricultural writers expressed their belief that life on farms
could be not only healthy and morally sound but as financially rewarding as other
career pursuits. At the same time, early conservationists such as George Perkins Marsh
began to warn that farmers were capable of destroying the very soil upon
which civilization depended. These years also saw increasing tension between
two visions of agrarian democracy, slave and free.*

3. The Machine in the Garden: The Rise of American Romanticism   105

*This section features writing from the first major "back to the land" move-
ment in the nation's history. Well-educated men and women disaffected with the
growing materialism of American culture tried to forge a more humane, integrated
way of life through closer contact with the land. Although these efforts often marked
a clear break from mainstream society, they also incorporated familiar agrarian ideals,
such as the belief in economic self-reliance as the basis for political independence.*

The movement was marked by a reverence for nature that would become central to twentieth-century versions of America's agrarian tradition. Yet in her childhood memoir, Louisa May Alcott injected some humorous skepticism of utopian dreams that would become equally familiar.

4. Agriculture in an Industrializing Nation, 1860–1910 149

The writing in this section reflects dramatic changes in American agriculture during the second half of the nineteenth century. As the economy became still more specialized and integrated, American farmers in many regions found themselves in deep financial trouble. The opening of vast territories of fertile soil drove production up faster than demand, leaving many farmers in debt and at the mercy of banks, railroads, and wholesalers. Economic pressure set the stage for the Populist revolt, which sought to restore agrarian moral principles to the governing of an increasingly urban nation.

5. Agrarians in an Industrial Nation, 1900–1945 199

The selections here focus on writing from the first half of the twentieth century. With the nation's available farmland settled and its urban population and industrial economy rapidly expanding, many worried about the ability of farmers to feed the

*nation. Some called for the most efficient, scientific, large-scale agriculture possible;
others, such as Liberty Hyde Bailey, envisioned an updated agrarian culture, founded
on values of community, conservation, and appreciation of nature, practicing
a more ecologically based "permanent agriculture." Although rural America was
perceived by many leaders to be falling behind modern standards of living and
education, these decades also saw a wide variety of agrarian movements
(including Jewish and Catholic versions), as some disaffected Americans
continued to look to the land for a more wholesome way of life.*

6. Southern Agrarianism, 1925–1940   251

*The writings here focus on the work of the so-called Nashville Agrarians, also
known as the Southern Agrarians, who published the collection of essays* I'll Take My
Stand *in 1930. These writers mounted a broad defense of the South's agrarian and
(as they saw it) humane traditions, in contrast to the culture of the increasingly
industrialized North. Their perspective was self-consciously shaped by the long
Southern agrarian tradition that went back to Thomas Jefferson and John Taylor.
Predictably, the Agrarians aroused intense skepticism from the likes of H. L. Mencken.
Deeply integrated into the Southern agricultural experience was the place of race
in the region, and so we also include some reflections from the black sharecropper
Ned Cobb on agricultural production in the Deep South.*

7. *Back to the Land Again, 1940–Present*   299

This section draws primarily on writings from postwar America. Long-term trends of land concentration have accelerated, leaving the nation with a heavily industrialized farming sector, a shrinking number of farmers, and a population increasingly unfamiliar with agriculture—developments often celebrated by mainstream agricultural leaders such as Earl Butz. But many in the modern era feel discomfort with urban and suburban life, a sense of alienation. The period has seen a revitalized back-to-the-land movement, strong growth in organic and sustainable farming, and a wealth of writers reflecting on what the changing nature of farming could mean for the nation.

Foreword

We moderns, with or without an agrarian disposition, need the essays in this book as touchstones for our thoughts and to inform our actions. And to have agrarian scholarship informed by direct experience on the landscape, as Edwin Hagenstein, Sara Gregg, and Brian Donahue have had—well, that is a big plus!

The change on the land covered by this collection, 1780 to the present, goes far beyond that of any other period since the beginning of agriculture. These pages represent the codification of events and ideas during the great transition from a world running on contemporary sunlight to one running on fossil carbon. The writers of these selections may sometimes be inspired by mere nostalgia, but more often they are by worry and a sense of loss. And crucially, the editors have noted the period's material, practical, social, political, and ecological consequences.

There have been other great transitions. The "big one," of course, was domestication of plants and animals, which provided our inventory of crops and livestock. But that occurred over millennia, not over just a quarter of one. Others involved changes in technology, such as the invention of the moldboard plow. But these pale in comparison to the explosive quantity of fossil energy recently applied to the landscape.

The industrial revolution also brought a dramatic increase in literacy. And it is not merely coincidental that in England, while the use of coal and iron were creating a new age, the natural history essay, beginning with Gilbert White, also began to flourish. This development led to the modern discipline of ecology. The two tracks of agriculture and ecology ran along more or less separately until the last century, which saw a serious attempt at a merger. But that is yet in early stages.

The "look to nature" as a source of practicality for humans has been around for millennia. Wendell Berry has spoken and written of this, from Job to Virgil and the *Georgics,* 36–29 BC, through Edmund Spenser, Shakespeare, Milton, and finally

Alexander Pope, who in his *Epistle to Burlington* counseled gardeners to "let Nature never be forgot" and to "Consult the Genius of the Place in all."

After the eighteenth century, Wendell notes, English poets lost this theme of a practical harmony between humans and nature. The Romantics made the human mind so central that, although they involved nature in their poetry, it was only as a "reservoir of symbols."

When the idea of nature as the standard or measure was codified again more than a century later, it was among agricultural writers with a scientific bent. Liberty Hyde Bailey, with *The Outlook to Nature* in 1905 and *The Holy Earth* in 1915, boldly advanced the notion that "a good part of agriculture is to learn how to adapt one's work to nature." A quarter-century later, in 1940, Sir Albert Howard wrote in *An Agricultural Testament* that we should farm as the forest does, for nature constitutes the "supreme farmer."

Wendell concluded that in the agrarian common culture, the look to nature proceeds by *succession*. Although there is little in the way of codification, it is easy to imagine that the idea of superior nature was ever present, especially among those who have livestock. The sow decides whether she will allow her pigs to suck, not the farmer. The show ring champion draft horse may be a product of our breeding, but it may not be inclined to "pull its weight" in the field. Wallace Stegner's "stickers" sought some sort of harmony with their places while the "boomers" were always looking for something better elsewhere. In the *formal* culture—meaning teachers and students—such thinking comes only now and then, and therefore, as a *series*. Finally there is the tradition of caretaking represented by Shakespeare's and Thomas Hardy's peasants. A practical harmony, at least among some farmers of every generation, seems certain to be actively sought.

Five years before Howard's book there began one extensive attempt to establish succession in the formal culture. In 1935, during the Great Depression and the dust bowl years, President Franklin D. Roosevelt selected Dr. Hugh H. Bennett to found what was later called the Soil Conservation Service (SCS). Bennett was affectionately called "Big Hugh," and *Farm Journal* honored him as "one of the few immortals of agricultural history." Bennett believed society benefited by saving the "back forty." The back forty was part of the entire system of forests and plains, rivers and power, lakes and cities. He traveled widely to make clear to all the inter-reliance of countryside and town.

He also gathered into service to save the soil the most able engineers, agronomists, nurserymen, biologists, foresters, soil surveyors, economists, accountants, clerks, stenographers, and technicians of many backgrounds. They felt an anxious urgency. As Wellington Brink put it, the SCS was "born with pride and loyalty and a sense of high destiny—an inner element that was to persist and spread and animate the organization and weld it together with a spirit altogether unique in modern government" ("Big Hugh's New Science," *The Land* 10, no. 3 [1951]). And because of the high caliber of the people employed, its good reputation grew fast.

The SCS program included four steps: science, farmer participation, publicity, and congressional relations. In his North Carolina accent, Bennett explained them many times.

1. "By science," he would say, "we tried to imitate nature as much as we could." This was one of his central ideas.
2. *The farmer-participation step was extremely successful.* SCS men would walk over a farm, field by field, with the owner. Tens of thousands came to the soil conservation demonstration meetings.
3. *Bennett tried to fully exploit publicity,* with journalists who knew good pictures, news angles, and feature possibilities when they saw them, and knew how to get them published. Papers picked up the material, it got back to Congressmen, and requests for more information poured in. Bennett was drawing on succession in the common culture.
4. *On Capital Hill Bennett succeeded,* even became a legend, because of his good sense, colorful personality, and charm. "When I appeared before a committee," he explained, "I never talked about correlations or replicas. But I did spread out a thick bath towel one day on a table before a committee, tipped the table back a bit, and poured a half pitcher of water on that towel. The towel absorbed most of the water, cutting its flow from the table to the rug.

   "I then lifted the towel and poured the rest of the pitcher on the smooth table top, watching it wash over the edge onto the rug. I didn't say anything right away, just stood there looking at the mess on the floor. Then I looked up at the committee and explained the towel represented well-covered, well-managed land that could absorb heavy, washing rains. And that the smooth table top represented bare, eroded land, with poor cover and management on it. They seemed to understand, because we got our appropriation" (Sanford Martin, "Some Impressions of Hugh H. Bennett, Father of Soil Conservation," *Better Crops with Plant Food,* summer 1959).

Bennett not only led the SCS to great effect but also helped establish a citizens' organization with like goals. It was called Friends of the Land. Friends of the Land featured in its quarterly journal well-known names in conservation: Liberty Hyde Bailey, cartoonist J. N. "Ding" Darling, ecologist Paul B. Sears, novelist Louis Bromfield, Aldo Leopold, E. B. White, Henry Wallace, and Gifford Pinchot (first head of the U.S. Forest Service). All wanted the organization to be more than a pressure group. They wanted members to develop a deep allegiance to soil conservation, always writing to connect people and land. The magazine cover boldly said, "A Society for the Conservation of Soil, Rain and Man." Leopold wrote that "health is the capacity of the land for self-renewal. Conservation is our effort to understand and preserve this capacity." Such agrarian values are expressed all through that fine old journal.

So we have a history of both the government and a citizens group working to support soil and water conservation, with nature as a model. Agrarian values were probably never higher, never more visible.

But the Friends of the Land was phased out in the late 1950s, and even earlier its publication clearly was sinking. How could an organization that held hundreds of meetings at summer camps and teachers institutes, at metropolitan centers and in the hinterland, fade away? No one called it cultish or fanatical. It was typically American, like the Farm Bureau or the Chamber of Commerce. But it died, and few know of it today.

The SCS certainly helped the country, most significantly in setting aside marginal land from tillage. And it survives today as the Natural Resource Conservation Service (NRCS). But neither group succeeded in preventing soil loss beyond natural replacement levels over most till agricultural land. Land still erodes, seriously so, and now both our land and waters are increasingly contaminated with agrichemicals. What might have helped re-establish in the formal culture a succession in looking to nature instead failed.

To partially explain this failure I will tell of a visit to a northwest Missouri farm one mid-May evening in 2010. Some fifty adults had gathered on the lawn of a farmstead. The evening air was cool, still, and insect-free, the mood convivial. We chatted in small groups and enjoyed drinks and nibbles on a flat part of a high lawn, which dropped sharply toward a field of about eighty acres. The top part of the field was fairly steep and had three parallel terraces standing out lush, green, and uncropped. The terraces were probably built according to NRCS specifications, the owner's cost likely shared by taxpayers. Below, most of the field sloped gently and was strewn with last year's corn stalks. Except for the green terraces, the rest appeared devoid of life. The field had been sprayed with an herbicide in preparation for planting. Below was a catchment pond built to slow the water before it entered a large gully through the woods. The pond was filled with mud. Even the relatively flat land had countless rills and larger gullies. Because of recent rain, tons of soil had been sent toward the Missouri, not ten miles away as the crow flies, and from there to the Mississippi and to the Gulf of Mexico. In the water and soil was nitrogen that would contribute to a New Jersey–sized dead zone in the gulf.

That eroding field was not the farmer's failure. It was a cultural failure. It represented the current gold standard, the best offering, of the U.S. Department of Agriculture, the land grant university system, and agribusiness to prevent soil erosion on annual grain fields.

That field is partly the product of the ten-thousand-year-old idea that came with the invention of agriculture, the idea that nature must be subdued or ignored. Our ancestors had to clear land of its original vegetation to plant those annual grains. Eventually, they tore out vast stretches of wild perennial polycultures to make way for annual monocultures. As at the convivial suppertime gathering last May in northwest Missouri, that old idea to subdue is part of who we are.

Also shaping our minds is the scientific revolution, born of the likes Copernicus, Galileo, Bacon, Descartes, and Newton. Central was the idea that the advancement of knowledge would come from being reductive. This led to a history of scientific inquiry devoted to placing priority on parts over the whole, rather than to acknowledging the interpenetration of parts and whole. Most of us still assume that breaking a problem down provides insight essential for a solution. There is nothing wrong with breaking a problem down, of course, so long as we remember that the world is not like the method. But too many of us forget.

Having been limited by our thinking that nature is to be subdued or ignored, and by the reductive methodology featured in science, we can plainly see that ideas do matter as we view that eighty-acre field. The field is eroded. Alien chemicals in the form of pesticides are in the water and soil. So is an overabundance of nitrogen headed for the Gulf of Mexico or wells in the surrounding area. All of this is the result of research featuring the molecular end of the spectrum. To help clarify, the hierarchy of structure runs from atoms to molecules to cells to tissues to organs to organisms to ecosystems and finally to the ecosphere. In the history of science we have increasingly looked to the smaller elements of that hierarchy. And so, given our desire to reduce soil erosion and control weeds, we have invented, unfortunately, no-till or minimum-till farming, which depends on an herbicide, a poisonous molecule. And this requires resistance in the desired crop induced by splicing in a foreign DNA molecule. In that brown field on that beautiful Missouri evening we saw the two-molecule approach to soil erosion and weed management.

Our problem comes not only from thinking reductively and subjugating or ignoring nature. All organisms need food, energy-rich carbon. Like bacteria on a petri dish with sugar, we all just go for it, humans measuring progress by yield per acre. With modern machinery and chemicals powered by and made from fossil fuels, we have increased yields beyond the imagination of a century ago.

Agrarians, especially those with animals, since 1750 have not had molecules on their mind. They have been forced to think several steps upward in the hierarchy of structure, to the organism level, measuring by the animals that provided them meat, milk, and traction. Agrarian thought would integrate organisms within the ecosystems that supported them. They thought in terms of what one might call symbiotic units. Agrarian thought on a broad scale is sure to have its time again. No matter that fewer than 2 percent of people in our country are on farms now. That will change as we approach the end of the fossil fuel interlude. For that we need to prepare, and this volume will help point the way. Will the romantics in the formal culture return to use nature only as "a reservoir of symbols"? I doubt it. They will see practical value in the lessons of nature. Agrarianism seems certain to become a succession in the formal culture.

We all know that not all problems in agriculture will go away just because we have an agrarian view. Farmers have a long history of revolt over land ownership, overproduction, and low prices. But we are beginning faintly to see that, for those who work the fields, justice has ecological implications. To be reminded of the consequences of agricultural injustice, witness the Roman Empire's decline.

Will our future depend on more science? Yes, but the science of the future will focus on the ecosystem concept and the creatures in it. The molecular end of the spectrum will need to be in the service of the higher levels. This shift will bring the science of ecology and evolutionary biology to the farm as never before.

Where is the core American constituency for embracing the new agrarianism? It is found mostly in clusters on the coasts, with only a sprinkling in the middle states; and focused still too narrowly, on vegetables and a few chickens grown locally and organically. Those are the allies, and they are beginning to draw their attention toward the Midwestern grain fields, which grow most of our calories on the bulk of our cropland, and where erosion and chemical contamination of land and water remain very serious. These coastal allies remain the primary hope for those resources that sustain our food supply. A new kind of agrarian is in the making, and this volume will be an indispensable help.

Wes Jackson

Acknowledgments

We are grateful for suggested readings and other practical advice from Professor Clinton Machann at Texas A&M, Lynne Heasley at Western Michigan University, and David Hickman at the University of California–Davis. As the manuscript came together and was submitted for review, Steven Stoll and Eric Freyfogle offered thoughtful critiques that served us well. We appreciate the great care both took in their reviews of early versions of the manuscript. Given the nature of the work, we also depended heavily on various libraries to help us gather readings from sometimes obscure sources. Leslie Perrin Wilson, curator of the Concord Public Library's Special Collection in Concord, Massachusetts, was helpful in this regard. Staff members at the Wayland Public Library, including Jan DeMeo, Andy Moore, Kathy Powers, and Sandy Raymond, also stand out for their helpfulness and their mastery of the necessary research aids. Alison Roberts of the Pendergrass Library in Knoxville provided timely help late in the editorial process. Jeff Cramer at the Thoreau Institute at Walden Woods and the staff of the Prints and Photographs Reading Room at the Library of Congress were also helpful in identifying images to illustrate the text.

The expense of producing this book has been borne in part by a Publication Subvention Grant funded by the Office of the Vice President for Research and Economic Development at Iowa State University. We are most grateful for its timely support.

We also thank family, friends, and colleagues for the interest they showed in the book and their encouragement, including Deborah Lee, Ana Nuncio, Hal Ober, Susan Stark, Marta Thornton, Lynne Weiss, and, especially, Helen Byers.

This book has been decades in the making, the product of a sense that the connections between agrarian ideals and American identity deserve more sustained study and discussion. The idea for the book came partly from a conversation with

Wes Jackson at the Land Institute. We would like to thank him and Wendell Berry for their suggestions about what to include in the collection. Both men have inspired us for many years through their writings and the example of their lives.

Finally, we thank our editor at Yale University Press, Jean Thomson Black, who gave her support to this project early on and guided it through development and production with skill and good cheer.

Introduction

"Those who labor in the earth are the chosen people of God, if ever he had a chosen people, whose breasts he has made his peculiar deposit for substantial and genuine virtue." So wrote Thomas Jefferson in 1781 in his *Notes on the State of Virginia*. "Generally," he continued, "the proportion which the aggregate of the other classes of citizens bears in any state to that of its husbandmen, is the proportion of its unsound to its healthy parts, and is a good-enough barometer whereby to measure its degree of corruption."

If Jefferson was right, then by his measure the state of America today is 98 percent rotten. To many Americans, Jefferson's claim that farmers are peculiarly virtuous and thus indispensable to a democratic republic might seem at best quaint, at worst preposterous. For more than two centuries the proportion of the nation's cultivators has been on an inexorable slide, while the proportion of citizens with the right to vote has steadily increased. Moreover, historians like to point out that farmers have been enthusiastic participants in America's growing market economy from the start, and therefore the "agrarian ideal" of the yeoman standing free from the corruption of commerce might be better termed the "agrarian myth." Any claim Jefferson himself might have had to agrarian virtue was confounded by the privileges he enjoyed as an owner of slaves. For many, far from being the foundation of democratic freedom, American agrarianism is deeply tainted with backwardness and racism.

And yet, myth or not, Jefferson's belief in the virtue of cultivating the earth was widely held in his own time (not least among American farmers themselves), and was fundamental to the nation's sense of its destiny. Although farmers certainly took part in the market economy, throughout the nineteenth century many continued to believe that government should protect the interests of the rural majority, allowing the nation to achieve a purpose higher than the naked pursuit of wealth. During the twentieth century, critics of unrestrained industrial growth looked to the land for visions of a more fulfilling way of life. Some of the most vigorous, incisive, and enduring criticisms of industrial society, and some of the most appealing alternative visions, have been essentially agrarian in nature.

American Georgics is a collection of representative agrarian writings of the past two centuries and more, from J. Hector St. John de Crèvecoeur to Wendell Berry. The collection reveals the great reach of the agrarian idea and its durability in American thinking—initially as an expression of mainstream rural culture, and later reborn as a dissenting vision of radical reform. Agrarianism begins with the understanding that the work we do, as individuals and as a society, is a critical factor in shaping our character. Agrarians continue to believe, with Jefferson, that there is something profoundly satisfying and valuable for human beings in working the earth; and that there is something equally important to the health of society in the way that the land is cultivated. These ideas were central to American culture (though increasingly beleaguered) in the nineteenth century. By the twentieth century, they became marginal to modern industrial culture but remained central to recurring critiques of it. No doubt, agrarian ideas have become easy to romanticize, or to manipulate on behalf of a narrow "farm interest," or to satirize and dismiss—yet they strike a deep chord for all of that. The editors of *American Georgics* are confessed agrarians themselves, though critical ones. This collection is intended to help those interested in agrarian thinking to grasp the origins of these ideas, the long journey they have made, and how contested they have always been.

The agrarian ideal preceded the European discovery of America, grew up with the country, moved to the city, and longs to return to its roots. Yet those roots are diverse, and the meaning of agrarianism in America has grown and changed, sometimes in contradictory ways. One root, going very deep in European peasant culture, is that the land should belong, by moral right, to those who labor upon it and improve it. Another, having its origins perhaps in higher culture, is the idea that people are ennobled by contact with the earth. Both of these ideas found vigorous expression in America. Leo Marx opened his 1964 book *The Machine in the Garden* with the claim that "the pastoral idea has been used to define the meaning of America ever since the age of discovery, and it has not yet lost its hold on the imagination." Europeans at the time were "dazzled by the prospect" of creating on this (to their eyes) unspoiled continent the kind of ideal rural society evoked in Virgil's *Georgics*. That prospect has continued to be a force in American life to the present day. Major strands of the environmental movement, from the defense of natural landscapes to the drive for local, sustainable agriculture, have drawn on these agrarian roots. For all Marx's fascination with pastoralism, in the end he declared the philosophy anachronistic and unworkable for an industrial society, believing it had about exhausted its useful possibilities. Yet even as Marx rather fondly kissed agrarianism good-bye, Wendell Berry was writing *A Place on Earth*. Soon another wave of young back-to-the-landers was washing out toward Mendocino and Vermont, learning to farm from the *Mother Earth News*. Predictably, the froth of that neoagrarian tide soon receded, but what remained on the land has proven remarkably durable. The agrarian ideal continues to thrive in the American imagination, refusing to surrender its grip on the soul of agriculture.

Somehow, agrarianism—which holds that a healthy proportion of the citizenry should work the land—has survived in the face of an overpowering demographic and economic current in the opposite direction. At the time of the War for Independence, the American population was overwhelmingly rural and engaged in farming, for both subsistence and market production. Today, by contrast, less than 2 percent of the American people farm, and the nature of farming has changed so radically as to be only a distant cousin to the husbandry of the late eighteenth century. Even though everyone still eats, taking part in the practice of growing food has less of a direct influence on people's lives than at any time in our history. Agrarians have steadfastly regarded society's drift in this direction not as a historical inevitability to be accepted, let alone celebrated, but as a cultural calamity to be prevented or reversed.

The outlines of a long-standing debate about the place of agriculture in the United States took shape during the founding era. Whereas Jefferson argued that the young nation would be best served by maintaining its agricultural character, Alexander Hamilton believed that fostering robust commercial and manufacturing sectors would provide the strongest foundation for its future. The debate wasn't quite so two-dimensional as that. The contestants understood that the nation could not be simply one or the other, either all agricultural or all industrial. Hamilton paid homage to agriculture's "strong claim to pre-eminence over every other kind of industry," which at least conceded the facts on the ground at the time. Jefferson, for his part, acknowledged that the American economy was bound to include strong elements of manufacturing and trade. To a European friend who had asked whether he would like to see a more commercial United States, Jefferson replied that were he to indulge his own "theory," the country would remain agricultural and stay aloof from Europe. "But," he continued, "this is theory only, a theory which the servants of America are not at liberty to follow. Our people have a decided taste for navigation and commerce. They take this from their mother country."

In spite of this apparent common ground, there was fundamental disagreement about what goals the young nation should pursue, and the outlines of that debate anticipated conflicts of the future. At the heart of the argument was the belief that the economic, political, and moral spheres were entwined in profound ways. Freehold agriculture was widely held to promote such virtues as moderation, self-discipline, prudence, and love of country. Agrarians believed that commerce and manufacturing tended to encourage dependence and dissipation. And yet by the early nineteenth century, American farmers were becoming as devoted as any other class to the benefits of economic growth. Could agrarian values even survive, let alone be sustained by, wholehearted engagement with the market economy? The qualities of the nation's citizenry would determine its fate. Given that, the nature of the economy was a matter of national survival as a democratic republic.

For all Jefferson's rhetorical skill and the consistent electoral victories of his party throughout the first half of the nineteenth century, Hamilton and his followers won the long-term battle and laid the institutional and legal foundations of the immensely vital American market economy. And in truth, Jefferson's loyalty to the westward ambitions of his own planter class allowed stridently conflicting agrarian cultures to expand in the North and South—American agrarianism was undermined as much by the spread of slavery as by the rise of industrialism. Yet Jefferson's concerns about the importance of agriculture to the broader society remain, and have given rise to searching questions ever since. To what degree are democracy and freehold agriculture interdependent? What are the social and cultural implications of the massive transformation of our economy away from decentralized farming? If our work shapes us in crucial ways, what is the impact on individuals and communities of the wholesale shift to industry and commerce? Agrarian writers still rail against commercialization and defend the role of agriculture in sustaining a healthy society. Such thinkers have been present throughout the history of our nation and, taken together, have composed an enormously provocative body of work. The purpose of this anthology is to document some of these agrarian arguments and present them in a way that builds historical context for contemporary readers.

Agrarianism stands as a set of political, economic, ecological, and social convictions rising from the period when agriculture was central to American life. To the classic agrarian, cultivation was unlike commerce, industry, civil service, or military life, offering a uniquely valuable realm of experience. These convictions are sometimes stoutly defended by the modern agrarian, but are often recast in ways that are meant to be compatible with and yet transform industrial society.

We take agriculture here in its root sense of tending and caring for the land, not just in its current economic meaning of an industry that produces commodities. The Latin root verb, *colere,* carried overtones of honoring the object of cultivation. Much the same sense abides in the ancient English term *husbandry.* In keeping with this, we define as agrarian those who speak to something beyond the purely economic value of American farming.

In choosing selections for this collection, we looked for writers who share a belief in agriculture's special value and have defended the place of agriculture in American life. When agriculture was central in the lives of most citizens, in what ways was it celebrated? As the position of farming eroded within the broader society, by what terms did its defenders argue for its protection or revival? As back-to-the-land movements blossomed, how did participants articulate what they hoped to find by working the soil? These are the kinds of questions our anthology addresses.

A number of intertwined themes within American history are documented by this collection, through successive generations and altered circumstances. The first is political: a recurring strain of agrarian thought is the belief that a sound footing of economic independence, planted in widespread ownership of land, is necessary to political independence. This was a central element in Jefferson's defense of the yeoman farmer and was still prominent in later thinkers, ranging from Helen and Scott Nearing to Victor Davis Hanson. But the politics of agrarianism also take us into realms of conflict, such as divisions over slavery, the Populist revolt against control of the government by plutocrats, and the future of the American family farm.

A second major theme of *American Georgics* is economic, revolving around the vexed relationship between farmers and the market. Motives of profit and material reward have been major driving forces in the rapid expansion of American agriculture, but at the same time have often undermined other goals of farming as a way of life. Closely related to this is an ecological question: Can American agriculture be improved by progressive scientific means alone, or does it require more radical reform to care for the land sustainably? While the terms *ecological* and *sustainable* may belong mostly to the twentieth century, the concerns they express go back to the early nineteenth century and are found repeatedly in agrarian writing. Finally, and again closely related, there is a social theme concerning the benefits of farm life to the human soul and its value to our communities and culture—a promise that has never been easy to reconcile with the economic reality of farming.

American Georgics is divided into seven sections. These sections follow a general chronological order, but there are some exceptions. There is some inevitable overlap, for instance, when a piece we've included is germane thematically to one period but actually published in a year covered by another section. Two of the book's sections stand somewhat apart from this general chronological flow. In the first case, section 3, we gathered selections that reflect the blossoming Romantic naturalism of the 1840s and 1850s as it began to reshape American agrarianism. In the second case, section 6, we selected writings from the Southern Agrarians of the mid-twentieth century. Both movements seemed to us to merit special attention for their seminal importance. Each has served as a touchstone for later thinking about the direction of American society and neither has lost relevance over time.

Readers will also find occasional pieces that express a contrary or antiagrarian point of view. These appear once per section and are meant to exemplify the context in which the ongoing arguments about agriculture and American society took place. In some cases, the antiagrarian piece was very much a response to agrarian writings, as with H. L. Mencken's comments on the Southern Agrarian manifesto *I'll Take My Stand*. In other cases, such as with selections from Alexander Hamilton, Edwin Nourse, and Earl Butz, the piece simply expresses a sharply different take—

quite often a more narrowly economic one—on the proper place of agriculture in American life.

Other writings included here don't fit neatly into any scheme of public argument. An excerpt from Willa Cather's *My Ántonia* is included because it is so fully infused with the sensibility Cather gained from her youth in rural Nebraska. Cather's piece is the only fiction represented in *American Georgics*, attributable mainly to the limits of space rather than any shortage of excellent novels and short stories to draw on. Plenty of American fiction embodies an agrarian spirit, including O. E. Rolvaag's *Giants in the Earth*, Sarah Orne Jewett's *Country of the Pointed Firs*, Caroline Gordon's *Green Centuries*, Allen Tate's *The Fathers*, Lois Phillips Hudson's *Bones of Plenty*, and Wendell Berry's tales of the Port William membership. We have also included a poem by Hayden Carruth. As is the case with fiction, there is a great deal of poetry with agrarian themes, including the poems that were a staple of agricultural periodicals for generations.

American Georgics aims to document agrarian thought as it evolved in the nation's history. The book does not attempt to tell the complete story of agriculture in the United States. That story is reflected here, but no anthology of this kind could adequately capture its diversity. Our full agricultural heritage includes the pre-Columbian Indian era, the experiences of Mexican American farmers and migrant workers, the efforts of Japanese American farmers in the West, and much more. No doubt many of these farm cultures have infused, and will continue to infuse, the American agrarian tradition in ways we have not recognized here. It remains to be seen whether the twenty-first century will be one in which industrial agriculture remains overwhelmingly dominant, or whether we will witness a revival of farming on a smaller scale, designed to serve a broader set of social and environmental values. In either case, the agrarian idea is so vigorous, and so durable, that we are confident it will continue to spring up, ever hopeful, with each new growing season.

AMERICAN GEORGICS

Life of George Washington—The Farmer. Painted by Junius Brutus Stearns, lithograph by Régnier, imprinted Lemercier, Paris, c. 1853. Courtesy Library of Congress, Washington, DC.

1. Shaping the Agrarian Republic, 1780–1825

The United States began as a nation of farmers. The vast majority of Americans worked the land, and for many of them farming represented a state of economic independence and self-determination that was as essential to their own sense of well-being as to the health of a democratic republic. But would the political institutions of the new nation be framed in ways that upheld these ideals? What would be the place of agriculture in the country's rapidly but fitfully growing economy? These were questions that animated not only the nation's leading thinkers and politicians but a few million American farmers.

The ideals of American agrarianism were first fully expounded by an enigmatic Frenchman turned English citizen and farmer in colonial New York. Toward the end of the War for Independence, when events in America had attracted a new level of interest from overseas observers, J. Hector St. John de Crèvecoeur published his *Letters from an American Farmer*. Adopting the fictitious persona of a native-born American yeoman writing to an English acquaintance, Crèvecoeur addressed European readers who held that cultivation was the most moral life for a man, and the foundation of economic and political strength for a nation. In America, said Crèvecoeur, this was not just a nice idea dreamed up by physiocrats who had never had to dirty their hands in the soil: it was really happening.

"What is an American?" asked Crèvecoeur. By way of answering, he offered a picture of an independent yeoman farmer. This freeholder's life was set in contrast with the downtrodden peasants and dispossessed urban poor of Europe. "Here the rewards of [the farmer's] industry follow with equal steps the progress of his labour. . . . Wives and children, who before in vain demanded of him a morsel of bread, now, fat and frolicsome, gladly help their father to clear those fields whence exuberant crops are to arise to feed and clothe them all." The enthusiastic reception that

9

the *Letters* received, both at home and abroad, suggests that Crèvecoeur's depiction dovetailed with the ideals of Americans and the hopes of many more who saw in the new country the dawn of a better future. Was it impossibly optimistic?

The first few letters offered the sunny, confident portrait of life in rural America for which Crèvecoeur is best known, but the book closed on a darker note. One of the last letters was an outcry against the cruelty of slavery in South Carolina, which Crèvecoeur saw as a harbinger of Old World corruption reaching the New. And Crèvecoeur ended with an anguished description of the upheavals of the Revolution, which had his narrator contemplating flight to the frontier to live among the Indians, but which in real life caused the author to flee in the other direction, back to England and France. Taken as a whole, Crèvecoeur's *Letters* posed the essential dilemma between the great promise of the agrarian dream in America and the conflicts and contradictions that would hinder it from ever being fully realized. In reality, how many Americans could become a freehold yeoman farmer?

Not Native Americans, clearly—their land was systematically expropriated to supply the burgeoning families of European yeomen. To the extent that any legal justification for this was thought necessary, the English claimants employed an essentially agrarian argument: that the Indians did not significantly cultivate and improve the land, and hence had no moral right to it. But even when Native groups did attempt to conform fully to English agrarian ways—as did the members of the Cherokee Nation, in the most dramatic example—ultimately the great bulk of their land passed to the invaders anyway. Consequently, the Native contribution to the way the land was worked—not least, many of the crops themselves—survives mainly as a ghostly, disembodied presence within the American agrarian tradition. Today, Indian communities in many regions provide both inspiring models of modern agrarian ways of life and particularly distressing examples of the difficult conditions afflicting rural America in general.

Not African slaves, obviously. Agrarian independence and felicity for some Americans, notably Southern planters like Thomas Jefferson, James Madison, and John Taylor of Caroline, rested on the labor of others who were brought here by force. Even Crèvecoeur, who denounced plantation slavery so bitterly, apparently owned a few slaves himself. And yet, African slaves and their descendants made enormous contributions to agrarian culture—in the cultivation of rice, for example, and the herding of cattle, and in the hunting, fishing, and foraging traditions that became an important part of Southern rural life. From an early date both slaves and freed blacks in America had agrarian traditions and dreams of their own, and although those aspirations were persistently thwarted, they did not die.

What of the role of women? Achieving the independent yeoman ideal of the colonial period depended on the closely integrated but clearly second-class place of women, who performed unremitting labor within the domestic sphere of the household economy. In very few cases could women actually (or even in effect) own and run farms, given cultural and legal barriers. Women in thriving yeoman households contributed more than half of the work and many enjoyed rewards and satisfactions from it, but because women's role in the public sphere was so severely restricted, we hardly ever hear their voices. We also know that whatever its satisfactions, the burden of women's labor in bearing and rearing children and in processing food, clothing, and nearly everything else to provide for their families, day in and day out—a woman's work truly was never done—was so exhausting that women were eager to find relief from it. Crèvecoeur's picture of his young wife bringing her knitting when she came to watch him plow is charming, but it smacks more of courtship than of housekeeping.

But for the class of people that Crèvecoeur describes—peasants, artisans, and laborers arriving from Europe, often with meager resources—his optimistic portrait was largely accurate: America truly was the "best poor man's country." As long as land remained cheap and abundant, through the colonial period and well into the nineteenth century, many immigrants were indeed able to obtain a freehold and achieve a "competence" and a large measure of economic independence. We can hardly exaggerate what a novel and positive thing this was, in contrast to the long history of European rural society. It was the real foundation of the American agrarian ideal, and it was no mere literary construction. But for how long and for how many would that opportunity persist? Should land always remain available to the poor and landless as a common moral right, even if that meant breaking up large private landholdings? That was the radical position termed "agrarianism" in Jefferson's time, and there is no evidence that even he was willing to go that far. In America, what we now call the agrarian ideal embraced the policy of making public land that had been wrested from the Native people easily available to white farmers as long as it lasted, but it held no long-term guarantee of access to land for all, or any meaningful limitation on the ability of private owners to accumulate and hold as much land as possible.

The greatest policy challenge facing American farmers after the War for Independence was how far to support a strong central government that would promote the growth of a commercial economy. Farmers deeply mistrusted the corrupting influence of government power, but some measure of political stability and economic growth was crucial to their own prosperity. If the Revolution brought jarring disruptions to Crèvecoeur and other Americans, the years immediately following were not much better. An economic downturn after the war meant hardship for many farmers, saddled with debts they were hard-pressed to repay. States slipped toward bankruptcy, and such national government as the country had under the Articles of Confederation proved impotent. The malaise of the mid-1780s led to a painful period of reassessment about prospects for the newly independent states. Many feared that the American experiment would be short-lived. "No Morn ever dawned more favorable than ours did—and no day was ever more clouded than the present," George Washington wrote to James Madison in 1786.

Fears for the future led to bitter debates about how best to secure self-government against external threats and internal disorder. The debates came to focus on how, or even whether, to build strong national political and economic institutions to address America's problems. If adopted, these institutions might enable a return to prosperity and a more stable future. On the other hand, those same institutions might undermine the liberties of rural citizens and communities and offer unfair economic advantage to commercial interests that would be in a position to profit from them. Proposals for a powerful central government reminded many Americans of why they had rebelled against Great Britain in the first place.

A few Americans, however, looked to England as a model for enabling the country's economic growth. The British had undertaken a financial revolution a century before that came to include the expansion of banking and credit, the development of stock markets, and currency reforms. In its wake came an increase in wealth, the beginnings of industrialization, and quickening urbanization. This revolution was facilitated by a strengthened central government and a powerful (and expensive) army and navy capable of pursuing ambitions of empire. Where this activist government aligned with private interests, corruption at times deformed policy. The deal making that led to the South Sea Bubble was just one notorious example.

The British model might help unlock the wealth of a fertile continent, but many Americans were reluctant to follow it. Not only had the states just fought a war against Britain, that war had been largely fueled by a body of thought that originated with British opponents of the new political and economic system. One famous indictment was the Irish poet Oliver Goldsmith's "The Deserted Village," with its ringing couplet remembered long after the rest of the poem was forgotten: "Ill fares the land, to hastening ills a prey, / Where wealth accumulates, and men decay." *Cato's Letters* was another leading example of this "country" thought decrying the corruption that accompanied the financial revolution and stressing the need for individuals to be vigilant in protecting their liberties from the government's growing power. *Cato's Letters* was widely read and embraced in America before the war. In his *Ideological Origins of the American Revolution,* Bernard Bailyn noted the close fit between *Cato's Letters* and the self-image of colonial farmers: "There the moral basis of a healthy, liberty-preserving polity seemed already to exist in the unsophisticated lives of independent, uncorrupted, landowning yeoman farmers who comprised so large a part of the colonial population." Mistrust of government was part of the bedrock of the American agrarian creed.

But the economic woes that followed the Revolution, along with civil disturbances like Shays's Rebellion, convinced many Americans that stronger centralized political institutions were needed, despite the warnings of traditional country arguments. Once the Constitution was ratified and the new government formed, with all its checks and balances, the nation faced the question of how to apply its new powers to find its economic way. Although only a few American leaders were inclined to promote reforms along British lines, they had an exceptionally able champion in Alexander Hamilton. Hamilton famously defended the British government over dinner one night with John Adams and Thomas Jefferson. As Jefferson recalled the event, Adams suggested that the British constitution would be the most perfect of all, except for the corruption that marred it. According to Jefferson, Hamilton answered, "Purge it of its corruption . . . & it would become an impracticable government: as it stands at present, with all it's [*sic*] supposed defects, it is the most perfect government which ever existed."

This was rank heresy to Jefferson, and Hamilton became a bête noire for the Virginian and the opposition that gathered around him during the presidency of George Washington. As secretary of the treasury, Hamilton pressed his program of reform, which he laid out in three great reports: the *Report on Public Credit*, the *Report on the Bank of the United States*, and the *Report on Manufactures*. And as secretary of state, Jefferson opposed Hamilton with all his might. Jefferson's opposition to the Hamilton program grew, in part, out of his agrarian beliefs. Though he never wrote in a sustained way about those beliefs, he did touch on them frequently enough in the *Notes on Virginia* and in his letters.

"Cultivators of the earth are the most valuable citizens. They are the most vigorous, the most independent, the most virtuous, & they are tied to their country & wedded to it's [*sic*] liberty & interests by the most lasting bonds," wrote Jefferson to John Jay in 1785. And, writing from Paris to James Madison in 1787, Jefferson added, "I think we shall be [virtuous], as long as agriculture is our principal object, which will be the case, while there remains vacant lands in any part of America. When we get piled upon one another in large cities, as in Europe, we shall become corrupt as in Europe," and, he later added, "go to eating one another as they do there."

Despite Jefferson's best efforts, Hamilton got much of the legislation he wanted. His program included assuming state debts at the federal level, setting up the Bank of the United States, founding the national mint, and imposing new taxes and tariffs with an efficient system to collect them. These reforms laid the foundation for rapid growth and diversification of the U.S. economy in the decades that followed. Although later administrations (including Jefferson's) managed to rescind parts of Hamilton's system, discarded elements often returned, as in the case of the national bank. Jefferson remarked in 1802, "When this government was first established, it was possible to have kept it going on true first principles, but the contracted, English, half-lettered ideas of Hamilton, destroyed that hope in the bud. We can pay off his debt in 15 years; but we can never get rid of his financial system."

Among the criticisms leveled against Hamilton's system was the charge that it fostered a culture obsessed with money. Modern readers will be struck, for instance, by his approval of the increased productivity in British factories that resulted from employing children. But Hamilton's biographer Forrest McDonald claims that Hamilton had higher purposes in mind than simple profit. "[Hamilton] believed that the greatest benefits of his system were spiritual—the enlargement of the scope of human freedom and the enrichment of opportunities for human endeavor."

This defense of Hamilton sets his achievement—and much of the United States' subsequent economic growth—in a properly broad context. But if we are to credit those enriched opportunities with the kind of spiritual value that McDonald suggests, a similarly broad range of costs should also be considered. For example, what some experienced as economic opportunity could as easily turn to economic coercion and ruin for others. The farmer content with his life in a stable agricultural society might well have felt unfairly burdened by the escalation of market forces, such as demands for prompt repayment of debts in cash.

Moreover, by giving economic effort and financial gain so central a role in American society, to this day the Hamiltonian worldview may discount those natural and social values that don't fully register in monetary metrics. Does the pursuit of profit foster the exploitation of land and people? Does it undermine community values? One hallmark of agrarian arguments since the founding of the Republic has been to try to weigh economic gain in the larger balance of overall well-being, and not to assume that if profit is put first, all that follows will be positive.

Arguments about the Constitution and subsequent economic reforms in America linked the economic, political, and moral aspects of the challenges facing the new states. Republics, it was assumed, depended on the actions—and so, at bottom, on the virtues—of their citizens. For a republic to survive, its citizens must be willing to put the good of the public above their own immediate self-interest. And it was widely assumed that the work people do shapes their character. The republican virtues, including modesty, self-discipline, sobriety, and frugality, grew most readily, according to common belief, among farmers. The corresponding vices—arrogance, opulence, cunning, avarice—were associated with the world of finance and the "stockjobbers" who gamed and profited from a speculative commercial economy.

Farmers may have regarded those who grew rich by commerce with suspicion, but they were not of one mind about commerce itself. Could they stand free of the market, and did they want to? One writer urged his readers not to be buffaloed into concentrating power in a central government. "We are told, that agriculture is without encouragement; trade is languishing; private faith and credit are disregarded, and public credit is prostrate. . . . But suffer me, my countrymen, to call attention to a serious and sober estimate of the situation in which you are placed. . . . Does not every man sit under his own fig-tree, having none to make him afraid?"

Perhaps—but most farmers would not be satisfied to sit unafraid beneath the fig tree forever. From the moment European farmers arrived in America they were linked to the expanding Atlantic economy—at the very least by a need for goods they could not easily produce themselves, but even more by a desire for increased material comforts and a prosperous life. The planter James Madison embraced the market economy and the new Constitution. He hoped that a stronger central government would ensure a market for American agricultural products abroad. He assumed that farmers would take advantage of a favorable international market by raising more produce and selling their surplus. Like his friend Thomas Jefferson, he believed that farmers would continue moving beyond the Appalachians and that their prospects were critical for the future of the nation. They would extend the dominance of agriculture in American society into the future as cultivation was extended across the continent, but only if they had an outlet for their products. If content with mere subsistence, they would become stagnant and slothful. Better markets for the produce of vigorously expanding American farms would be critical in preserving American self-government.

From the beginning, American farmers and agrarian thinkers have wrestled with a deep ambivalence about their place in the market economy and the degree to which they could avoid the corruption of commerce yet still enjoy the increased prosperity it offered. This was painfully true for planters such as Madison and Jefferson, who lived as gentlemen upon the commodities produced from vast landholdings by the forced labor of their slaves. But the tension between farming as a means to wealth and farming as a way of life would also arise for Northern yeomen as the nation grew. This conflict was never far beneath the surface of debates about how to frame a new, agrarian nation.

J. HECTOR ST. JOHN DE CRÈVECOEUR

J. Hector St. John de Crèvecoeur was born Jean Michel Guillaume de Crèvecoeur near Caen, France, in 1735. He led a remarkably varied and adventurous life. As a youth he moved alone to England for a time, then in 1755 he came to New France, where he fought as a minor officer in the French and Indian War. After the war he resigned his commission and traveled in the American colonies, spending some years exploring the interior as an Indian trader. By 1769 he had become a naturalized English citizen, settled on a farm near present-day Goshen, New York, and changed his name to reflect his new identity. Crèvecoeur apparently farmed there quietly for the next several years, raising his family.

The years of the American Revolution threw Crèvecoeur's life into turmoil. Suspected by local patriots of Tory sympathies and wanting to see his ailing father, in 1779 he set out on a journey to France with one son, leaving the rest of his family behind on the farm. Along the way he was thrown in prison by the British forces occupying New York City, now suspected of being a rebel spy. Upon his release in 1780 he sailed to England and France, where Letters from an American Farmer *was published to great acclaim in both countries. Crèvecoeur returned to America in 1783, only to find that his farm had been burned and his wife was dead, though he was eventually reunited with his children. He served with great success as French consul to New York. In time he left America and returned to France, where he survived the French Revolution and lived out his remaining years at his old family estate.*

The few years he spent farming in New York seem to have been the happiest of his life and gave rise to Letters from an American Farmer. *In form, the book is a series of letters from a modestly prosperous Pennsylvania farmer to a European friend seeking a better understanding of life across the ocean. Much of the* Letters *was shaped by Crève-coeur's experiences, though it is not a verbatim account of Crèvecoeur's life on his own farm—for example, for the purposes of the book he made himself the son of an American frontiersman, rather than an immigrant. In the opening chapters, the* Letters *succeeds in articulating the agrarian ideals of the emerging nation of farmers—a pre-Revolutionary English America just discovering its national identity.*

Later chapters, however, were not so flattering. In one, Crèvecoeur wrote about the savagery of the backwoods, whose hunters seemed to have slipped back to a more primitive existence. Another chapter presented American slavery in it cruelest aspects, which Crèvecoeur cast as the first appearance in America of the European condition of greed, corruption, and enormous disparities of wealth, where "men eat men." Finally, the Letters *closed with the author's anguished description of the havoc and destruction that arrived with the Revolution. The book's trajectory, from wholesome rural simplicity to the horror of slavery to the turmoil of war, in a remarkable way foreshadowed the national experience of succeeding generations: from the youthful exuberance of agrarian expansion to the disaster of the Civil War.*

From *Letters from an American Farmer* (1782)

LETTER II. ON THE SITUATION, FEELINGS, AND PLEASURES, OF AN AMERICAN FARMER

As you are the first enlightened European I had ever the pleasure of being acquainted with, you will not be surprised that I should, according to your earnest desire and my promise, appear anxious of preserving your friendship and correspondence. By your accounts, I observe a material difference subsists between your husbandry, modes, and customs, and ours. Everything is local. Could we enjoy the advantages of the English farmer, we should be much happier, indeed, but this wish, like many others, implies a contradiction; and, could the English farmer have some of those privileges we possess, they would be the first of their class in the world. Good and evil, I see, are to be found in all societies, and it is in vain to seek for any spot where those ingredients are not mixed. I therefore rest satisfied, and thank God that my lot is to be an American farmer, instead of a Russian boor or an Hungarian peasant. I thank you kindly for the idea, however dreadful, which you have given me of their lot and condition. Your observations have confirmed me in the justness of my ideas, and I am happier now than I thought myself before. It is strange that misery, when viewed in others, should become to us a sort of real good; though I am far from rejoicing to hear that there are in the world men so thoroughly wretched. They are no doubt as harmless, industrious, and willing to work, as we are. Hard is their fate to be thus condemned to a slavery worse than that of our negroes. Yet, when young, I entertained some thoughts of selling my farm. I thought it afforded but a dull repetition of the same labours and pleasures. I thought the former tedious and heavy: the latter few and insipid. But, when I came to consider myself as divested of my farm, I then found the world so wide, and every place so full, that I began to fear lest there would be no room for me. My farm, my house, my barn, presented to my imagination, objects from which I adduced quite new ideas: they were more forcible than before. Why should not I find myself happy, said I, where my father was before? He left me no good books it is true; he gave me no other education than the art of reading and writing: but he left me a good farm and his experience: he left me free from debts, and no kind of difficulties to struggle with—I married; and this perfectly reconciled me to my situation. My wife rendered my house all at once cheerful and pleasing: it no longer appeared gloomy and solitary as before. When I went to work in my fields, I worked with more alacrity and sprightliness. I felt that I did not work for myself alone, and this encouraged me much. My wife would often come with her knitting in her hand, and sit under the shady tree, praising the straightness of my furrows and the docility of my horses. This swelled my heart and made everything light and pleasant, and I regretted that I had not married before.

I felt myself happy in my new situation, and where is that station which can confer a more substantial system of felicity than that of an American farmer, possessing freedom of action, freedom of thoughts, ruled by a mode of government which requires but little from us? I owe nothing but a pepper-corn to my country, a small tribute to my king, with loyalty and due respect. I know no other landlord than the Lord of all land, to whom I owe the most sincere gratitude. My father left me three hundred and seventy-one acres of land, forty-seven of which are good timothy meadow, an excellent orchard, a good house, and a substantial barn. It is my duty to think how happy I am that he lived to build and pay for all these improvements. What are the labours which I have to undergo? What are my fatigues when compared to his, who had every thing to do, from the first tree he felled to the finishing of his house? Every year I kill from 1500 to 2000 weight of pork, 1200 of beef, half a dozen of good wethers in harvest; of fowls my wife has always a great stock; what can I wish more? My negroes are tolerably faithful and healthy. By a long series of industry and honest dealings, my father left behind him the name of a good man. I have but to tread his paths to be happy and a good man like him. I know enough of the law to regulate my little concerns with propriety, nor do I dread its power. These are the grand outlines of my situation, but as I can feel much more than I am able to express, I hardly know how to proceed.

When my first son was born, the whole train of my ideas was suddenly altered. Never was there a charm that acted so quickly and powerfully. I ceased to ramble in imagination through the wide world. My excursions, since, have not exceeded the bounds of my farm; and all my principal pleasures are now centered within its scanty limits: but, at the same time, there is not an operation belonging to it in which I do not find some food for useful reflections. This is the reason, I suppose, that, when you were here, you used, in your refined style, to denominate me the farmer of feelings. How rude must those feelings be in him who daily holds the ax or the plough! How much more refined, on the contrary, those of the European, whose mind is improved by education, example, books, and by every acquired advantage! Those feelings, however, I will delineate as well as I can, agreeably to your earnest request.

When I contemplate my wife, by my fire-side, while she either spins, knits, darns, or suckles our child, I cannot describe the various emotions of love, of gratitude, of conscious pride, which thrill in my heart, and often overflow in involuntary tears. I feel the necessity, the sweet pleasure, of acting my part, the part of a husband and father, with an attention and propriety which may entitle me to my good fortune. It is true these pleasing images vanish with the smoke of my pipe, but, though they disappear from my mind, the impression they have made on my heart is indelible. When I play with the infant, my warm imagination runs forward, and eagerly anticipates his future temper and constitution. I would willingly open the book of fate, and know in which page his destiny is delineated. Alas! where is the father, who, in those moments of paternal extasy, can delineate one half of the thoughts which dilate his heart? I am sure I cannot. Then again I fear for the health of those who are become so dear to me, and in their sicknesses, I severely pay for the joys I experienced while they were well. Whenever I go abroad it is always involuntary. I never return home without feeling some pleasing emotion, which I often suppress as useless and foolish. The instant I enter on my own land, the bright idea of property, of exclusive right, of independence, exalt my mind. Precious soil, I say to myself, by what singular custom of law is it that thou wast made to constitute the riches of the freeholder? What should we American farmers be without the distinct possession of that soil? It feeds us, it clothes, us: from it we draw even a great exuberancy, our best meat, our richest drink; the very honey of our bees comes from this privileged spot. No wonder we should thus cherish its possession: no wonder that so many Europeans, who have never been able to say that such portion of land was theirs, cross the Atlantic to realize that happiness! This formerly rude soil has been converted by my father into a pleasant farm, and, in return, it has established all our rights. On it is founded our rank, our freedom, our power, as citizens; our importance as inhabitants of such a district. These images, I must confess, I always behold with pleasure, and extend them as far as my imagination can reach; for this is what may be called the true and the only philosophy of an American farmer.

Pray do not laugh in thus seeing an artless countryman tracing himself through the simple modifications of his life. Remember that you have required it, therefore with candour, though with diffidence, I endeavour to follow the thread of my feelings, but I cannot tell you all. Often, when I plough my low ground, I place my little boy on a chair which screws to the beam of the plough. Its motion and that of the horses please him: he is perfectly happy, and begins to chat. As I lean over the handle, various are the thoughts which crowd into my mind. I am now doing for

Plan of an American New Cleared Farm. From P. Campbell, *Travels in the Interior Inhabited Parts of North America in the Years 1791 and 1792* (1793). Courtesy Library of Congress, Washington, DC.

him, I say, what my father formerly did for me: may God enable him to live that he may perform the same operations for the same purposes when I am worn out and old! I relieve his mother of some trouble while I have him with me; the odoriferous furrow exhilarates his spirits, and seems to do the child a great deal of good, for he looks more blooming since I have adopted that practice. Can more pleasure, more dignity, be added to that primary occupation? The father, thus ploughing with his child, and to feed his family, is inferior only to the emperor of China ploughing as an example to his kingdom. In the evening, when I return home through my low grounds, I am astonished at the myriads of insects which I perceive dancing in the beams of the setting sun. I was before scarcely acquainted with their existence; they are so small that it is difficult to distinguish them: they are carefully improving this short evening space, not daring to expose themselves to the blaze of our meridian sun. I never see an egg brought on my table but I feel penetrated with the wonderful change it would have undergone but for my gluttony. It might have been a gentle useful hen leading her chicken with a care and vigilance which speaks shame to many women. A cock, perhaps, arrayed with the most majestic plumes, tender to its mate, bold, courageous, endowed with an astonishing instinct, with thoughts, with memory, and every distinguishing characteristic of the reason of man! I never see my trees drop their leaves and their fruit in the autumn, and bud again in the spring, without wonder. The sagacity of those animals, which have long been the tenants of my farm, astonish me: some of them seem to surpass even men in memory and sagacity. I could tell you singular instances of that kind. What then is this instinct which we so debase, and of which we are taught to entertain so diminutive an idea? My bees, above any other tenants of my farm, attract my attention and respect. I am astonished to see that nothing exists but what has its enemy; one species pursues and lives upon the other. Unfortunately our kingbirds are the destroyers of those industrious insects; but on the other hand, these birds preserve our fields from the depredation of crows which they pursue on the wing with great vigilance and astonishing dexterity. . . .

LETTER III. WHAT IS AN AMERICAN?

I wish I could be acquainted with the feeling and thoughts which must agitate the heart and present themselves to the mind of an enlightened Englishman, when he first lands on this continent. He must greatly rejoice that he lived at a time to see this fair country discovered and settled. He must necessarily feel a share of national pride when he views the chain of settlements which embellish these extended shores. When he says to himself, this is the work of my countrymen, who, when convulsed by factions, afflicted by a variety of miseries and wants, restless and impatient, took refuge here. They brought along with them their national genius, to which they principally owe what liberty they enjoy and what subsistence they possess. Here he sees the industry of his native country displayed in a new manner, and traces, in their works, the embryos of all the arts, sciences, and ingenuity, which flourish in Europe. Here he beholds fair cities, substantial villages, extensive fields, an immense country filled with decent houses, good roads, orchards, meadows, and bridges, where, a hundred years ago, all was wild, woody, and uncultivated! What a train of pleasing ideas this fair spectacle must suggest! It is a prospect which must inspire a good citizen with the most heartfelt pleasure! The difficulty consists in the manner of viewing so extensive a scene. He is arrived on a new continent: a modern society offers itself to his contemplation, different from what he had hitherto seen. It is not composed, as in Europe, of great lords who possess every thing, and of a herd of people who have nothing. Here are no aristocratical families, no courts, no kings, no bishops, no ecclesiastical dominion, no invisible power giving to a few a very visible one, no great manufactures employing thousands, no great refinements of luxury. The rich and the poor are not so far removed from each other as they are in Europe. Some few towns excepted, we are all tillers of the earth, from Nova Scotia to West

Florida. We are a people of cultivators, scattered over an immense territory, communicating with each other by means of good roads and navigable rivers, united by the silken bands of mild government, all respecting the laws, without dreading their power, because they are equitable. We are all animated with the spirit of an industry which is unfettered and unrestrained, because each person works for himself. If he travels through our rural districts, he views not the hostile castle and the haughty mansion contrasted with the clay-built hut and miserable cabin, where cattle and men help to keep each other warm, and dwell in meanness, smoke, and indigence. A pleasing uniformity of decent competence appears throughout our habitations. The meanest of our log-houses is a dry and comfortable habitation. Lawyer or merchant are the fairest titles our towns afford: that of a farmer is the only appellation of the rural inhabitants of our country. It must take some time ere he can reconcile himself to our dictionary, which is but short in words of dignity and names of honour. There, on a Sunday, he sees a congregation of respectable farmers and their wives, all clad in neat homespun, well mounted, or riding in their own humble waggons. There is not among them an esquire, saving the unlettered magistrate. There he sees a parson as simple as his flock, a farmer who does not riot on the labour of others. We have no princes, for whom we toil, starve, and bleed. We are the most perfect society now existing in the world. Here man is free as he ought to be; nor is this pleasing equality so transitory as many others are. Many ages will not see the shores of our great lakes replenished with inland nations, nor the unknown bounds of North America entirely peopled. Who can tell how far it extends? Who can tell the millions of men whom it will feed and contain? for no European foot has, as yet, travelled half the extent of this mighty continent.

The next wish of this traveller will be to know whence came all these people? They are a mixture of English, Scotch, Irish, French, Dutch, Germans, and Swedes. From this promiscuous breed, that race now called Americans, have arisen. The Eastern provinces must indeed be excepted, as being the unmixed descendents of Englishmen. I have heard many wish that they had been more intermixed also: for my part, I am no wisher, and think it much better as it has happened. They exhibit a most conspicuous figure in this great and variegated picture. They too enter for a great share in the pleasing perspective displayed in these thirteen provinces. I know it is fashionable to reflect on them, but I respect them for what they have done; for the accuracy and wisdom with which they have settled their territory; for the decency of their manners; for their early love of letters; their antient college, the first in this hemisphere; for their industry; which to me, who am but a farmer, is the criterion of every thing. There never was a people, situated as they are, who, with so ungrateful a soil, have done more in so short a time. Do you think that the monarchical ingredients, which are more prevalent in other governments, have purged them from all foul stains? Their histories assert the contrary.

In this great American asylum, the poor of Europe have by some means met together, and in consequence of various causes. To what purpose should they ask one another what countrymen they are? Alas, two thirds of them had no country. Can a wretch, who wanders about, who works and starves, whose life is a continual scene of sore affliction or pinching penury; can that man call England or any other kingdom his country? A country that had no bread for him; whose fields procured him no harvest; who met with nothing but the frowns of the rich, the severity of the laws, with jails and punishments; who owned not a single foot of the extensive surface of this planet. No! Urged by a variety of motives here they came. Every thing has tended to regenerate them. New laws, a new mode of living, a new social system. Here they are become men; in Europe they were as so many useless plants, wanting vegetative mould and refreshing showers. They withered; and were mowed down by want, hunger, and war; but now, by the power of transplantation, like all other plants, they have taken root and flourished! Formerly they were not numbered in any civil lists of their country, except in those of the poor: here they rank as citizens. By what invisible power has this surprising metamorphosis been performed? By that of the laws and that of their industry. The laws, the indulgent laws, protect them as they arrive, stamping on them the symbol of adoption: they receive ample rewards for their labours: these accumulated rewards procure them lands: those lands confer on them the title of freemen, and to that title every benefit is affixed which men can possibly require. This is the great operation daily performed by our laws. Whence proceed these laws? From our government. Whence that government? It is derived from the original genius and strong desire of the people ratified and confirmed by the crown. This is the great chain which links us all; this is the picture which every province exhibits, Nova Scotia excepted. There the crown has done all. Either there were no people who had genius, or it was not much attended to. The consequence is, that the province is very thinly inhabited indeed. The power of the crown, in conjunction with the musketoes, has prevented men from settling there. Yet some parts of it flourished once, and it contained a mild harmless set of people. But for the fault of a few leaders, the whole was banished. The greatest political error, the crown ever committed in America, was, to cut off men from a country which wanted nothing but men.

What attachment can a poor European emigrant have for a country where he had nothing? The knowledge of the language, the love of a few kindred as poor as himself, were the only cords that tied him. His country is now that which gives him land, bread, protection, and consequence. *Ubi panis ibi patria* [Where there is bread, there is the homeland] is the motto of all emigrants. What then is the American, this new man? He is neither an European, nor the descendent of an European: hence that strange mixture of blood, which you will find in no other country. I could point out to you a family, whose grandfather was an Englishman, whose wife was Dutch, whose son married a French woman, and whose present four sons have now four wives of different nations. He is an American, who, leaving behind him all his antient prejudices and manners, receives new ones from the new mode of life he has embraced, the new government he obeys, and the new rank he holds. He becomes an American by being received in the broad lap of our great *alma mater*.

Here individuals of all nations are melted into a new race of men, whose labours and posterity will one day cause great changes in the world. Americans are the western pilgrims, who are carrying along with them that great mass of arts, sciences, vigour, and industry, which began long since in the east. They will finish the great circle. The Americans were once scattered all over Europe. Here they are incorporated into one of the finest systems of population which has ever appeared, and which will hereafter become distinct by the power of the different climates they inhabit. The American ought therefore to love this country much better than that wherein either he or his forefathers were born. Here the rewards of his industry follow, with equal steps, the progress of his labour. His labour is founded on the basis of nature, *self-interest:* can it want a stronger allurement? Wives and children, who before in vain demanded of him a morsel of bread, now, fat and frolicksome, gladly help their father to clear those fields whence exuberant crops are to arise, to feed and clothe them all, without any part being claimed, either by a despotic prince, a rich abbot, or a mighty lord. Here religion demands but little of him; a small voluntary salary to the minister, and gratitude to God: can he refuse these? The American is a new man, who acts upon new principles; he must therefore entertain new ideas and form new opinions. From involuntary idleness, servile dependence, penury, and useless labour, he has passed to toils of a very different nature, rewarded by ample subsistence.—This is an American.

British America is divided into many provinces, forming a large association, scattered along a coast 1500 miles extent and about 200 wide. This society I would fain examine, at least such as it appears in the middle provinces; if it does not afford that variety of tinges and gradations which may be observed in Europe, we have colours peculiar to ourselves. For instance, it is natural to conceive that those who live near the sea must be very different from those who live in the woods: the intermediate space will afford a separate and distinct class.

Men are like plants. The goodness and flavour of the fruit proceeds from the peculiar soil and exposition in which they grow. We are nothing but what we derive from the air we breathe, the climate we inhabit, the government we obey, the system of religion we profess, and the nature of our employment. Here you will find but few crimes; these have acquired as yet no root among us. I wish I were able to trace all my ideas. If my ignorance prevents me from describing them properly, I hope I shall be able to delineate a few of the outlines, which is all I propose.

Those, who live near the sea, feed more on fish than on flesh, and often encounter that boisterous element. This renders them more bold and enterprising: this leads them to neglect the confined occupations of the land. They see and converse with a variety of people. Their intercourse with mankind becomes extensive. The sea inspires them with a love of traffic, a desire of transporting produce from one place to another; and leads them to a variety of resources, which supply the place of labour. Those who inhabit the middle settlements, by far the most numerous, must be very different. The simple cultivation of the earth purifies them; but the indulgences of the government, the soft remonstrances of religion, the rank of independent free-holders, must necessarily inspire them with sentiments very little known in Europe among people of the same class. What do I say? Europe has no such class of men. The early knowledge they acquire, the early bargains they make, give them a great degree of sagacity. As freemen they will be litigious. Pride and obstinacy are often the cause of law-suits; the nature of our laws and governments may be another. As citizens, it is easy to imagine that they will carefully read the newspapers, enter into every political disquisition, freely blame, or censure, governors and others. As farmers, they will be careful and anxious to get as much as they can, because what they get is their own. As northern men, they will love the cheerful cup. As Christians, religion curbs them not in their opinions: the general indulgence leaves every one to think for themselves in spiritual matters. The law inspects our actions; our thoughts are left to God. Industry, good living, selfishness, litigiousness, country politics, the pride of freemen, religious indifference, are their characteristics. If you recede still farther from the sea, you will come into more modern settlements: they exhibit the same strong lineaments in a ruder appearance. Religion seems to have still less influence, and their manners are less improved.

Now we arrive near the great woods, near the last inhabited districts. There men seem to be placed still farther beyond the reach of government, which in some measure, leaves them to themselves. How can it pervade every corner, as they were driven there by misfortunes, necessity of beginnings, desire of acquiring large tracks [sic] of land, idleness, frequent want of oeconomy, antient debts. The re-union of such people does not afford a very pleasing spectacle. When discord, want of unity and friendship, when either drunkenness or idleness, prevail in such remote districts, contention, inactivity, and wretchedness, must ensue. There are not the same remedies to these evils as in a long-established community. The few magistrates they have are, in general, little better than the rest. They are often in a perfect state of war; that of man against man; sometimes decided by blows, sometimes by means of the law: that of man against every wild inhabitant of these venerable woods, of which they are come to dispossess them. There men appear to be no better than carnivorous animals, of a superior rank, living on the flesh of wild animals when they can catch them, and, when they are not able, they subsist on grain. He, who would wish to see America in its proper light, and to have a true idea of its feeble beginnings and barbarous rudiments, must visit our extended line of frontiers, where the last settlers dwell, and where he may see the first labours of settlement, the mode of clearing the earth, in all their different appearances. Where men are wholly left dependent on their native tempers and on the spur of uncertain industry, which often fails when not sanctified by the efficacy of a few moral rules. There, remote from the power of example and check of shame, many families exhibit the most hideous parts of our society. They are a kind of forlorn hope, preceding, by ten or twelve years, the most respectable army of veterans which come after them. In that space, prosperity will polish some, vice and the law will drive off the rest, who, uniting again with others like themselves, will recede still farther, making room for more industrious people, who will finish their improvements, convert the log-house into a convenient habitation, and, rejoicing that the first heavy labours are finished, will change, in a few years, that hitherto-barbarous country into a fine, fertile, well-regulated, district. Such is our progress, such is the march of the Europeans toward the interior parts of this continent. In all societies there are off-casts. This impure part serves as our precursors or pioneers. My father himself was one of that class; but he came upon honest principles, and was therefore one of the few who held fast. By good conduct and temperance he transmitted to me his fair inheritance, when not above one in fourteen of his contemporaries had the same good fortune.

Forty years ago this smiling country was thus inhabited. It is now purged. A general decency of manners prevails throughout, and such has been the fate of our best countries.

LETTER IX. DESCRIPTION OF CHARLES-TOWN; THOUGHTS ON SLAVERY, ETC.

Charles-Town is in the north what Lima is in the south; both are capitals of the richest provinces of their respective hemispheres; you may therefore conjecture, that both cities must exhibit the appearances necessarily resulting from riches. Peru abounding in gold, Lima is filled with inhabitants, who enjoy all those gradations of pleasure, refinement, and luxury, which proceed from wealth. Carolina produces commodities, more valuable perhaps than gold, because they are gained by greater industry; it exhibits also on our northern stage a display of riches and luxury, inferior indeed to the former, but far superior to what are to be seen in our northern towns. Its situation is admirable; being built at the confluence of two large rivers, which receive, in their course, a great number of inferior streams; all navigable, in the spring, for flat boats. Here the produce of this extensive territory concentres; here, therefore, is the seat of the most valuable exportation; their wharfs, their docks, their magazines, are extremely convenient to facilitate this great commercial business. The inhabitants are the gayest in America; it is called the centre of our beau monde, and is always filled with the richest planters in the province, who resort hither in quest of health and pleasure. Here is always to be seen a great number of valetudinarians from the West-Indies, seeking for the renovation of health, exhausted by the debilitating nature of their sun, air, and modes of living. Many of these West-Indians have I seen, at thirty, loaded with the infirmities of old age; for, nothing is more common, in those countries of wealth, than for persons to lose the abilities of enjoying the comforts of life at a time when we northern men just begin to taste the fruits of our labour and prudence. The round of pleasure, and the expences of those citizens tables, are much superior to what you would imagine: indeed the growth of this town and province have been astonishingly rapid. It is pity that the narrowness of the neck, on which it stands, prevents it from increasing, and which is the reason why houses are so dear. The heat of the climate, which is sometimes very great in the interior parts of the country, is always temperate in Charles-Town, though, sometimes, when they have no sea breezes, the sun is too powerful. The climate renders excesses of all kinds very dangerous, particularly those of the table; and yet, insensible or fearless of danger, they live on, and enjoy a short and a merry life: the rays of their sun seem to urge them irresistibly to dissipation and pleasure: on the contrary, the women, from being abstemious, reach to a longer period of life, and seldom die without having had several husbands. An European at his first arrival must be greatly surprised when he sees the elegance of their houses, their sumptuous furniture, as well as the magnificence of their tables; can he imagine himself in a country, the establishment of which is so recent? . . .

While all is joy, festivity, and happiness, in Charles-Town, would you imagine that scenes of misery overspread in the country? Their ears, by habit, are become deaf, their hearts are hardened; they neither see, hear, nor feel for, the woes of their poor slaves, from whose painful labours all their wealth proceeds. Here the horrors of slavery, the hardship of incessant toils, are unseen; and no one thinks with compassion of those showers of sweat and of tears which from the bodies of Africans daily drop, and moisten the ground they till. The cracks of the whip, urging these miserable beings to excessive labour, are far too distant from the gay capital to be heard. The chosen race eat, drink, and live happy, while the unfortunate one grubs up the ground, raises indigo, or husks the rice: exposed to a sun full as scorching as their native one without the support of good food, without the cordials of any cheering liquor. This great contrast has often afforded me subjects of the most afflicting meditations. On the one side, behold a people enjoying all that life affords most bewitching and pleasurable, without labour, without fatigue, hardly subjected to the trouble of wishing. With gold, dug from Peruvian mountains, they order vessels to the coasts of Guinea; by virtue of that gold, wars, murders, and devastations, are committed in some harmless, peaceable, African neighborhood, where dwelt innocent people, who even knew not but that all men were black. The daughter torn from her weeping mother, the child from the wretched parents, the wife from the loving husband; whole families swept away, and brought, through storms and tempests, to this rich metropolis! There, arranged like horses at a fair, they are branded like cattle, and then driven to toil, to starve, and to languish, for a few years, on the different plantations of these citizens. And for whom must they work? For persons they know not, and who have no other power over them than that of violence; no other right than what this accursed metal has given them! Strange order of things! O Nature, where art thou?—Are not these blacks thy children as well as we? On the other side, nothing is to be seen but the most diffusive misery and wretchedness, unrelieved even in thought or wish! Day after day they drudge on without any prospect of ever reaping for themselves; they are obliged to devote their lives, their limbs, their will, and every vital exertion, to swell the wealth of masters, who look not upon them with half the kindness and affection with which they consider their dogs and horses. Kindness and affection are not the portion of those who till the earth, who carry burdens, who convert the logs into useful boards. This reward, simple and natural as one would conceive it, would border on humanity; and planters must have none of it!

ALEXANDER HAMILTON

Alexander Hamilton was born in 1755, the illegitimate son of a Scottish immigrant to the West Indies. He was raised by his mother, who ran a small store in St. Croix. Having caught the eye of several business and civic leaders, Hamilton was sent in 1772 to New York for his education. Within a few years the young man was writing pamphlets on behalf of the rights of the colonies and serving under General Washington in the Continental Army. He fought at Brandywine and Germantown, wintered at Valley Forge, and finally, in 1781, fought at Yorktown. For much of the war he worked closely with Washington and earned the general's trust.

That trust served Hamilton well during Washington's first administration. From his position as secretary of the treasury, Hamilton put in place structures that helped set in motion the dynamic growth that would characterize the American economy. The central points of these policies were set out in Hamilton's three major reports, published between 1789 and 1791.

The first two, focusing on the nation's debt and plans for a national bank, were aimed at providing the country with a reformed financial system. The third, his Report on Manufactures, *gave a fuller picture of the kind of society that might be built on the new financial foundations. Hamilton called for the government to use tariffs and bounties to foster rapid industrial growth in America, building a more powerful, diverse economy. The report confirmed for critics how far Hamilton wanted to take the United States from its agrarian base and helped spur the formation of the Democratic-Republican Party in opposition to his plans.*

From *Report on the Subject of Manufactures* (1791)

The Secretary of the Treasury in obedience to the order of ye House of Representatives, of the 15th day of January 1790, has applied his attention, at as early a period as his other duties would permit, to the subject of Manufactures; and particularly to the means of promoting such as will tend to render the United States, independent on foreign nations, for military and other essential supplies. And he there upon respectfully submits the following Report. . . .

It ought readily to be conceded, that the cultivation of the earth—as the primary and most certain source of national supply—as the immediate and chief source of subsistence to man—as the principal source of those materials which constitute the nutriment of other kinds of labor—as including a state most favourable to the freedom and independence of the human mind—one, perhaps, most conducive to the multiplication of the human species—has *intrinsically a strong claim to pre-eminence over every other kind of industry.*

But, that it has a title to any thing like an exclusive predilection, in any country, ought to be admitted with great caution. . . .

One of the arguments made use of, in support of the idea may be pronounced both quaint and superficial. It amounts to this—That in the production of the soil, nature cooperates with man; and that the effect of their joint labour must be greater than that of the labour of man alone.

This however, is far from being a necessary inference. It is very conceivable, that the labor of man alone laid out upon a work, requiring great skill and art to bring it to perfection, may be more productive, *in value,* than the labour of nature and man combined, when directed towards more simple operations and objects: And when it is recollected to what an extent the Agency of nature, in the application of the mechanical powers, is made auxiliary to the prosecution of manufactures, the suggestion, which has been noticed, loses even the appearance of plausibility.

It might also be observed, with contrary view, that the labour employed in Agriculture is in a great measure periodical and occasional, depending on seasons, liable to various and long intermissions; while that occupied in many manufactures is constant and regular, extending through the year, embracing in some instances night as well as day. It is also probable, that there are among the cultivators of land more examples of remissness, than among artificers. The farmer, from the peculiar fertility of his land, or some other favorable circumstance, may frequently obtain a livelihood, even with a considerable degree of carelessness in the mode of cultivation; but the artisan can with difficulty effect the same object, without exerting himself pretty equally with all those, who are engaged in the same pursuit. And if it may likewise be assumed as a fact, that manufactures open a wider field to exertions of ingenuity than agriculture, it would not be a strained conjecture, that the labour employed in the former, being at once more *constant,* more uniform and more ingenious, than that which is employed in the latter, will be found at the same time more productive.

But it is not meant to lay stress on observations of this nature—they ought only to serve as a counterbalance to those of a similar complexion. Circumstances so vague and general, as well as so abstract, can afford little instruction in a matter of this kind. . . .

The foregoing suggestions *are not designed to inculcate an opinion that manufacturing industry is more productive than that of Agriculture.* They are intended rather to shew that the reverse of this proposition is not ascertained; that the general arguments which are brought to establish it are not satisfactory; and consequently that a supposition of the superior productiveness of Tillage ought to be no obstacle to listening to any substantial inducements to the encouragement of manufactures, which may be otherwise perceived to exist, through an apprehension, that they may have a tendency to divert labour from a more to a less profitable employment.

It is extremely probable, that on a full and accurate development of the matter, on the ground of fact and calculation, it would be discovered that there is no material difference between the aggregate productiveness of the one, and of the other kind of industry; and that the propriety of the encouragements, which may in any case be proposed to be given to either ought to be determined upon considerations irrelative to any comparison of that nature. . . .

It is now proper to proceed a step further, and to enumerate the principal circumstances, from which it may be inferred—That manufacturing establishments not only occasion a positive augmentation of the Produce and Revenue of the Society, but that they contribute essentially to rendering them greater than they could possibly be, without such establishments. These circumstances are—

1. The division of Labour.
2. An extension of the use of Machinery.
3. Additional employment to classes of the community not ordinarily engaged in the business.
4. The promoting of emigration from foreign Countries.
5. The furnishing greater scope for the diversity of talents and dispositions which discriminate men from each other.
6. The affording a more ample and various field for enterprise.
7. The creating in some instances a new, and securing in all, a more certain and steady demand for the surplus produce of the soil.

Each of these circumstances has a considerable influence upon the total mass of industrious effort in a community. Together, they add to it a degree of energy and effect, which are not easily conceived. Some comments upon each of them, in the order in which they have been stated, may serve to explain their importance.

I. AS TO THE DIVISION OF LABOUR.

It has justly been observed, that there is scarcely any thing of greater moment in the oeconomy of a nation, than the proper division of labour. The separation of occupations causes each to be carried to a much greater perfection, than it could possible acquire, if they were blended. This arises principally from three circumstances.

1st—The greater skill and dexterity naturally resulting from a constant and undivided application to a single object. It is evident, that these properties must increase, in proportion to the separation and simplification of objects and the steadiness of the attention devoted to each; and must be less, in proportion to the complication of objects, and the number among which the attention is distracted.

2nd. The oeconomy of time—by avoiding the loss of it, incident to a frequent transition from one operation to another of a different nature. This depends on various circumstances—the transition itself—the orderly disposition of the implements, machines and materials employed in the operation to be relinquished—the preparatory steps to the commencement of a new one—the interruption of the impulse, which the mind of the workman acquires, from being engaged in a particular operation—the distractions hesitations and reluctances, which attend the passage from one kind of business to another.

3rd. An extension of the use of Machinery. A man occupied on a single object will have it more in his power, and will be more naturally led to exert his imagination in devising methods to facilitate and abrige labour, than if he were perplexed by a variety of independent and dissimilar operations. Besides this, the fabrication of Machines, in numerous instances, becoming itself a distinct trade, the Artist who follows it, has all the advantages which have been enumerated, for improvement in his particular art; and in both ways the invention and application of machinery are extended.

And from these causes united, the mere separation of the occupation of the cultivator, from that of the Artificer, has the effect of augmenting the productive powers of labour, and with them, the total mass of the produce or revenue of a Country. In this single view of the subject, therefore, the utility of Artificers or Manufacturers, towards promoting an increase of productive industry, is apparent.

II. As to an extension of the use of Machinery a point which though partly anticipated requires to be placed in one or two additional lights.

The employment of Machinery forms an item of great importance in the general mass of national industry. 'Tis an artificial force brought in aid of the natural force of man; and, to all the purposes of labour, is an increase of hands; an accession of strength, *unencumbered too by the expence of maintaining the laborer.* May it not therefore be fairly inferred, that those occupations, which give greatest scope to the use of this auxiliary, contribute most to the general Stock of industrious effort, and, in consequence, to the general product of industry?

It shall be taken for granted, and the truth of the position referred to observation, that manufacturing pursuits are susceptible in a greater degree of the application of machinery, than those of Agriculture. If so all the difference is lost to a community, which, instead of manufacturing for itself, procures the fabrics requisite to its supply from other Countries. The substitution of foreign for domestic manufactures is a transfer to foreign nations of the advantages accruing from the employment of Machinery, in the modes in which it is capable of being employed, with most utility and to the greatest extent.

The Cotton Mill invented in England, within the last twenty years, is a signal illustration of the general proposition, which has been just advanced. In consequence of it, all the different processes for spinning Cotton are performed by means of Machines, which are put in motion by water, and attended chiefly by women and Children; and by a smaller number of persons, in the whole, than are requisite in the ordinary mode of spinning. And it is an advantage of great moment that the operations of this mill continue with convenience, during the night, as well as through the day. The prodigious affect of such a Machine is easily conceived. To this invention is to be attributed essentially the immense progress, which has been so suddenly made in Great Britain in the various fabrics of Cotton.

III. As to the additional employment of classes of the community, not ordinarily engaged in the particular business.

This is not among the least valuable of the means, by which manufacturing institutions contribute to augment the general stock of industry and production. In places where those institutions prevail, besides the persons regularly engaged in them, they afford occasional and extra employment to industrious individuals and families, who are willing to devote the leisure resulting from the intermissions of their ordinary pursuits to collateral labours, as a resource of multiplying their acquisitions or their enjoyments. The husbandman himself experiences a new source of profit and support from the encreased industry of his wife and daughters; invited and stimulated by the demands of the neighboring manufactories.

Besides this advantage of occasional employment to classes having different occupations, there is another of a nature allied to it and of a similar tendency. This is—the employment of persons who would otherwise be idle (and in many cases a burthen on the community), either from the byass of temper, habit, infirmity of body, or some other cause, indisposing, or disqualifying them for the toils of the Country. It is worthy of particular remark, that, in general, women and Children are rendered more useful by manufacturing establishments, than they would otherwise be. Of the number of persons employed in the Cotton Manufactories of Great Britain, it is computed that 4/7 nearly are women and children; of whom the greatest proportion are children and many of them of a very tender age.

And thus it appears to be one of the attributes of manufactures, and one of no small consequence, to give occasion to the exertion of a greater quantity of Industry, even by the *same number* of persons, where they happen to prevail, than would exist, if there were no such establishments.

IV. As to the promoting of emigration from foreign Countries.

Men reluctantly quit one course of occupation and livelihood for another, unless invited to it by very apparent and proximate advantages. Many, who would go from one country to another, if they had a prospect of continuing with more benefit the callings, to which they have been educated, will often not be tempted to change their situation, by the hope of doing better, in some other way. Manufacturers, who listening to the powerful invitations of a better price for their fabrics, or their labour, of greater cheapness of provisions and raw materials, of an exemption from the chief part of the taxes burthens and restraints, which they endure in the old world, of greater personal independence and consequence, under the operation of a more equal government, and of what is far more precious than mere religious toleration—a perfect equality of religious privileges; would probably flock from Europe to the United States to pursue their own trades or professions, if they were once made sensible of the advantages they would enjoy, and were inspired with an assurance of encouragement and employment, will, with difficulty, be induced to transplant themselves, with a view to becoming Cultivators of Land.

If it be true then, that it is the interest of the United States to open every possible avenue to emigration from abroad, it affords a weighty argument for the encouragement of manufactures; which for the reasons just assigned, will have the strongest tendency to multiply the inducements to it. . . .

V. As to the furnishing greater scope for the diversity of talents and
dispositions, which discriminate men from each other.

This is a much more powerful means of augmenting the fund of national Industry than may at first sight appear. It is a just observation, that minds of the strongest and most active powers for their proper objects fall below mediocrity and labour without effect, if confined to uncongenial pursuits. And it is thence to be inferred, that the results of human exertion may be immensely increased by diversifying its objects. When all the different kinds of industry obtain in a community, each individual can find his proper element, and can call into activity the whole vigour of his nature. And the community is benefitted by the services of its respective members, in the manner, in which each can serve it with most effect.

If there be anything in a remark often to be met with—namely that there is, in the genius of the people of this country, a peculiar aptitude for mechanic improvements, it would operate as a forcible reason for giving opportunities to the exercise of that species of talent, by the propagation of manufactures.

VI. As to the affording a more ample and various field for enterprise.

This also is of greater consequence in the general scale of national exertion, than might perhaps on a superficial view be supposed, and has effects not altogether dissimilar from those of the circumstance last noticed. To cherish and stimulate the activity of the human mind, by multiplying the objects of enterprise, is not among the least considerable of the expedients, by which the wealth of a nation may be promoted. Even things in themselves not positively advantageous, sometimes become so, by their tendency to provoke exertion. Every new scene, which is opened to the busy nature of man to rouse and exert itself, is the addition of a new energy to the general stock of effort.

The spirit of enterprise, useful and prolific as it is, must necessarily be contracted or expanded in proportion to the simplicity or variety of the occupations and productions, which are to be found in a Society. It must be less in a nation of mere cultivators, than in a nation of cultivators and merchants; less in a nations of cultivators and merchants, than in a nation of cultivators, artificers and merchants.

VII. As to the creating, in some instances, a new, and securing in all
a more certain and steady demand, for the surplus produce of the soil.

This is among the most important of the circumstances which have been indicated. It is a principal mean, by which the establishment of manufactures contributes to an augmentation of the produce or revenue of a country, and has an immediate and direct relation to the prosperity of Agriculture.

It is evident, that the exertions of the husbandman will be steady or fluctuating, vigorous or feeble, in proportion to the steadiness or fluctuation, adequateness, or inadequateness of the markets on which he must depend, for the vent of the surplus, which may be produced by his labour; and that such surplus in the ordinary course of things will be greater or less in the same proportion.

For the purpose of this vent, a domestic market is greatly to be preferred to a foreign one; because it is in the nature of things, far more to be relied upon.

The National Gazette

Alexander Hamilton's growing power and the ambitious nature of his national project prompted the formation of a political opposition, which coalesced around Thomas Jefferson in the early 1790s and that evolved into the Democratic-Republican Party. Key to the new party's success was finding a way to present to a broad audience its basic beliefs as applied to matters of current policy. The National Gazette *came into being to meet this need.*

Jefferson and his allies were searching for a way to counter the influence of John Fenno's Gazette of the United States, *which generally mirrored the views of Federalist leaders. James Madison heard that an old Princeton friend, staunchly republican Philip Freneau, was preparing to found and edit a newspaper. Madison and Jefferson lured Freneau to Philadelphia with a sinecure in Jefferson's State Department and financial backing for a publishing venture that Freneau could lead. He became the founding editor of the* National Gazette, *the first issue of which came out at the end of October 1791. The* Gazette *lasted almost exactly two years.*

After a relatively nonpartisan start, early in 1792 the Gazette *began taking an increasingly critical and satirical stance toward Federalist policy after the publication of Hamilton's* Report on Manufactures. *A number of Democratic-Republican leaders published articles in the paper, including Madison, George Logan, and John Taylor. The* National Gazette *focused its indignation on the threat of corruption in the economic system that was taking shape in the country. These concerns were not always identical with agrarian interests. After all, many farmers, in New England and elsewhere, supported the Federalists. Still, the arguments presented in the* Gazette *frequently used agrarian images and themes to further the Democratic-Republican cause, as seen in the two selections that follow.*

From "An Old Prophecy" (1792)

About the year 1792, the people of the United States shall offer a curious phenomenon to the philosophic eye of the world—A whole nation, and that too a republic, in the morning of their glory, smitten with the love of gold!

And the philosophers and patriots of Europe shall hear thereof, and like the penitent nobles of Nineveh, shall cover themselves with sackcloth and ashes, and send messengers to enquire into the cause thereof.

And these messengers shall return for answer, that the United States possess many wise and virtuous citizens who love their country, and individually despise gold; among others, one that presides over their treasury department; but that this great minister seems not to be so skilful in the science of human nature as his genius and philanthropy deserve—hence all his schemes and plans have tended and tended only to meliorate the pockets, and not the heads and hearts of the people—that he has talked to them so much of imposts, and of funds, and of banks, and of manufactures, that they are considered as the cardinal virtues of the union.—Hence liberty, independence, philosophy, and genius have been struck out from the American vocabulary, and the hieroglyphic of money inserted in their stead, as a symbol of every thing worthy the estimation of man.

That gangrene thus begun at the heart of their constitution, though spreading too fast towards the extremities by the activity of its own virus, is yet about to be inoculated into the distant limbs, and this, as is pretended, will save the body from mortification and from death!

That poverty, though dignified by the purest virtue, is every where considered as the sum of all evil: hence all the ordinary walks of industry begin to be loathed, and are about to be abandoned for the *golden dreams of speculation*; and notwithstanding the wonted superiority of agriculture in the opinion of the wise and good, the phrase *"low voice of the unmonied farmer,"* is here considered as the emphatic description of misery and contempt.—That elevation, therefore, of soul, which, during their unequal conflict with Britain, raised them to a sublimity of character, which excited the envy and admiration of the world, has yielded to the contagion of avarice, and they are now sunk into all the meanness of stock jobbing.

They are also become liars, and hypocrites, and calumniators of the little virtue and merit which is left among them, for they say, that the Secretary of the Treasury and those who adopted his plans in Congress, intended thereby *"a monied aristocracy, and a government of influence;"* whereas their crime was, a too good opinion of the sense and virtue of their countrymen, whose excessive avarice, they did *not foresee*, was like the grave, which will never cry enough—Yet these diabolical ravens beset their state governors, and demand that they should follow the examples of the government, which, in the same breath, they have traduced. . . .

That they had sought in the usual haunts of virtue and of merit for those heroes and legislators, who had conducted them through the tempests of anarchy, treason, and war, and had given to their country a name among the nations; but were told that here they were not to be found; that the wealthy part thereof enjoyed the smiles of the community, and might be seen in the paths of excisemen, brokers, and stock jobbers; that the rest, frowned out of society on account of their poverty, and driven for lack of bread into the wilderness, had fallen in war by the hand of the savages, or had associated their humble dwelling with those of the tenants of the forest, as neighbours who, at least, would not insult their indigence.

That every discoverable symptom gave mournful intimation of a nation on the verge of infamy and ruin, by reason of prosperity, without an adequate share of wisdom and virtue!

A letter to the editor (1792)

To *the* Editor *of the* National Gazette.
Sir,

Being a real friend to manufactures, I am not disposed to discuss the power of Congress on that subject, or to throw obstacles in the way of their assuming it, if not delegated by the constitution. I am, however, so far from joining in opinion with those who are in favor of taxing agriculture by way of bounty to manufactures, that I think their learned arguments ought to be directly faced about and made to look towards the opposite conclusion, *to wit*, that manufactures ought to be taxed in order to raise premiums for improving and maturing agriculture; and this not so much with a view to the immediate advantage of agriculture, as to the solid and ultimate prosperity of manufactures. The following reasons, it is conceived, must overcome the most obstinate prejudices in this case.

First, all writers and all experience agree, that population [growth] is more rapid in the country among tillers of the earth, than in towns, the chief abode of manufacturers—Secondly, that an abundant population can alone support a flourishing state, either of agriculture or of manufactures—Thirdly, that hands for manufactures are to be drawn from that surplus of labourers which is found on the soil—Fourthly, to nurse and multiply labourers of the soil, is, therefore, the true and obvious means of providing the hands wanted for manufactures—Fifthly, cheapness of food and of materials; both of which are the fruit of the soil, the very life and soul of manufactures; and this cheapness will be promoted by filling up the vacant country with labourers, and stimulating by bounties their skill and industry—Sixthly, as population is the great reservoir from which manufacturers are to be derived, so immigration is one of the streams that may help to fill up the reservoir. Now, in every just view, it is better to invite foreigners from the country places, than from the towns of Europe, because they will bring with them equal, if not more, industry, and certainly less vice. And as to manufactures, they must gain as much from the introduction of tillers as of artizans; since the former by increasing the stock of hands on the soil, will enable it the faster to supply hands for manufactures; and, *by increasing the quantity and keeping down the price of provisions and materials,* will give immediate nourishment to that valuable branch of industry.

It will be asked, perhaps, where will be the justice, where the respect for the rights of property, where the equal protection to the free choice of our occupations and the free use of our faculties, thus to take money from the pocket of the manufacturers and give it to the farmer?

This question would be attended with its difficulties, were it not for two considerations: the one, that our opponents must first answer it themselves by shewing the justice of taking the money of the farmer, and giving it to the manufacturer; the other, that this is not a question of justice, but of mere policy, and being discussed in that light only, it is impertinent to view it in any other.

Should this answer not be satisfactory, and the justice, right, and equality of the measure still be insisted on as no less indispensable than its policy, I will endeavour to avoid the difficulty by another theory, derived not so much from my own principles or contrivance, as from the ingenuity of a friend who has been so kind as to communicate it to me.

This theory proposes, that instead of taxing agriculture in favour of manufactures, or manufactures in behalf of agriculture, all professions, trades, occupations, and employments whatsoever shall be reciprocally taxed and reciprocally bountied by a comprehensive provision of the government for that purpose. Thus let the manufacturer be taxed, and the tax paid as a bounty to the farmer; and the account be balanced by a tax levied on the farmer, and paid back as a bounty to the manufacturer. So again, a bounty may be taxed on the maker of hemp and given to the maker of cotton, and a like bounty in turn be levied on the maker of cotton and bestowed on the maker of hemp. Or, in a more circular way, by taxes skillfully adjusted, the maker of hemp, cotton or wheat, after receiving a bounty drawn from some other occupation, may pay a bounty to the grazer, he to the tanner, he to the shoemaker, he to the saddler, he to the coach maker, he to the ironmonger, he to the smith, he to the stocking weaver, he to the cotton manufacturer, &c. &c. quite round to the point from which you set out. In this manner every body will receive bounties. It will be a lottery where every ticket will draw a prize, and every adventurer consequently be pleased.

To the objection, that the prizes received by the whole society cannot exceed the sums paid by the whole, and that as great deductions must be made in the bargain for the expence of managing the scheme, every class of citizens instead of being gainers, must in fact be necessarily losers; to this objection, I say, my friend who is as dextrous in defending as he is ingenious in forming his thoughts, has a double reply ready. To those who contend for any other form of premiums and bounties, he repeats what has been already been said; that as they must either fall into evident partiality and injustice, or proceed in the same magical circle, they have no right to start the objection. To those, not under this embarrassment, he remarks, that in all lotteries and like schemes, where prizes and premiums are to be obtained, the fund for paying them must be taxed on the adventurers, and particularly, that the deductions and drawbacks not only make a part of the scheme, but constitute its very essence; the scheme being set on foot for the sake not of those who are to share the prizes and premiums, but of those who are to have the benefit of the drawbacks and profits. The former is quite a secondary, collateral, and incidental matter, well enough to amuse the calculations and hopes of the adventurers, but by no means the direct or primary object of the projectors. Considered in this point of view, every thing is smooth and square: for *note* (says he) the curious and charming effects of a universal system of bounties, supported by a system of taxes well digested and disguised for such an application.

First, having the appearance of equality, it silences, or, at least, softens the noisy declaimers against unequal principles of legislation. Secondly, it will diffuse good humour among all the superficial and inconsiderate part of the community, who receiving the bounty immediately and palpably, and paying the tax, they know not when and feel not how, forget that the bounty is taken out of their own pockets, and are ready to imagine it the product of some sleight-of-hand in the government beyond their comprehension, or not worth their enquiry. Thirdly, the deductions and drawbacks in such a case may fairly be estimated at 10 or 15 per cent. On the sum paid by the people, the whole of which becomes a fund for salaries and perquisites to collectors, receivers, treasurers, commissioners, managers, &c. &c. enabling the government to reward a greater number of its active friends, and encreasing its means of obtaining a willing obedience to all its measures, or of compelling obedience where its measures may produce an unwillingness. Fourthly, of no less value is the advantage that must accrue to the government from a *proper dispensation of the bounties.* No better opportunity can be conceived for rewarding political merit of all sorts, for extending the salutary influence of power in every direction; and for throwing a decent veil over the jobs and schemes in which the members of all well regulated governments have the immemorial privilege of sharing; but which in our new-fangled republic, and in these censorious times, it will be prudent to hide from the public eye. Fifthly, the last advantage to be noticed, out of the infinity that remains, is the solid ground it affords to the government for enlarging its whole system of taxation; for nothing can be more just and reasonable, or which is the same thing, can be more speciously so called, than that the government should reap where it has sown; that it should gather fruit from the tree it has planted; or, to speak plainly and without a metaphor, that it should excise every article as fast as its bounties have brought it to sufficient maturity for the operation.

Against this reasoning one objection only is foreseen by an ingenious theorist, *to wit,* that the whole of the advantages contemplated are to fall to the government; whereas, the interest of the people is the true object for which every public measure ought to be calculated. But this he treats with the greatest levity, as a pitiful quibble. He insists that the distinction is fanciful and inadmissible, and appeals with the utmost confidence to various numbers of *the Gazette of the United States,* where it has been demonstrated over and over again, that the government is the people and the people the government: that they are physically, morally, numerically, identically, and indivisibly one and the same; so that the more power the government assumes, the more freedom the people enjoy; and that every shilling which the members of the government put into their own pockets, is a shilling put into the pockets of their constituents.
—A Citizen
Philadelphia, Sept. 24

JOHN TAYLOR OF CAROLINE

John Taylor of Caroline County, Virginia, was born in 1753 into a family with strong ties to the Virginia elite. From his early years, when he was raised in large part by his uncle Edmund Pendleton and studied at an academy alongside James Madison, Taylor was thoroughly familiar with those who would make up Virginia's political leadership during and after the Revolution.

Portrait of John Taylor of Caroline County, Virginia, c. 1850.
Reproduced by permission of The Library of Virginia,
Richmond.

Taylor's own political career included positions in the Virginia Assembly and in the
U.S. Senate, but he devoted the greater share of his energy to writing on political matters.
His books include the daunting An Inquiry into the Principles and Policy of the Gov-
ernment of the United States, Tyranny Unmasked, *and* Construction Construed and
Constitutions Vindicated. *He wrote about the fundamental questions of the founding*
era, paying special attention to the intersection of economic interests and political power.

Taylor earned a reputation as one of the strictest republicans of his day. He was con-
cerned by the prospect of a stronger central government during the debate over ratification
and even more troubled by the loose construction of the Constitution that prevailed in
the years after the new government formed. Through misuse of these new powers under
Hamilton's program, Taylor believed, the government was encouraging the rise of a "paper
aristocracy" of bankers, capitalists, and speculators whose interests were at odds with those
of ordinary citizens and agriculturalists. "Land, being incapable of an artificial multipli-
cation, cannot by increasing its quantity, strengthen its influence—with paper the case is
different. Land cannot in interest be at enmity with the public good—paper is often so."

In his attacks on the paper economy and in his defense of agriculture, Taylor was
not finding fault with the new economy for its materialism. Rather, he was attacking

its abstraction and its corrupting influence, defending traditional virtues rooted in the very material realities of the farm. By encouraging physical health, rewarding curiosity about nature, and requiring steady habits, agriculture is "the best architect of the complete man," as Taylor put it. In his conservative humanism, Taylor stands at the head of a Southern agrarian tradition. Yet Taylor was also a slaveholder and a stout defender of the institution, holding the frankly racist belief that slavery, so long as it was conducted with proper humane treatment, improved the moral condition and happiness of both slave and master. For Taylor and many others of his class, liberty for the planter and servitude for the bondsman was not a contradiction but a double blessing. When Taylor speaks of "enslaved countries," he is not referring to his own.

Taylor's Arator essays first appeared in book form in 1813. At the time, agricultural interests in Virginia were under a heavy set of pressures, ranging from a weak tobacco market to soil erosion, and Arator was aimed at reversing the fortunes of the region's farming classes. Much of Arator is devoted to giving advice on farming practices, but it is also a political call to arms for the nation's farmers and planters.

From *Arator* (1813)

AUTHOR'S PREFACE

The essays above the signature *Arator,* were not preceded by an explanation of the motives by which they were dictated, or of the ends designed to be effected, because they were originally published in a newspaper without any anticipation of the form they have assumed. But being now stampt by the public acceptation with some degree of value, an account of these motives and ends may extend their efficiency, and promote the examination of subjects, so important and inexhaustible.

A conviction that the prosperity of our country depended upon a competent share of agricultural and political knowledge, and that an ignorance of either, would defeat the benefits naturally flowing from a proficiency in both, produced these essays, and also a larger book, entitled *An Inquiry into the Principles and Policy of the Government of the United States* for the ends of advancing practical improvements in one science, and of preserving those we had already made in the other.

Agriculture and politics are primary causes of our wealth and liberty. Both contain internal good principles, but both are liable to practical deterioration. If one is vitiated in practice, poverty, if the other, oppression ensues. If the agriculture is good and the government bad, we may have wealth and slavery. If the government is good and the agriculture bad, liberty and poverty. To secure both wealth and liberty, an intimate knowledge of the good principles comprised in both, and a strict accordance in practice with those principles, must be indispensably necessary.

Hence results the propriety of awakening the people to the good principles of agriculture, and of keeping them awake to those of our form of government. Without commemoration, the latter might be lost, and without enquiry, the former could never be found. Both subjects are vitally important to the success of the singular experiment now making by the United States upon the theatre of the world, and contain the only means of redeeming their pledge to mankind.

Both are treated of in the two books abovementioned. *Arator* is chiefly confined to agriculture, but it contains a few political observations. The *Inquiry*, to politics; but it labours to explain the true interest of the agricultural class. The affinity between the subjects, caused them to be intermingled. But the author never imagined himself able to tear the bandage of habit from the eyes of prejudice, nor to squeeze the tincture of corruption from the heart of avarice. Without aspiring to moral heroism, or to the renown of new inventions, he only attempts to extract good agricultural practices from his own experience, and good political measures from the wise and faithful archives of revolutionary patriotism, to increase the skill of his countrymen, in making good crops and in securing them for their own benefit.

That private industry combined with political fraud, may make a nation wealthy and miserable, is exemplified in England. That one interest or class of men, may reap oppression from carrying their occupation to great perfection, appears in the state of the manufacturers of the same country. And a similar fate awaits the agricultural class of this, although it could be driven by English coercion, even up to English perfection, unless it retains the American political principles alone competent to defeat the arts, under which the productive classes of mankind are universally groaning. . . .

The author had another reason for uniting the subjects of agriculture and politics. He considered agriculture as the guardian of liberty, as well as the mother of wealth. So long as the principles of our government are uncorrupted, and the sovereignty of majorities remains, she must occupy the highest political station, and owe to society the most sacred political duty. It is as incumbent upon her to learn how to protect defenceless minor interests, as to defend herself. And whilst the security for her patriotism "that she cannot find any body worth plundering" exists, she should take care not to betray her wards, by suffering herself to be made poor, either by a deficiency of skill and industry, or by legal spoliations; because wealth is power. She must be rich to be powerful, and she must be powerful to discharge faithfully the sacred obligation she owes to society, by constituting the majority. If her wealth is transferred, her power will go with it, and an irresistible political influence will be acquired by a minority, which can find some body worth plundering.

The Pleasures of Agriculture

In free countries, are more, and in enslaved, fewer, than the pleasures of most other employments. The reason of it is, that agriculture both from its nature, and also as being generally the employment of a great portion of a nation, cannot be united with power, considered as an exclusive interest. It must of course be enslaved, wherever despotism exists, and its masters will enjoy more pleasures in that case, than it can ever reach. On the contrary, where power is not an exclusive, but a general interest, agriculture can employ its own energies for the attainment of its own happiness.

Under a free government it has before it the inexhaustible sources of human pleasure, of fitting ideas to substances, and substances to ideas; and of a constant rotation of hope and fruition.

The novelty, frequency and exactness of accommodations between our ideas and operations, constitutes the most exquisite source of mental pleasure. Agriculture feeds it with endless supplies in the natures of soils, plants, climates, manures, instruments of culture and domestic animals. Their combinations are inexhaustible, the novelty of results is endless, discrimination and adaption are never idle, and an unsatiated interest receives gratification in quick succession.

Benevolence is so closely associated with this interest, that its exertion in numberless instances, is necessary to foster it. Liberality in supplying its labourers with the comforts of life, is the best sponsor for the prosperity of agriculture, and the practice of almost every moral virtue is amply remunerated in this world, whilst it is also the best surety for attaining the blessings of the next. Poetry, in allowing more virtue to agriculture, than to any other profession, has abandoned her privilege of fiction, and yielded to the natural moral effect of the absence of temptation. The same fact is commemorated by religion, upon an occasion the most solemn, within the scope of the human imagination. At the awful day of judgment, the discrimination of the good from the wicked, is not made by the criterion of sects or of dogmas, but by one which constitutes the daily employment and the great end of agriculture. The judge upon this occasion has by anticipation pronounced, that to feed the hungry, clothe the naked, and give drink to the thirsty, are the passports to future happiness; and the divine intelligence which selected an agricultural state as a paradise for its first favourites, has here again prescribed the agricultural virtues as the means for the admission of their posterity into heaven.

With the pleasures of religion, agriculture unites those of patriotism, and among the worthy competitors for pre-eminence in the practice of this cardinal virtue, a profound author assigns a high station to him who has made two blades of grass grow instead of one; an idea capable of a signal amplification, by a comparison between a system of agriculture which doubles the fertility of a country, and a successful war which doubles its territory. By the first the territory itself is also substantially doubled, without wasting the lives, the wealth, or the liberty of the nation which has thus subdued sterility, and drawn prosperity from a willing source. By the second, the blood pretended to be enriched, is spilt; the wealth pretended to be increased, is wasted; the liberty said to be secured, is immolated to the patriotism of a victorious army; and desolation in every form is made to stalk in the glittering garb of false glory, throughout some neighbouring country. Moral law decides the preference with undeviating consistency, in assigning to the nation, which elects true patriotism, the recompense of truth, and to the electors of the false, the expiation of error. To the respective agents, the same law assigns the remorses of a conquerour, and the quiet conscience of the agriculturalist.

The capacity of agriculture for affording luxuries to the body, is not less conspicuous than its capacity for affording luxuries to the mind; it being a science singularly possessing the double qualities of feeding with unbounded liberality, both the moral appetites of the one, and the physical wants of the other. It can even feed a morbid love of money, whilst it is habituating us to the practice of virtue; and whilst it provides for the wants of the philosopher, it affords him ample room for the most curious and yet useful researches. In short, by the exercise it gives both to the body and to the mind, it secures health and vigour to both; and by combining a thorough knowledge of the real affairs of life, with a necessity for investigating the arcana of nature, and the strongest invitations to the practice of morality, it becomes the best architect of a complete man.

If this eulogy should succeed in awakening the attention of men of science to a skilful practice of agriculture, they will become models for individuals, and guardians for national happiness. The discoveries of the learned will be practiced by the ignorant; and a system which sheds happiness, plenty and virtue all around, will be gradually substituted for one, which fosters vice, breeds want, and begets misery.

Politicians (who ought to know the most, and generally know the least, of a science in which the United States are more deeply interested than in any other) will appear, of more practical knowledge, or at least of better theoretical instruction; and the hopeless habit of confiding our greatest interest to people most ignorant of it, will be abandoned.

The errors of politicians ignorant of agriculture, or their projects designed to oppress it, can only rob it of its pleasures, and consign it to contempt and misery. This revolution of its natural state, is invariably effected by war, armies, heavy taxes, or exclusive privileges. In two cases alone, have nations ever gained any thing by war. Those of repelling invasion and emigrating into a more fruitful territory. In every other case, the industrious of all professions suffer by war, the effects of which in its modern form, are precisely the same to the victorious and the vanquished nation. The least evil to be apprehended from victorious armies, is a permanent system of heavy taxation, than which, nothing can more vitally wound or kill the pleasures of agriculture. Of the same stamp, are exclusive privileges in every form; and to pillage or steal under the sanction of the statute books, is no less fatal to the happiness of agriculture, than the hierarchical tyranny over the soul, under the pretended sanction of God, or the feudal tyranny over the body, under the equally fraudulent pretence of defending the nation. In a climate and soil, where good culture never fails to beget plenty, where bad cannot produce famine, begirt by nature against the risque of invasion, and favoured by accident with the power of self government, agriculture can only lose its happiness by the folly or fraud of statesmen, or by its own ignorance.

JAMES MADISON

James Madison was born in 1751 near Port Royal, Virginia, soon moving to Montpelier in Orange County, where he would make his home the rest of his life. He studied at the College of New Jersey (now Princeton), graduating in 1771. Returning to Virginia, Madison took his place as one of Orange County's leading supporters of the Revolution, rising to the rank of colonel in the county militia.

During the years of the Revolution, Madison took part in Virginia's nascent state government. He was elected to the Virginia House of Delegates in 1776, where he met Thomas Jefferson. He was appointed to the council of state and was elected to represent Virginia at the Continental Congress beginning in 1780. But Madison hardly had the temperament of a revolutionary, as he showed in his work shaping the Constitution. There Madison articulated the way competing interests would check one another, producing policy that would resist radical or revolutionary measures.

While he thought it healthy to have a broad array of interests represented in the government, Madison did worry about what he called the "distribution of citizens" within the Republic. When Hamilton's Report on Manufactures *appeared, he warned in a* National Gazette *essay against "experiments by power" that might foster the growth of manufacturing to the detriment of society as a whole. In that essay he echoed both Jefferson and broader agrarian sentiment: "The class of citizens who provide at once their own food and their own raiment, may be viewed as the most truly independent and happy. They are more; they are the best basis of public liberty and the strongest bulwark of public safety. It follows, that the greater the proportion of this class to the whole society, the more free, the more independent, and the more happy must be the society itself."*

Madison retained his interest in the moral aspects of the American economy in the years after his presidency. The following selection is taken from his address to the Albemarle, Virginia, Agricultural Society, one of many such societies that sprang up in the early nineteenth century to advance reforms in agricultural practices. Madison lays out his understanding of agriculture's essential role in civilization and the threat of losing an agricultural basis for society. On the one hand, Madison cautioned that the backwoods might tempt citizens to the freer—but less fruitful and civilized—life of the herder or hunter. On the other, he feared the vitiation of the crowded, unnatural urbanism of Europe. American civilization, he believed, would thrive best in the middle landscape of an improved and refined agriculture.

From "An Address Delivered before the Albemarle, Va., Agricultural Society" (1818)

It having pleased the Society to name me for their presiding member I feel it a duty, on my first appearing among you, to repeat my acknowledgements for that honorary distinction; with the assurances of my sincere desire to promote the success of an establishment, which has in view so valuable an object, as that of improving the agriculture of our Country.

The faculty of cultivating the earth, and of rearing animals by which food is increased beyond the spontaneous supplies of nature, belongs to man alone. No other terrestrial Being has recd. a higher gift than an instinct like that of the Beaver or the ant, which merely hoards, for future use, the food spontaneously furnished by nature. As this peculiar faculty gives to man a pre-eminence over irrational animals, so it is the use made of it by some, and the neglect of it by other communities, that distinguishes them from each other in the most important features of the human character.

The contrast between the enlightened & refined nations on some parts of the earth, and the rude & wretched tribes on others, has its foundation in this distinction. Civilisation is never seen without agriculture. Nor has agriculture ever prevailed, where the civilized arts did not make their appearance.

But closely as agriculture and Civilization are allied, they do not keep pace with one another. There is probably a much higher state of agriculture in China than in many other Countries far more advanced in the improvements of Civilized life. It is surely no small reproach to the latter, that with so great a superiority in science, and in the fuller possession of all auxiliary arts, they should suffer themselves to be outstripped in the very art by which both are essentially distinguished from the Brute Creation.

It must not be inferred however from the capacities and the motives of man for an artificial increase of the productions of the earth, that the transition from the Hunter or even the Herdsman state to the Agricultural, is a matter of course. The first steps in this transition are attended with difficulty; and what is more with disinclination.

Without a knowledge of the metals, and the implements made of them, the process of opening & stirring the soil is not an easy operation; though one perhaps not requiring more effort & contrivance than produced the instruments used by savages in war & in the chase.

And that there is a disinclination in human nature, to exchange the savage for the Civilized State, can not be questioned. We need not look for proofs beyond our own neighbourhood. The Indian Tribes have ever shown an aversion to the change. Neither the persuasive examples of plenty & comfort derived from the culture of the earth by their white brethren, nor the lessons & specimens of tillage placed in the midst of them and seconded by actual sufferings from a deficient & precarious subsistence, have converted them from their strong propensities & habitual pursuits. In the same spirit they always betray an anxious disposition to return to their pristine life, after being weaned from it by time, and apparently moulded by intellectual and moral instruction, into the habits and tastes of an agricultural people. A still more conclusive evidence of the biass of human nature, is seen in the familiar fact that our own people nursed & reared in those habits & tastes, easily slide into those of the savages & are rarely reclaimed to Civilized Society with their own consent.

Had the Europeans, on their arrival, found this Continent destituted of human inhabitants whose dangerous neighbourhood kept them in a compact and agricultural State, and more especially if their communication with the countries they left had been discontinued, they might have spread themselves into the forests where game & fruits wd. have abounded; and gradually forgetting the arts, no longer necessary to their immediate wants, have degenerated into Savage Tribes. . . .

The bent of human nature may be traced on the chart of our own Country. The manufacturer readily exchanges the loom for the plow, in oposition often to his own interest as that of the public. The Cultivator, in situations presenting an option, prefers to the labors of the field, the more easy employment of rearing a herd; & as the game of the forest is approached, the resort to the hunting life shows its powerful attractions. Where do we behold a retrograde march; the hunter becoming the herdsman, the latter a follower of the plow, & the last repairing to the manufactory or the workshop?

Such indeed are the charms of that personal independence which belongs to the uncivilized State, and such the disrelish and contempt of the monotonous labor of tillage, compared with the exciting occupations of the chase, or with the indolence enjoyed by those who subsist on the mere bounties of nature or on their migratory flocks, that a voluntary relinquishment of these latter modes of life is little to be expected. We certainly perceive nothing in the character of our Savage neighbors from which it could be inferred that even the germs of agriculture presented in their spots of maiz, and of a few other cultivated plants, wd. ever be developed into the extent implied by an agricultural life. To that little resource combined with the game furnished by the forest & by the lake or the stream, their population and their habits are adjusted. There may be said in fact to be a plenum of the former; because it is fully commensurate to their food; and this cannot be increased witht. a change of habits, which being founded in natural propensities do not change of themselves.

The first introduction of Agriculture among a Savage people appears accordingly never to have taken place without some extraordinary interposition. Where it has not been obtruded by Colonies transplanted from Agricultural Countries, as from Phoenicia and Egypt into Greece, and from Greece herself among her savage neighbors, the revolution has proceeded from some individual, whose singular endowments and supernatural pretensions had given him a sufficient ascendancy for the purpose. All these Great Reformers, in ancient times, were regarded as more than men, and ultimately worshiped as Gods. A very remarkable example of Modern date is found in the Revolution from the Savage to the Agricultural State said to have been brought about by Manco Capac among the Peruvians, to whom he represented himself as the offspring of the Sun.

Agriculture once effectually commenced, may proceed of itself under impulses of its own creation. The mouths fed by it increasing, and the supplies of nature decreasing, necessity becomes a spur to industry; which finds another spur in the advantages incident to the acquisition of property in the civilized State. And thus, a progressive agriculture, and progressive population naturally ensue.

But altho no determinate limit presents itself, to the increase of agricultural food, & to a population commensurate with it, other than the limited productiveness of the Earth itself, we can scarcely be warranted in supposing that all productive powers of its surface can be made subservient to the use of man, in exclusion of all the plants and animals not entering into his stock of sustenance; that all the elements & combinations of elements, in the earth, the atmosphere, and the water, which now support such various & such numerous descriptions of created beings animate and inanimate could be withdrawn from that general destination and appropriated to the exclusive support and increase of the human part of the creation; so that the whole habitable earth shd. be as full of people as the spots most crowded now are, or might be made; & as destitute as those spots are of the plants & animals not used by man.

The supposition can not well be reconciled with that symmetry in the face of nature, which derives new beauty from every insight that can be gained into it. It is forbidden also by the principles and laws which operate in various departments of her economy falling within the scope of common observation, as well as within that of philosophical researches. . . .

Animals including man & plants may be regarded as the most important parts of the terrestrial Creation. They are pre-eminent in their attributes; and all nature teems with their varieties and their multitudes, visible & invisible. To all of them the Atmosphere is the breath of life. When deprived of it they all equally perish. But it answers their purpose by virtue of its appropriate constitution and character. What are These?

The Atmosphere is not a simple but a compound body. In its least compound State, it is understood to contain, besides what is called vital air, others noxious in themselves, yet without a portion of which the vital air becomes noxious. But the Atmosphere in its natural State and in its ordinary communication with the organized world, comprizes various ingredients, or modifications of ingredients, derived from the use made of it by the existing variety of animals & plants. The expirations & perspirations; the putrifactions & dissolutions; the effluvia & transpirations of these are continually charging the atmosphere with a heterogeneous variety & immense quantity of matter, which together must contribute to the character which fits it for its destined purpose of supporting the life and health of organized beings. Is it unreasonable to suppose, that if instead of the actual composition & character of the animal & vegetable creation, to which the atmosphere is now accommodated, such a composition & character of that creation were substituted as a result from a reduction of the whole to man and a few kinds of animals & plants; is the supposition unreasonable that the change might essentially affect the aptitude of the Atmosphere for the functions required of it, and that so great an innovation might be found in this respect, not to accord with the order & economy of nature?

The relation of the animal part and the vegetable part of the creation to each other thru the medium of the atmosphere comes in aid of the reflection suggested by the general relation between the Atmosphere & both. It seems to be now well understood, that the atmosphere when respired by animals, becomes unfitted for their further use; and fitted for the absorption of vegetables; and that when evolved by the latter, it is refitted for the respiration of the former; an interchange being thus kept up by which this breath of life is recd. by each in a wholesome state, in return for it in an unwholesome one.

May it not be inferred from this happy arrangement and beautiful feature in the economy of nature, that if the whole class of animals were extinguished, the use of the Atmosphere by the vegetable class alone would exhaust it of its life supporting property; in like manner that if the whole class of vegetables were extinguished, the use of it by the animal class alone would soon deprive it of its fitness for their support. And if such would be the effect of an entire destruction of either class in relation to the other, the inference seems to press itself upon us, that so vast a change in the proportions of each class to the other, and in the species composing the respective classes, as that in question, might not be compatible with the continued existence and health of the remaining species of the 2 classes.

Other views of the economy of nature coincide with the preceding. There is a known tendency in all organized beings to multiply beyond the degree necessary to keep up their actual numbers. It is a wise provision of nature, 1. To guard agst. the failure of the species, 2. To afford in the surplus, food for animals whether subsisting on vegetables, or on other animals which subsist on vegetables. Nature has been equally provident in guarding agst. an excessive multiplication of any one species, which might too far encroach on others, by subjecting each when unduly multiplying itself to be arrested in its progress, by the effect of the multiplication itself, 1. in producing a deficiency of food, and where that does not happen, 2. in producing a state of the atmosphere unfavorable to life & health. All animals as well as plants sicken & die when too much crowded. So our domestic animals of every sort, when in that situation or where a scarcity of food cannot be the cause. (In the case of certain insects which multiply with peculiar rapidity, and visit particular regions in such dense swarms, the destruction is almost instantaneous.)

To the same laws mankind are equally subject. An increase not consistent with the general plan of nature arrests itself. According to the degree in which the number thrown together exceeds the due proportion of Space and of atmosphere, disease and mortality ensue. It was the viciated air alone which put out human life in the crowded Hole at Calcutta. In a space somewhat enlarged, the effect would have been slower but not less certain. In all confined situations from the Dungeon to the crowded Workhouses; and from these to the compact population of overgrown Citizens, the atmospheres become in corresponding degrees, unfitted by reiterated use for sustaining human life & health. Were the atmosphere breathed in Cities not diluted by fresh supplies from the surrounding Country, the mortality would soon become general. Were the surrounding Country thickly peopled and not refreshed in like manner, the decay of health, tho' a later, wd. be a necessary consequence. And were the whole habitable earth covered with a dense population and without any other resource of a purer air than in the exhaustible one covering the aqueous parts of the globe, wasteful maladies might be looked for that would thin the nos. into a healthy proportion.

WILLIAM COBBETT

William Cobbett (1763–1835) was born in Farnham, a Surrey County village about thirty-five miles southeast of London. He came from a family of farm laborers and small businessmen—his father kept a public house. Though Cobbett left Farnham by the age of twenty, his years there left him with a lifelong love of the English countryside and rural life.

Cobbett emigrated to the United States in 1792, living first in Wilmington, Delaware, before moving to Philadelphia. He worked as a teacher for a time but found his calling during the agitation set off by Joseph Priestley's arrival in New York. Priestley, a scientist and political thinker, was a prominent critic of British society and supporter of the French Revolution. When he came to America in 1794, his admirers celebrated with public welcome ceremonies, denunciations of Great Britain, and tributes to the French Revolution. This outpouring touched a nerve with Cobbett, who responded with a biting attack on Priestly and a defense of his homeland.

Before long Cobbett, using the penname Peter Porcupine, was writing widely read polemics aimed at Jefferson, Madison, Paine, and other republican leaders. In 1796 he opened a bookshop where portraits of George III and other British leaders were displayed as a goad to Philadelphia's Democratic-Republicans. After a libelous attack on the medi-

cal practices of republican Benjamin Rush, Cobbett fled to England in 1800 rather than pay a $5,000 fine.

Back home, Cobbett's political sympathies evolved as he watched the accelerating industrialization and urbanization around him. He came to denounce not only the British government but the landed aristocracy, which he had once seen as a pillar of England's virtues. Increasingly he championed the cause of all workers, though his deepest sympathies remained with rural laborers. Cobbett's political sensibilities changed to the point that in 1819 he had the bones of Thomas Paine disinterred and brought to England, where he planned to honor his onetime foe with a mausoleum (though the bones were mislaid and no memorial built).

Cobbett had returned to America in 1817 under pressure stemming from his political writing, which the British government believed was fueling the Luddite movement. For two years he lived and farmed on Long Island, during which time he wrote his Journal of a Year's Residence in the United States of America, *from which the selection that follows is taken. It amounts to a report to English readers on conditions and customs among America's farmers and farmhands. Given Cobbett's political aims, his praise of farming in America should also be read as an indirect attack on British domestic policies.*

From *Journal of a Year's Residence in America* (1819)

Chapter IX. Prices of Land, Labour, Food and Raiment.

Land is of various prices, of course. But, as I am, in this Chapter, addressing myself to *English Farmers,* I am not speaking of the price either of land in the *wildernesses,* or of land in the immediate vicinage of great cities. The wilderness price is two or three dollars an acre; the city price four or five hundred. The land at the same distance from New York that Chelsea is from London, is of higher price than the land at Chelsea. The surprizing growth of these cities, and the brilliant prospect before them, give value to every thing that is situated in or near them.

It is my intention, however, to speak only of *farming land.* This, too, is, of course, affected in its value by the circumstance of distance from market; but, the reader will make his own calculations in this matter. A farm, then, on this Island, any where not nearer than thirty miles of, and not more distant than sixty miles from, New York, with a good farm-house, barn, stables, sheds, and styes; the land fenced into fields with posts and rails, the wood-land being in the proportion of one to ten of the arable land, and there being on the farm a pretty good orchard; such a farm, if the land be in a good state, and of an average quality, is worth *sixty dollars an acre, or thirteen pounds sterling;* of course, a farm of a hundred acres would cost one thousand three hundred pounds. The rich lands on the *necks* and *bays,* where

A Design to Represent the Beginning and Completion of an American Settlement or Farm.
"Painted by Paul Sandby, from a Design made by his Excellency Governor Pownal."
Engraving by James Peake. From *Scenographia Americana; or, A Collection of Views in
North America and in the West Indies* (1793). Courtesy Library of Congress, Washington, DC.

there are *meadows* and surprizingly productive orchards, and where there is *water
carriage,* are worth, in some cases, three times this price. But, what I have said will
be sufficient to enable the reader to form a pretty correct judgment on the subject.
In New Jersey, in Pennsylvania, every where the price differs with the circumstances
of water carriage, quality of land, and distance from market.

When I say a good farm-house, I mean a house *a great deal better* than the *general
run* of farm-houses in England. More neatly finished on the inside. More in a *parlour*
sort of style; though *round about* the house, things do not look so neat and tight as in
England. Even in Pennsylvania, and amongst the Quakers too, there is a sort of out-
of-doors slovenliness, which is never hardly seen in England. You see bits of wood,
timber, boards, chips, lying about, here and there, and pigs and cattle trampling
about in a sort of confusion, which would make an English farmer fret himself to
death; but which is here seen with great placidness. The out-buildings, except the
barns, and except in the finest counties of Pennsylvania, are not so numerous, or

so capacious, as in England, in proportion to the size of the farms. The reason is, that the *weather is so dry.* Cattle need not covering a twentieth part so much as in England, except hogs, who must be *warm* as well as dry. However, these share with the rest, and very little covering they get.

Labour is the great article of expence upon a farm; yet it is not nearly so great as in England, in proportion to the amount of the produce of a farm, especially if the poor-rates be, in both cases, included. However, speaking of the positive wages, a *good* farm-labourer has *twenty-five pounds sterling a year* and his board and lodging; and a *good* day-labourer has, upon an average, *a dollar a day.* A woman servant, in a farm-house, has from forty to fifty dollars a year, or eleven pounds sterling. These are the average of the wages throughout the country. But, then, mind, the farmer has nothing (for, really, it is not worth mentioning) to pay in *poor-rates;* which in England, must always be added to the wages that a farmer pays; and, sometimes, they far exceed the wages.

It is, too, of importance to know, *what sort* of labourers these American are; for, though a labourer is a labourer, still there is some difference in them; and, these Americans are *the best that ever I saw.* They mow *four acres of oats, wheat, rye,* or *barley* in a day, and, with a cradle, lay it so smooth in the swarths, that it is tied up in sheaves with the greatest neatness and ease. They mow *two acres and a half of grass* in a day, and they do the work well. And the crops, upon an average, are all, except the wheat, *as heavy* as in England. The English farmer will want nothing more than these facts to convince him, that the labour, after all, is not so *very dear.*

The causes of these performances, so far beyond those in England, is first, the men are *tall* and well built; they are *bony* rather than *fleshy;* and they *live,* as to food, as well as man can live. And, secondly, they have been *educated* to do much in a day. The farmer here generally is at the *head* of his "*boys,*" as they, in the kind language of the country, are called. Here is the best of examples. My old and beloved friend, Mr. James Paul, used, at the age of nearly *sixty* to go at *the head of his mowers,* though his fine farm was his own, and though he might, in other respects, be called a rich man; and, I have heard, that Mr. Elias Hicks, the famous Quaker Preacher, who lives about nine miles from this spot, has this year, at *seventy* years of age, cradled down four acres of rye in a day. I wish some of the *preachers* of other descriptions, especially our fat parsons in England, would think a little of this, and would betake themselves to "work with their hands the things which be good, that they may have to give to him who needeth," and not go on any longer gormandizing and swilling upon the labour of those who need.

Besides the great quantity of work performed by the American labourer, his *skill,* the *versatility* of his talent, is a great thing. Every man can use an *ax,* a *saw,* and a *hammer.* Scarcely one who cannot do any job at rough carpentering, and mend a plough or a waggon. Very few indeed, who cannot kill and dress pigs and sheep, and many of them Oxen and Calves. Every farmer is a *neat* butcher, a butcher for *market;* and, of course, "the boys" must learn. This is a great convenience. It makes you so independent as to a main part of the means of housekeeping. All are *ploughmen.* In short, a good labourer here, can do *any thing* that is to be done upon a farm.

The operations necessary in miniature cultivation they are very awkward at. The *gardens are ploughed* in general. An American labourer uses a *spade* in a very awkward manner. They *poke the earth about* as if they had no eyes; and toil and muck themselves half to death to dig as much ground in a day as a Surrey man would dig in about an hour of hard work. *Banking, hedging,* they know nothing about. They have no idea of the use of a *bill-hook,* which is so adroitly used in the coppices of Hampshire and Sussex. An *ax* is their tool, and with that tool, at *cutting down* trees or *cutting them up,* they will do *ten times* as much in a day as any other men that I ever saw. Set one of these men on upon a wood of timber trees, and his slaughter will astonish you. A neighbour of mine tells a story of an Irishman, who promised he could *do any thing,* and whom, therefore, to begin with, the employer sent into the wood to cut down a load of wood to burn. He staid a long while away with the team, and the farmer went to him fearing some accident had happened. "What are you about all this time?" said the farmer. The man was hacking away at a hickory tree, but had not got it half down; and that was all he had done. An American, black or white, would have had half a dozen trees cut down, cut up into lengths, put upon the carriage, and brought home in the time.

So that our men, who come from England, must not expect, that, in these *common labours* of the country, they are to surpass, or even equal these *"Yankees,"* who, of all men that I ever saw, are the most *active* and the most *hardy.* They skip over a fence like a greyhound. They will catch you a pig in an open field by *racing* him down; and they are afraid of nothing. This was the sort of stuff that filled the *frigates* of DECATUR, HULL, and BRAINBRIDGE [Bainbridge]. No wonder that they triumphed when opposed to poor pressed creatures, worn out by length of service and ill-usage, and encouraged by no hope of fair-play. . . .

This is the stuff that stands between the rascals, called the Holy Alliance, and the slavery of the whole civilized world. This is the stuff that gives us Englishmen an asylum; that gives us time to breathe; that enables us to deal our tyrants blows, which, without the existence of this stuff, they never would receive. This America, this scene of happiness under a free government, is the beam in the eye, the thorn in the side, the worm in the vitals, of every despot upon the face of the earth.

An American labourer is not regulated, as to time, by *clocks* and *watches.* The *sun,* who seldom hides his face, tells him when to begin in the morning and when to leave off at night. He has a dollar, a *whole dollar* for his work; but then it is the work of a *whole day.* Here is no dispute about *hours.* Hours "were made for *slaves,*" is an old saying; and, really, they seem here to act upon it as a practical maxim. This is a *great thing* in agricultural affairs. It prevents so many disputes. It removes so great a cause of disagreement. The American labourers, like the tavern-keepers, are never *servile,* but always *civil.* Neither *boobishness* nor *meanness* mark their character. They never *creep* and *fawn,* and they are never *rude.* Employed about your house as day-labourers, they never come to interlope for victuals or drink. They have no idea of such a thing; Their pride would restrain them if their plenty did not; and, thus would it be with all labourers, in all countries, were they left to enjoy the fair produce of their labour. Full pocket or empty pocket, these American labourers are always the *same men;* no saucy cunning in the one case, and no base crawling in the other. This, too, arises from the free institutions or government. A man has a voice *because he is a man,* and not because he is the *possessor of money.* And, shall I *never* see our English labourers in this happy state?

Let those English farmers, who love to see a poor wretched labourer stand trembling before them with his hat off, and who think no more of him than of a dog, remain where they are; or, go off, on the cavalry horses, to the devil at once, if they wish to avoid the tax-gatherer; for, they would, here, meet with so many mortifications, that they would, to a certainty, hang themselves in a month.

There are some, and even many, farmers, who *do not work themselves in the fields.* But, they all *attend* to the thing, and are all equally civil to their working people. They manage affairs very judiciously. Little talking. Orders plainly given in few words, and in a decided tone. This is their only secret.

The *cattle* and *implements* used in husbandry are cheaper than in England; that is to say, *lower priced.* The wear and tear not nearly half so much as upon a farm in England of the same size. The climate, the soil, the gentleness and docility of horses and oxen, the lightness of the waggons and carts, the lightness and toughness of the *wood* of which husbandry implements are made, the simplicity of the harness, and, above all, the ingenuity and handiness of the workmen in *repairing,* and in *making shift;* all these make the implements a matter of very little note. Where horses are kept, the *shoing* of them is the most serious kind of expence.

The first business of a farmer is, here, and ought to be every where, to *live well;* to live in ease and plenty; to *"keep hospitality,"* as the old English saying was. To *save money* is a good secondary consideration; but, any English farmer, who is a good farmer there, may, if he will bring his industry and care with him, and be *sure* to leave his pride and insolence (if he have any) along with his anxiety, behind him, live in ease and plenty here, and keep hospitality, and save a great parcel of money, too. If he have the Jack-Daw taste for heaping little round things together in a hole, or chest, he may follow his taste. I have often thought of my good neighbour, John Gater, who, if he were here, with his pretty clipped hedges, his garden-looking fields, and his neat homesteads, would have visitors from far and near; and, while every one would admire and praise, no soul would envy him his possessions. Mr. Gater would soon have all these things. The hedges only want planting; and he would feel so comfortably to know that the Botley Parson could never again poke his nose into his sheepfold or his pig-stye. However, let me hope, rather, that the destruction of the Borough-tyranny, will soon make England a country, fit for an honest and industrious man to live in. Let me hope, that a relief from grinding taxation will soon relieve men of their fears of dying in poverty, and will, thereby, restore to England the "*hospitality,*" for which she was once famed, but which now really exists no where but in America.

Edward Hicks, *The Cornell Farm*, 1848. Inscribed "An Indian summer view of the Farm & Stock OF JAMES C. CORNELL of Northampton Bucks county Pennsylvania. That took the Premium in the Agricultural society, October the 12, 1848. Painted by E. Hicks in the 69th year of his age." Gift of Edgar William and Bernice Chrysler Garbisch. Image courtesy of the Board of Trustees, National Gallery of Art, Washington, DC.

2. A Nation of Farmers: The Promise and Peril of American Agriculture, 1825–1860

In an 1817 letter to François de Marbois, Thomas Jefferson reflected on the health of the American experiment in republican government. He compared the fate of the French and American revolutions over the intervening years. "Our lot has been happier. When you witnessed our first struggles in the war of independence, you little calculated, more than we did, on the rapid growth and prosperity of this country," Jefferson commented. He went on to predict that America's good fortune would "proceed successfully for ages to come" and remarked that his hopes depended on "the enlargement of the resources of life going hand in hand with the enlargement of territory."

By the eve of the Civil War, America was indeed an enlarged nation of farmers. Its population remained three-quarters rural, and agriculture was still the largest sector of its burgeoning economy. Farm settlement had expanded with astounding speed across the Mississippi River and beyond, to the verge of the Great Plains—and leapt past the western mountains to the Oregon country and California. A popular agrarian rhetoric extolling the virtues of farm life helped to impel this advance, as did the nation's apparently limitless supply of superbly arable land, sold at a low cost by the federal government. In little more than three generations after independence was gained, half the continent had been settled by American cultivators. On the surface, it appeared that Jefferson's dream was being fulfilled.

But beneath the surface this great tide of agricultural settlement was being driven by powerful currents that threatened the future of the agrarian Republic. That threat did not come in any simple way from the competing rise of cities and factories. Many farmers welcomed the vigorous new manufacturing branch of industry as a boon to agriculture. Textile mills and industrial laborers provided new markets; improved transportation moved produce to those markets; new implements eased the labor of working the land. More than a few prosperous farmers owned railroad stocks. But this widening of agricultural horizons took place as part of a larger transformation of the American economy and culture that would ultimately prove very difficult to reconcile with agrarian values, including the health of the land itself. And rapid westward migration was driven by more than a growing population and economy: it was also a deadly race between two conflicting visions of agrarianism—Free-Soil and Slave—that would soon tear the nation of farmers apart.

During the first half of the nineteenth century, both the expansion of agricultural settlement and the surge of industrial growth were manifestations of a profound underlying change in American society that historians call the "market revolution." In brief, much more of what the nation produced began to flow through the cash economy, and on ever-wider scales of exchange. More of what people consumed came not from the household or the local community but was purchased from afar—and the volume of consumption rapidly increased. Along with this sweeping economic change came a corresponding adoption of middle-class values as more Americans strove for individual prosperity in place of seeking a "comfortable subsistence." This transformation has been celebrated as a capitalist "takeoff," and indeed it generated great material wealth for American society, especially once the new powers of the industrial revolution had been unleashed. But at the same time, the development of a commercial economy brought wrenching change to increasing numbers of citizens who had not previously participated fully in the market, for along with opportunity came risk. The market revolution, according to historian Melvyn Stokes, "undermined the security, dignity, and personal autonomy of subsistence life, substituting the pressures of the competitive marketplace for the characteristic independence and neighborliness of the subsistence economy."

Of course American farmers had always produced at least partly for market, some more than others—a Virginia tobacco planter more than a Massachusetts yeoman, for example. But such production had often been balanced with, or even subordinate to, a larger degree of family household independence. Now, as many farmers took on middle-class aspirations and sought to purchase more consumer goods, they began to devote their energy to growing a narrower range of cash crops. They invested in buildings and equipment to increase their productivity, and they took on debt—often borrowing against the value of their land. With this came prosperity in good times but danger and despair in bad times—when crops failed or markets were glutted or the wider economy crashed. Establishing a successful farm now required capital, access to credit, and reliance on strangers to ship crops to distant customers. Farmers were no longer in control of their own destiny—the economic independence that lay at the heart of the Jeffersonian ideal was in peril. And as wealth came to be concentrated in the hands of an emerging capitalist class, small independent farmers began to find that their grip on the levers of legal and political power was less secure than republican ideology had promised—government often acted to facilitate the growth of new enterprises but left farmers to look out for themselves.

These dramatic changes taking hold in American society provoked resistance. The most politically effective was the rise of Andrew Jackson, whose populist movement reached back to the anti-Federalist sentiment and Jeffersonian agrarianism of the founding era. Speaking in his Fourth Annual Message to Congress, Jackson used rhetoric that would have sounded familiar fifty years earlier: "The wealth and strength of a country are its population, and the best part of that population are cultivators of the soil. Independent farmers are every where the basis of society and true friends of liberty." Those words came in defense of his policy of encouraging rapid settlement of the nation's public lands. Jackson was wildly popular in the old Southwest for his support of territorial expansion, cheap land for farmers, and leniency for settlers in debt.

Jackson shared with Jefferson and with John Taylor a deep distrust of banks and the speculative economy, and like them feared the formation of an aristocracy of money spawned by commerce. In cobbling together his political base Jackson sided not only with farmers but with the "mechanical and laboring classes" as well, in opposition to the new capitalist class. In New York City, workers were beginning to organize and demand improvements in wages and working conditions. Some in the workers' movement, such as George Henry Evans, also called for land reform, seeing the vast lands of the public domain as a refuge for the urban poor and a safety valve for the broader society. Here was the old agrarian boast of the superiority of the American freehold farmer over the downtrodden European landless laborer, now being adopted by the champions of impoverished wageworkers accumulating in *American* cities. And here also appeared one of the most intractable challenges for agrarians of every era that followed: to make common cause with the labor movement and the aspirations of urban Americans.

Evans's call for free land and an end to speculation was taken up by others, most famously *New York Tribune* editor Horace Greeley, and eventually resulted in the passage of the Homestead Act in 1862. In fact, most of those who went west to establish farms (both before and after the Homestead Act) were not the urban poor but rather migrants from rural areas in both Europe and America, where families continued to produce children more rapidly than the local economy could absorb them. Those who moved to the frontier mostly came not from cities but from farms, and most came with some capital, however limited. Many had both an abiding faith in freehold farming as a way of life and a strong desire to pursue economic opportunity—it was never just one or the other. Most believed that hard work on good land and shrewdness in managing their affairs, while it might never make them rich, ought to secure an abundant living and a good home. But the star to which the agrarian wagon had now been hitched was the market, and the question for American farmers to this day has remained: Would that meteor lift them out of the mud or just drag them through it?

For all that the feverish westward movement of the frontier seemed to capture the spirit of the agrarian age, in reality rapid expansion was a mixed blessing for agrarianism at best. Abundant land held great promise for America, but on what terms would that land be held? Along with the cession of Florida by Spain and the acquisition of territory from various Indian nations, Jefferson's purchase of Louisiana in 1803 had provided an apparently inexhaustible supply of land for the nation. "The glory of Jefferson was complete," proclaimed Missouri senator Thomas Hart Benton to his colleagues. "He had found the Mississippi the boundary, and he made it the centre of the Republic. He re-united the two halves of the Great Valley, and laid the foundation for the largest empire of freemen that Time or Earth ever beheld."

Seven states were added to the Union between 1803 and 1821, including Benton's home state on that western bank of the Great Valley. Before too many years, American settlers were flocking not only to the lands drained by the Mississippi but also to the Willamette and the Sacramento. In spite of Benton's enthusiastic promotion of Jefferson's democratic empire of small farmers, that empire was not entirely one of freemen—and therein lay agrarianism's greatest contradiction of all. For even as stock and grain farmers from New England and Pennsylvania swept through western New York and over the Allegheny Mountains to Ohio and the rest of the old Northwest, a parallel stream of planters from Virginia and the Carolinas was striking west to rich lands in Tennessee, Alabama, and Mississippi, and with them went their slaves. As every schoolchild knows, the invention of the cotton gin (together with the adoption of Mexican short-staple cotton, which throve on cooler uplands) threw open the doors for the creation of a vast empire of cotton in the Southern states. Economic historians have long recognized that above any other single commodity, cotton drove the antebellum economy of the United States. The cotton market had its occasional dips, but for the most part the boom rolled on decade after decade, answering skyrocketing demand in the textile mills of Europe and New England. And as cotton boomed in the South, from the flat, fertile farms of the Midwest flowed a bonanza of wheat, corn, whiskey, beef, and pork: down the Mississippi by flatboat and steamboat to help provision Southern plantations; east by the Great Lakes and the Erie Canal, and later by the railroad, to feed the burgeoning Eastern mill towns and port cities; and across the Atlantic to crowded Europe. But even as all the sections of the nation were drawing together into one thriving economy, they were moving farther and farther apart in culture and ideology. This was not simply a clash between an industrial North and an agricultural South: it was also a clash between two starkly different agrarian visions—and agrarian realities.

Cotton may have saved slavery from extinction and instead drove the South's "peculiar institution" to staggering new heights. At the turn of the century, many Americans—Thomas Jefferson among them—had allowed themselves to believe that plantation production of commodities such as tobacco and rice had reached its peak, and that the intractable and divisive issue of slavery could somehow be confined to a small corner of the older Southern states as a nation of yeomen farmers grew beyond it. Cotton put paid to all of that. Even seaboard states such as Virginia that couldn't grow cotton could still produce slaves and ship them in chains to the frontier. Plantation owners convinced themselves that black servitude in a free nation was not a shameful contradiction but a good thing: good for the welfare of the less than fully human African race and essential to the economic independence and liberty of their white owners. Northern farmers watched in alarm as a repellant social system that most of them had simply hoped to be able to ignore instead exploded across the South, showing tremendous economic vigor and no apparent natural boundaries. Thus, the rapid expansion of farms across the Midwest and South was not the result of demographic growth and the lure of fertile land alone: it was also driven by a fiercely competitive market economy and by an increasingly desperate contest between North and South to bring slave and free states into the Republic and their representatives into Congress. America was becoming not one agrarian nation racing westward, but two.

In an idiosyncratic defense of slavery, *Cannibals All,* Virginian George Fitzhugh attacked the North and its free-labor economy: "Why have you . . . Mormons, and anti-renters, and 'vote myself a farm' men, Millerites and Spiritual Rappers, and Shakers, and Widow Wakemenites, and Agrarians, and Grahamites, and a thousand other superstitious and infidel Isms at the North? Why is there faith in nothing, speculation about everything? Why is this unsettled, half-demented state of the human mind coextensive in time and space with free society?" Apologists for slavery like Fitzhugh delighted in pointing out the abysmal working and living conditions of wage earners in Northern factories and cities. His disgust with the effects of the North's free-labor system was so strong that he advocated the implementation of some form of paternalistic servitude for Northern workers. The cotton economy was profitable, and in an unvarnished market economy, with abundant fertile land stretching across the continent, who was to say that farming with slaves might not prove more profitable than independent yeoman farming that employed family labor or expensive wage labor, no matter what the crop? As Roger Kennedy points out in *Mr. Jefferson's Lost Cause,* a tragedy unfolded during the first decades of the nineteenth century. Jeffersonian rhetoric about the value of the yeoman to the Republic and the ideals enshrined in the Declaration of Independence collided with the resurgent economic and political power of Jefferson's own planter class, and the planters won. Although the Southern yeomanry—Frank Owsley's "plain folk of the old South"—never disappeared, the slave interest increasingly dominated agriculture across the South—and threatened to spread beyond it. In response, not only abolitionists such as George Washington Julian but Northern Free-Soilers in general viewed the slaveocracy with growing hostility and dismay. The possibility of a Republic ruled by united agrarian interests was disappearing long before farmers became a national minority.

The headlong expansion of American agriculture onto the fresh and fertile soils of the interior, driven by the growth of a powerful market economy, served to undermine the foundations of the agrarian Republic in yet another way. Abundant natural capital, next to free for the taking—forests, game, the soil itself—fostered an extractive agriculture that quickly began degrading the very land that supported it. This was apparently in spite of the best advice of agricultural leaders. Reformers took up the cause of agricultural improvement that had begun with the likes of Washington, Jefferson, and Madison. Agricultural societies and county fairs began to appear around the turn of the century and gathered momentum through the 1820s and 1830s. The movement toward improved agriculture was supported by an increasingly active agricultural press. Jesse Buel, Edmund Ruffin, and many others published regional agricultural journals full of the latest farming techniques, and these periodicals were widely read by practical farmers. Ruffin was noteworthy for his writings promoting the use of calcareous amendments such as lime and marl to replenish exhausted and acidic Southern soils. Due to Ruffin's research on these matters Avery Craven, his biographer, named him the father of American soil chemistry. Unfortunately, and tellingly, Ruffin achieved greater fame as a defiant defender of slavery.

The ambitions of agricultural improvers such as Ruffin and Buel, who focused on nourishing the soil of the home farm, were undermined by the rush to the West as well as by the imperatives of the market economy. Improvers called for scientific, labor-intensive farming that would restore and enhance the fertility of the soil. But how could farming that relied on careful manuring and legume rotations compete with cheap, fresh land on the frontier? How could a farmer who needed to maximize the output of his slaves, or pay high wages to hired laborers, afford to farm in ways that took good care of the soil? The reformers shied away from these questions, and instead promoted employing the latest scientific methods to boost the land's productivity and succeed in the market—indeed, only farmers who were committed to a businesslike approach would have the wherewithal to employ new equipment, new breeds of livestock, new fertilizers. Farmers in the older settled districts invested in new buildings and tools and experimented with a wide range of new crops and methods, and some of them did find ways to succeed, at least for a time. But specialized production for the market often profited by the most rapid exploitation of available resources more than by their conservation. In many cases the complex, ecologically diverse elements of an older agrarian economy relying on local production and exchange were stripped away in the drive to increase production of a few cash crops.

The land often paid the price for increased production. In Virginia, the very adoption of plowing and small grain culture that Washington and Jefferson advocated led to massive erosion from the Piedmont slopes into Chesapeake Bay. In Vermont, George Perkins Marsh watched the merino sheep boom sweep the forest from the hills. Another fundamental problem of American agriculture had appeared: clearly, farmers who wanted their farms to succeed in the long run needed to take good care of their land, and perhaps most of them wished to. But if they devoted themselves to the long run, they still had to survive in the short run. Agricultural improvers did not often notice this contradiction in their doctrine. It is notable that Marsh, who is widely regarded as the father of American conservation, did not blame the unbridled demands of the market for causing environmental destruction. Instead, like most improvers he placed blame on the hidebound traditions of his farming neighbors. Forward-looking "mechanics, merchants and professional men make in the end the best farmers," Marsh declared. He equated business and industry with science and progress, which bore the promise of understanding the land more precisely and treating it with more care. Conservation, from its birth, was often profoundly antiagrarian. But this was not always true: the naturalist Wilson Flagg foresaw the future relationship of industrial agriculture to farmers and the land in a very different way.

The first half of the nineteenth century saw an exuberant expansion of American agriculture. But by the time of the Civil War, America was a nation of farmers set in an economic, political, and environmental framework very different than anything Thomas Jefferson had imagined. The pivotal political figure of the age, Abraham Lincoln, came from Whig politics. The popular image of Lincoln the rail-splitter and son of the frontier is true, but shouldn't be taken to mean that Lincoln held any special regard for the life of the yeoman. He was restless on his father's farm and sought out town life as soon as he could. While addressing an agricultural fair in Wisconsin, Lincoln eschewed the familiar blandishments in praise of farming. "I presume I am not expected to employ the time assigned me, in the mere flattery of the farmers, as a class. My opinion of them is that, in proportion to numbers, they are neither better nor worse than other people. . . . And I believe there really are more attempts at flattering them than any other; the reason of which I cannot perceive, unless it be that they cast more votes than any other." Lincoln advocated a vigorous program of internal improvements, economic development, and a strong market economy. Free soil and free labor under such conditions were "the just and generous and prosperous system, which opens the way for all—gives hope to all, and energy, and progress, and improvement of condition to all."

By the end of Thomas Jefferson's life in 1825, just a few years after his letter to François de Marbois, even his hopes for his country were faltering. When Jefferson felt the warning of the "fire bell in the night" from the Missouri Compromise of 1820 fill him with dread of oncoming sectional conflict, he was also hearing the death knell for his dream of an agrarian Republic. By the time of Abraham Lincoln and the turning of the Civil War at Gettysburg, America was a profoundly different place than it had been when Jefferson drafted the Declaration four score and seven years earlier. It wasn't simply that the industrial North defeated the agricultural South in that terrible conflict. By then, the lives of Americans north and south, in city and in countryside, had been utterly transformed by new economic and political forces and a new set of values, within which the older vision of agrarian virtue and independence would find an increasingly uncertain and beleaguered place.

JESSE BUEL

Jesse Buel was born in 1778, in Coventry, Connecticut, son to Elias, a major in the War for Independence. After the war, the Buels moved to Rutland, Vermont, where Jesse was apprenticed to a local printer. He mastered the trade quickly, and over the next twenty-seven years founded a series of newspapers in upstate New York. During his years as a publisher of these papers, Buel became engaged in and concerned with agricultural matters. In 1821 he cashed out of his publishing interests and bought eighty-five acres of farmland west of Albany. Having little practical experience of farming, Buel learned what he could from agricultural reformers and experimented with crop rotation, deep plowing, systematic manuring, and other methods the reformers encouraged. In time, he turned his farm into a model operation.

As his farm thrived, Buel also returned to publishing, this time with the Cultivator, *a journal devoted to reforming the agricultural practices of ordinary farmers. The* Cultivator *was a prominent example of the burgeoning agricultural press of the 1830s, 1840s, and 1850s. Buel also wrote books, including* The Farmer's Companion *of 1839.*

Buel shared with earlier agrarians a belief in the importance of agriculture in sustaining the virtues necessary for the health of a republic. "Commerce and manufactures may give temporary consequence to a state," he wrote, "but these are always a precarious dependence. They are effeminating and corrupting; and unless backed by a prosperous agricultural population, they engender the elements of speedy decay and ruin." However, his writings, especially on the possibility of improvements in the comfort and attractiveness of farm life, show a defensiveness and concern for the future of the vocation. In the developing market economy, Jesse Buel felt the need to sell farming as a worthy career for ambitious young men and women.

From *The Farmer's Companion* (1839)

CHAPTER I. THE IMPORTANCE OF AGRICULTURE TO A NATION

There is no business of life which so highly conduces to the prosperity of a nation, and to the happiness of its entire population, as that of cultivating the soil. Agriculture may be regarded, says the great Sully, as the breasts from which the state derives support and nourishment. Agriculture is truly our nursing mother, which gives food, and growth, and wealth, and moral health and character, to our country. It may be considered the great wheel which moves all the machinery of society; and that whatever gives to this a new impulse, communicates a corresponding impetus to the thousand minor wheels of interest which it propels and regulates. While the other classes of the community are directly dependant upon agriculture, for a regular and sufficient supply of the means of subsistence, the agriculturist is able to supply all the absolute wants of life from his own labors; though he derives most of his pleasures and profits from an interchange of the products of labor with the other classes of society. Agriculture is called the parent of arts, not only because it was the first art practised by man, but because the other arts are its legitimate offspring, and cannot continue long to exist without it. It is the great business of civilized life, and gives employment to a vast majority of almost every people. . . .

But agriculture is beneficial to a state, in proportion as its labors are encouraged, enlightened, and honored—for in that proportion does it add to national and individual wealth and happiness.

Agriculture feeds all. Were agriculture to be neglected, population would diminish, because the necessaries of life would be wanting. Did it not supply more than is necessary for its own wants, every other art would not only be at a stand, but every science, and every kind of mental improvement, would be neglected. Manufactures and commerce originally owed their existence to agriculture. Agriculture furnishes, in a great measure, raw materials and subsistence for the one, and commodities for barter and exchange for the other. In proportion as these raw materials and commodities are multiplied, by the intelligence and industry of the farmer, and the consequent improvement of the soil, in the same proportion are manufactures and commerce benefited—not only in being furnished with more abundant supplies, but in the increased demand for their fabrics and merchandise. The more agriculture produces, the more she sells—the more she buys; and the business and comfort of society are mainly influenced and controlled by the results of her labors.

Agriculture, directly or indirectly, pays the burdens of our taxes and our tolls, which support the government, and sustain our internal improvements; and the more abundant her means, the greater will be her contributions. The farmer who manages his business ignorantly and slothfully, and who produces from it only just enough for the subsistence of his family, pays no tolls on the transit of his produce, and but a small tax upon the nominal value of his lands. Instruct his mind, and awaken him to industry, by the hope of distinction and reward, so that he triples the products of his labor, the value of his lands is increased in a corresponding ratio, his comforts are multiplied, his mind disinthralled, and two thirds of his products go to augment the business and tolls of our canals and roads. If such a change in the situation of one farm, would add one hundred dollars to the wealth, and one dollar to the tolls of the state, what an astonishing aggregate would be produced, both in capital and in revenue, by a similar improvement upon 250,000 farms, the assumed number in the State of New York. The capital would be augmented two millions, and the revenue two hundred and fifty thousand dollars per annum.

Agriculture is the principal source of our wealth. It furnishes more productive labor, the legitimate source of wealth, than all the other employments in society combined. The more it is enlightened by science, the more abundant will be its products; the more elevated its character, the stronger the incitements to pursue it. Whatever, therefore, tends to enlighten the agriculturist, tends to increase the wealth of the state, and the means for the successful prosecution of the other arts, and the sciences, now indispensable to their profitable management.

Agriculturists are the guardians of our freedom. They are the fountains of political power. If the fountains become impure, the stream will be defiled. If the agriculturist is slothful, and ignorant, and poor, he will be spiritless and servile. If he is enlightened, industrious, and in prosperous circumstances, he will be independent in mind, jealous of his rights, and watchful for the public good. His welfare is identified with the welfare of the state. He is virtually fixed to the soil; and has, therefore, a paramount interest, as well as a giant power, to defend it, from the encroachment of foreign or domestic foes. If his country suffers, he must suffer; if she prospers, he too may expect to prosper. Hence, whatever tends to improve the intellectual condition of the farmer, and to elevate him above venal temptation, essentially contributes to the good order of society at large and to the perpetuity of our country's freedom.

Agriculture is the parent of physical and moral health to the state—it is the salt which preserves from moral corruption. Not only are her labors useful in administering to our wants, and in dispensing the blessing of abundance to others, but she is constantly exercising a salutary influence upon the moral and physical health of the state, and in perpetuating the republican habits and good order of society. While rural labor is the great source of physical health and constitutional vigor to our population, it interposes the most formidable barrier to the demoralizing influence of luxury and vice. We seldom hear of civil commotions, of crimes, or of hereditary disease, among those who are steadily engaged in the business of agriculture. Men who are satisfied with the abundant and certain resources of their own labor, and their own farms, are not willing to jeopard these enjoyments, by promoting popular tumult, or tolerating crime. The more we promote the interest of the agriculturist, by developing the powers of his mind, and elevating his moral views, the more we shall promote the virtue and happiness of society.

The facts which are here submitted must afford ample proof, that agriculture is all-important to us as a nation; and that our prosperity in manufactures, in commerce, and in the other pursuits of life, will depend, in a great measure, upon the returns which the soil makes to agricultural labor. It therefore becomes the interest of every class, to cherish, to encourage, to enlighten, to honor, and to reward those who engage in agricultural pursuits. Our independence was won by our yeomanry, and it can only be preserved by them.

The Farm Yard. From *Prang's Aids for Object Teaching* (1874). Courtesy Library of Congress, Washington, DC.

CHAPTER IV. AGRICULTURE CONSIDERED AS AN EMPLOYMENT

Every provident parent is anxious to see his children settled in some business of life, that promises to confer wealth and respectability; and every young man, who aims to arrive at future and honorable distinction, is anxious to select that employment which is most likely to realize his wishes. It is with a view to enable both parent and son to act wisely in this matter, that we propose to point out some of the advantages which agriculture holds out to those who embark in its pursuits.

We propose to consider agricultural employment under the following heads:—

1. As a means of obtaining wealth;
2. As promotive of health, and the useful developement of the mind;
3. As a means of individual happiness, the great pursuit of life;
4. As a means of enabling us to fulfil the high objects of our being;—of performing the duties which we owe to our families, our country, and our God.

1. As a Means of obtaining Wealth

Adequate to our wants, and to all the beneficial purposes of life, agriculture certainly holds a pre-eminent rank. With that industry and prudence, which Providence seems to have made essential to human happiness, and that knowledge which we all have the means of acquiring, its gains are certain, substantial, and sufficient—sufficient for ourselves, for the good of our children, and the healthful tone of society. It does not, we admit, afford that prospect of rapid gain, which some other employments hold out to cupidity, and which too often distract and bewilder the mind, and unsettle for life the steady business habits of early manhood; yet neither does it, on the other hand, involve the risks, to fortune and to morals—to health and to happiness—with which the schemers and speculators of the day, who would live by the labor of others, seem ever to be environed. Great wealth begets great care and anxiety, and is too apt to engender habits unfriendly alike to the possessor and to society. Wealth that comes without labor, is often wasted without thought; but that which is acquired by toil and industry is preserved with care, and expended with judgement. The farmer, therefore, who secures an annual and increasing income by his industry, though it be small in the outset, is much more likely to become ultimately rich, not only in dollars and cents, but in all the substantial elements of happiness, than the man of almost any other profession in life.

We have shown that farm lands have been made to produce an annual income of thirty dollars an acre; and have said, that by good husbandry they may certainly be made to produce a nett income of fourteen dollars an acre. Now, if a farmer, upon a hundred acres of land, can save fourteen hundred dollars a year, to buy superfluities for his family, educate his children, and to add to his capital, he must, at the end of twenty years, be either a rich man or an improvident one; and if improvident, he will probably remain poor, be his employment what it may. But suppose the nett income of a farm should be but half, or a quarter of the sum we have assumed—that is, $7, or $3.50, an acre;—even this income, prudently managed, will in a few years place the possessor in independent circumstances.

2. As promotive of Health and the Developement of the Mind

The grand requisites to health, or rather for the prevention of diseases, are declared by Dr. Johnson, one of the highest medical authorities of the age, to be—*exercise in the open air—temperance in our living—moderation in our pleasures and enjoyments—restraint on our passions—limitation to our desires, and limitation to our ambition.*

What employment is there in life, so highly favorable to all the benign influences of exercise—so conducive to repose and tranquillity of mind—and which has so few temptations to intemperate enjoyments—as that of agriculture. And the only ambition which is likely to obtrude upon the farmer, and this is in no wise, we believe, prejudicial to the health either of his body or his mind—is the ambition of increasing the prolific properties of the soil, whereby he may benefit himself and society. Political ambition, which, like a cancer, is apt to prey upon and corrupt the mortal upon whom it fixes its fangs, abides not upon the farm; at least it should not abide there—for that farmer must be either weak or unfortunate who is willing to give up the certain and tranquil pleasures of a rural home, for the vexing, precarious, and corrupting cares and responsibilities of political eminence, otherwise than as duty may require it at his hands. "Horticulture and agriculture are better fitted for the promotion of health and of sound morals," says an eminent medical author [Dr. Caldwell, Transylvania College, Kentucky], "than any other human occupation." The business of agriculture is one of exercise in its most approved forms. It brings into healthful action the entire muscular system; and when exercised with prudence, as all employments should be, it insures appetite, digestion, sleep, a sound constitution, and a contented mind. "The declaration is as trite as it is true, that exercise promotes virtue, and subdues the storms of passions" [Dr. Harris of Philadelphia].

Although the garden and the farm may be made to furnish a great many delicacies and luxuries for the table, yet these delicacies and luxuries are such as conduce alike to health and to rational pleasure. It is a remark of St. Pierre [French social philosopher], that every country and every clime furnishes, within itself, the food which is best fitted for the wants of the animals which dwell in it. The same remark, with a trifling modification, will apply to the farm. The products of the farm and garden *do* constitute the best food for the farmer; and there is no class who can indulge in a greater variety of native products, or enjoy them in a higher state of freshness and perfection, than those who grow them. And upon the farm, and among an intelligent rural population, the pleasures of social intercourse are not curtailed by the cold formalities, nor taxed by the extravagant folly, of the town and city. The agriculturist relies upon his own resources—upon his industry and the blessing of Providence, for the enjoyments of life. His farm and his family are the special objects of his care, and his ambition is to obtain good crops, a good name and reputation in society, and to deserve them, by a liberal and kind deportment to all around him. He is exempt from a crowd of evils—of rivalships and jealousies—of corroding cares and feverish anxieties—which not unfrequently hang around other professions, mar the pleasures of life, and undermine health. He should hate no one; for he should dread no rivals. If his neighbor's field is more productive than his own, he borrows a useful lesson. If his own field is the most productive, it affords him pleasure to benefit his neighbor by his example. He learns to identify his own, with the prosperity of his neighborhood and of his country. . . .

And what an expansive field is ever before the eye of the agriculturist, for study, for reflection, for usefulness, for the enjoyment of rational happiness! The book of Nature, replete with the teachings of Divine Wisdom, always lies open before him!

The elements are subservient to his use; the vegetable and animal kingdoms are subject to his control! And the natural laws which govern them all, and which exert a controlling influence upon his prosperity and happiness, are constantly developing to his mind new harmonies, new beauties, perfect order, and profound wisdom, in the works of Nature which surround him. Nor need he, in these studies of usefulness, be restricted to his own personal observation. He may call to his aid, both in the prosecution of his business, and the improvement of his intellectual faculties, the counsels of eminent men of every age and every country, who have left for our use the record of their experience and their wisdom. And we say it without qualification, that there are few professions in the community, which give more leisure for general reading, or whose employments embrace a greater scope of *useful* reading, than the business of agriculture. The artisan is generally obliged to employ his winter evenings in labor; and those engaged in the liberal professions, and in mercantile business, are not only accustomed to do the like, but their study is in a measure restricted to their particular calling. The agriculturist, on the contrary, may devote his evenings, or most of them, to study—to the improvement of his mind—to the acquisition of useful knowledge. He may devote three hours out of the twenty-four to study, without infringing upon his necessary business, or fatiguing his mind, or impairing his health. This is allowing eight hours for sleep, ten for labor, and three for contingencies. What profession is there, which, if well conducted, gives a larger portion of time to the acquisition of general knowledge? . . .

Labor is in no wise incompatible with study; but, on the contrary, it is necessary, or exercise is necessary, to the development of the faculties of the mind; and where study and labor are directed to the same object, as they may be in agriculture, they tend particularly to stimulate, and to give pleasure and profit to each other. Many of the most eminent and useful men in the improvement of society have been such as have prosecuted their studies while daily laboring in their professional business. Among those, of our country, who have been distinguished for public usefulness, we may name Franklin, Rittenhouse, Fulton, Sherman, &c, who were all hard-working men, and who greatly improved their minds, while they daily labored with their hands.

3. As a Means of Individual Happiness

One of our good and great men [Chancellor Livingston, in his address to the Society of Agriculture and the Arts] has said—"If happiness is to be found upon earth, it must certainly be sought in the indulgence of those benign emotions which spring from rural cares and rural labors." "As Cicero," he continues, "sums up all human knowledge in the character of a perfect orator, so we might, with much more propriety, claim every virtue, and embrace every science, where we draw that of an accomplished farmer. He is the legislator of an extensive family; and not only man, but the brute creation are subject to his laws. He is the magistrate, who expounds and carries these laws into operation. He is the physician, who heals the wounds, and cures the diseases, of his various patients. He is the divine, who studies and enforces the precepts of reason. And he is the grand almoner of the Creator, who is continually dispensing his bounties not only to his fellow-mortals, but to the fowls of the air, and to the beasts of the field."

Though there are many ways and devices by which men endeavor to obtain wealth and happiness, there is perhaps no employment in which these are obtained with so much certainty—few which apparently better fulfil the beneficent designs of the Creator—than that assigned to our first parents—the cultivation of the soil. It has, to be sure, like all other avocations, its cares and its toils—its thorns;—yet its cares and its toils often turn out to be substantial blessings; and, unlike most other avocations, it has more of the roses than the thorns of life. "Agriculture," wrote Socrates, "is an employment most worthy of the application of man, the most ancient, and the most suitable to his nature; it is the common nurse of all persons, in every age and condition of life; it is the source of health, strength, plenty, and riches, and of a thousand sober delights and honest pleasures. It is the mistress and school of sobriety, temperance, justice, religion, and, in short, of all the virtues, civil and military."

4. As a Means of enabling us to fulfil the Temporal Duties of Life

These duties consist, first, in providing honestly for ourselves and families; secondly, in helping our neighbor; and, thirdly, in promoting the good of society at large. It is the due performance of these duties that gives worth and dignity to the human character,—that makes the *good* man,—that renders him useful and respected, and that constitutes the temporal elements of human happiness. Every virtue has its reward, and every vice a punishment, in one form or another, even here, to say nothing of a hereafter. The indolent man, who provides not for himself and his own, but lives upon the labor of others, becomes a dependent upon the sympathies or charities of the world, and is a stranger to the high and manly feelings that flow from conscious independence. He who cares not for the welfare of his neighbor, or seeks not to promote it, is a stranger to the best feelings of humanity—he is a misanthrope in practice, if not in heart. And he who feels not his obligations to society, for the protection and security it affords him, in the enjoyment of life, liberty, and property—and who does not use a portion of his means and his influence, from a high sense of duty, to promote the common weal—to maintain order, law, and a tone of moral health in society,—is not a good citizen, whatever may be his pretensions to talents or to wealth.

Now, agricultural employment, in the first place, enables us to provide by our industry for all the first wants, and for most of the substantial comforts of life;—to superintend and assist in the education of our children; to form their habits, restrain their bad passions and propensities, and to start them in life in a course of industry and usefulness.

In the second place, the condition of the agriculturist enables him to help his neighbor, and promote his welfare, in a variety of ways—by his counsel, by pecuniary aid, and particularly by his example. In the city, individual example is limited in its influence, or lost in the crowd, except in very eminent individuals; but in the country, it becomes conspicuous to all; and the good farmer is sure of benefiting those around him, not only by the improvements which he introduces upon his farm, but his exemplary deportment in life.

In the third place, no one is better fitted than the farmer, to appreciate his high obligations to society,—no one has a stronger interest in performing them. He enjoys the fruits of his labor in peace and quietude, because the laws protect him. He participates in all public improvements, as they tend to enhance the value of his farm and his products. He rejoices in the prosperity of other professions, as they are his customers. He sees constantly around him the works of Creative Wisdom; he sees that they are all governed by immutable laws—and that he is himself subject to these laws; and his employments, his reflections, and a conscious sense of duty, impel in him a desire to aid in carrying out the great and beneficent designs of the Lawgiver.

George Perkins Marsh

Born in 1801 to a leading family in Woodstock, Vermont, George Perkins Marsh grew up watching rampant clearing for sheep pastures in the mountains around his home reach its greatest extent. Marsh witnessed the results of these changes as destructive floods swept more frequently over the best fields of his father's farm along the Quechee River. A brilliant scholar, Marsh graduated from Dartmouth College in 1820. Although he eventually gained recognition as a linguist and philologist, for two decades he cast about for a suitable way to make a living—as a schoolteacher in Woodstock and then as a lawyer and businessman in Burlington. Marsh was unsatisfied as a lawyer, and his business ventures—as banker, owner of a large sheep farm and woolen mill, and railroad promoter, so characteristic of the spirit of his time and place—all suffered reversals also characteristic of such risky enterprises. When Jacksonian Democrats in Congress were able to sink the wool tariff in the late 1830s, Marsh's mill also went under.

Vermont's Whig Party sent Marsh to the House of Representatives in Washington in 1843, and he remained in office until 1849. He was a staunch protectionist and abolitionist and a fierce opponent of the Mexican War—in fact, Marsh was skeptical about American expansionism in general, which he felt would undermine the agriculture and culture of the East. While in Congress, he was instrumental in establishing the Smithsonian Institu-

tion. In 1849 President Taylor appointed Marsh as the U.S. minister to Turkey, and he found his true calling as a diplomat. In 1861 President Lincoln appointed him minister to Italy, where he remained until he died in 1882.

Marsh's observations of the anciently deforested and heavily grazed lands of the Mediterranean, combined with the sweeping land changes he had seen happen in his native Vermont, led him to publish Man and Nature *in 1864. Marsh argued that misuse of the earth could undermine civilizations, unless people were able to become careful stewards of the land over which they had gained such power. The book was widely read and is justly regarded as one of the seminal texts of the American conservation movement. In his 1847 speech to the Rutland Agricultural Society, excerpted here, Marsh offered unusually faint praise for the progress of agriculture but joined a few other perceptive reformers of the era in issuing a warning about "improvements" improvidently carried too far.*

From "Address to the Agricultural Society of Rutland County" (1847)

Although the Association, which I have the honor to address, is styled an Agricultural Society, its influence is not designed to be limited to the encouragement and improvement of the culture of the soil, but its objects are threefold, and embrace as well the toils of the herdsman and the mechanic as the labors of the ploughman. I shall, therefore, not be expected to confine my remarks within a narrower range than your sphere of operations, and while I shall make no attempt to lay down minute practical rules for the conduct or economy of either of these great branches of productive industry, I shall endeavor briefly to illustrate the importance of them all, considered as means and instruments of civilization and social progress, and shall suggest, in a general way, some improvements, the promotion of which seems to me an object well worthy the zealous efforts of the agricultural associations of Vermont. . . .

It is little to the credit of our agriculturists, that the greatest progress in these and other modern improvements should have been made by persons not bred to agricultural pursuits, and it has often been said mechanics, merchants and professional men make in the end the best farmers. If there is any truth in this opinion, it is probably because these persons, commencing their new calling at a period of life when judgment is mature, tied down by habit to no blind routine of antiquated practice, and ridden by no nightmare of hereditary prejudice in regard to particular modes of cultivation, are conscious of the necessity of observation and reflection, in an occupation, the successful pursuit of which requires so much of both, and feel themselves at liberty to select such processes as are commended by the results of actual experience, or accord with the known laws of vegetable physiology. Under such circumstances, a judicious man, encouraged by the stimulus of novelty, would be likely to study the subject with earnestness, and to profit by his own errors, as well as by the experience of others.

There are certain other improvements connected with agriculture, to which I desire to draw your special attention. One of these is the introduction of a better economy in the management of our forest lands. The increasing value of timber and fuel ought to teach us, that trees are no longer what they were in our fathers' time, an incumbrance. We have undoubtedly already a larger proportion of cleared land in Vermont than would be required, with proper culture, for the support of a much greater population than we now possess, and every additional acre both lessens our means for thorough husbandry, by disproportionately extending its area, and deprives succeeding generations of what, though comparatively worthless to us, would be of great value to them. The functions of the forest, besides supplying timber and fuel, are very various. The conducting powers of trees render them highly useful in restoring the disturbed equilibrium of the electric fluid; they are of great value in sheltering and protecting more tender vegetables against the destructive effects of bleak or parching winds, and the annual deposit of the foliage of deciduous trees, and the decomposition of their decaying trunks, form an accumulation of vegetable mould, which gives the greatest fertility to the often originally barren soils on which they grow, and enriches lower grounds by the wash from rains and the melting of snows. The inconveniences resulting from a want of foresight in the economy of the forest are already severely felt in many parts of New England, and even in some of the older towns in Vermont. Steep hill sides and rocky ledges are well suited to the permanent growth of wood, but when in the rage for improvement they are improvidently stripped of this protection, the action of sun and wind and rain soon deprives them of their thin coating of vegetable mould, and this, when exhausted, cannot be restored by ordinary husbandry. They remain therefore barren and unsightly blots, producing neither grain nor grass, and yielding no crop but a harvest of noxious weeds, to infest with their scattered seeds the richer arable grounds below. But this is by no means the only evil resulting from the injudicious destruction of the woods. Forests serve as reservoirs and equalizers of humidity. In wet seasons, the decayed leaves and spongy soil of woodlands retain a large proportion of the falling rains, and give back the moisture in time of drought, by evaporation or through the medium of springs. They thus both check the sudden flow of water from the surface into the streams and low grounds, and prevent the droughts of summer from parching our pastures and drying up the rivulets which water them. On the other hand, where too large a proportion of the surface is bared of wood, the action of the summer sun and wind scorches the hills which are no longer shaded or sheltered by trees, the springs and rivulets that found their supply in the bibulous soil of the forest disappear, and the farmer is obliged to surrender his meadows to his cattle, which can no longer find food in his pastures, and sometimes even to drive them miles for water. Again, the vernal and autumnal rains, and the melting snows of winter, no longer intercepted and absorbed by the leaves or the open soil of the woods, but falling everywhere upon a comparatively hard and even surface, flow swiftly over the smooth ground, washing away the vegetable mould as they seek their natural outlets, fill every ravine with a torrent, and convert every river into an ocean. The suddenness and violence of our freshets increases in proportion as the soil is cleared; bridges are washed away, meadows swept of their crops and fences, and covered with barren sand, or themselves abraded by the fury of the current, and there is reason to fear the valleys of many of our streams will soon be converted from smiling meadows into broad wastes of shingle and gravel and pebbles, deserts in summer, and seas in autumn and spring. The changes, which these causes have wrought in the physical geography of Vermont, within a single generation, are too striking to have escaped the attention of any observing person, and every middle-aged man who revisits his birth-place after a few years of absence, looks upon another landscape than that which formed the theatre of his youthful toils and pleasures. The signs of artificial improvement are mingled with the tokens of improvident waste, and the bald and barren hills, the dry beds of the smaller streams, the ravines furrowed out by the torrents of spring, and the diminished thread of interval that skirts the widened channel of the rivers, seem sad substitutes for the pleasant groves and brooks and broad meadows of his ancient paternal domain. If the present value of timber and land will not justify the artificial re-planting of grounds injudiciously cleared, at least nature ought to be allowed to reclothe them with a spontaneous growth of wood, and in our future husbandry a more careful selection should be made of land for permanent improvement. It has long been a practice in many parts of Europe, as well as in our older settlements, to cut the forests reserved for timber and fuel at stated intervals. It is quite time that this practice should be introduced among us. After the first felling of the original forest it is indeed a long time before its place is supplied, because the roots of old and full grown trees seldom throw up shoots, but when the second growth is once established, it may be cut with great advantage, at periods of about twenty-five years, and yields a material, in every respect but size, far superior to the wood of the primitive tree. In many European countries, the economy of the forest is regulated by law; but here, where public opinion determines, or rather in practice constitutes law, we can only appeal to an enlightened self-interest to introduce the reforms, check the abuses, and preserve us from an increase of the evils I have mentioned.

There is a branch of rural industry hitherto not much attended to among us, but to the social and economical importance of which we are beginning to be somewhat awake. I refer to the agreeable and profitable art of horticulture. The neglect of this art is probably to be ascribed to the opinion, that the products of the garden and the fruityard are to be regarded rather as condiments or garnishings than as nutritious food, as something calculated to tickle the palate, not to strengthen the system; as belonging in short to the department of ornament, not to that of utility. This is an

View of Mount Peg from Mount Tom, c. 1890. Reproduced by permission of
the Woodstock Historical Society, Woodstock, VT.

unfortunate error. The tendency of our cold climate is to create an inordinate appetite for animal food, and we habitually consume much too large a proportion of
that stimulating aliment. This, when we compare the relative cost of a given quantity
of nutritive matter obtained from animals and vegetables, seems very indifferent
economy, and considerations of health most clearly indicate the expediency of increasing the proportion of our fruit and vegetable diet. We cannot in this latitude
expect to rival the pomona or more favored climes, but in most situations, we may,
with little labor or expense, rear such a variety of fruits as to supply our tables with
a succession of delicious and healthful viands throughout the entire year. It is for
us a happy circumstance, that most fruits attain their highest perfection near the
northern limit of their growth. And though the fig and the peach cannot be naturalized among us, we may, to say nothing of the smaller fruits, successfully cultivate
the finer varieties of the apple, the pear and even the grape.

Another mode of rural improvement may be fitly mentioned in connection with
this last. I refer to the introduction of a better style of domestic architecture, which
shall combine convenience, warmth, and reasonable embellishment. A well arranged

and well proportioned building costs no more than a misshapen disjointed structure, and commodity and comfort may be had at as cheap a rate as inconvenience and confusion. Neither is a little expenditure in ornament thrown away. The paint which embellishes tends also to preserve, and the shade trees not only furnish a protection against the exhausting heats of summer, but they serve, if thickly planted, to break the fury of the blasts of winter, and in the end they furnish a better material for fuel or mechanical uses than the spontaneous forest growth. The habit of domestic order, comfort, and neatness will be found to have a very favorable influence in the manner in which the outdoor operations of husbandry are conducted. A farmer, whose house is neatly and tastefully constructed and arranged, will never be a slovenly agriculturist. The order of his dwelling and his courtyard will extend to his stables, his barns, his granaries, and his fields. His beasts will be well lodged and cared for, his meadows free from stumps and briars, and bushes, and the strength of his fences will secure him against the trespasses of his thriftless neighbor's unruly cattle. Another consideration, which most strongly recommends attention to order and comfort and beauty in domestic and rural arrangements is that all these tend to foster a sentiment, of which the enterprising and adventurous Yankee has in general, far too little—I mean a feeling of attachment to his home, and by a natural association, to the institutions of his native New England. To make our homes in themselves desirable is the most effectual means of compensating for that rude climate which gives us three winters each year—two Southern, with a Siberian intercalated between—and of arming our children against the tempting attractions of the milder sky and less laborious life of the South, and the seductions of the boasted greatness and exaggerated fertility of the West. A son of Vermont who has enjoyed, beneath the paternal roof the blessings of a comfortable and well ordered home, and whose eye has been trained to appreciate the charms of rural beauty, which his own hands have helped perhaps to embellish, will find little to please in the slovenly husbandry, the rickety dwellings, and the wasteful economy of the Southern planter, little to admire in the tame monotony of a boundless prairie, and little to entice in the rude domestic arrangements, the coarse fare and the coarser manners of the Western squatter. A youth will not readily abandon the orchard he has dressed, the flowering shrubs which he has aided his sisters to rear, the fruit or shade tree planted on the day of his birth, and whose thrifty growth he has regarded with as much pride as his own increase of stature and who that has been taught to gaze with admiring eye on the unrivalled landscapes unfolded from our every hill, where lake, and island, and mountain and rock, and well-tilled field, and evergreen wood, and purling brook, and cheerful home of man are presented at due distance and in fairest proportion, would exchange such scenes as these, for the miry sloughs, the puny groves, the slimy streams, which alone diversify the dead uniformity of Wisconsin and Illinois!

I have now shown, I hope, rather by suggestion than by argument, that the profession of agriculture in this age and land is an honorable, and in its true spirit, an elevated and an enlightened calling. I have adverted to its importance as an instrument of primary civilization, endeavored to indicate its present position as an art, and hinted at its future hopes and encouragements. It only remains for me to say a word on that other great branch of industry, the promotion of which is one of your leading objects, the ARTS OF CONVERSATION, namely, or as they are more generally called, the MECHANIC ARTS.

Although these arts are practiced to some extent by the rudest savages, as I said at the outset, yet they do not in general attain to any considerable degree of perfection, until agriculture has made great advances, and as they are the last of the industrial arts to be fully developed, so are they the ultimate material means, by which the power, and wealth, and refinement of social man are carried to their highest pitch. The distinction between agriculture and proper mechanic art may be thus stated. The one avails itself of the organic forces of nature, for the purpose of simple reproduction and multiplication; the other employs the more powerful inorganic forces for the conversion of natural forms into artificial shapes. The mechanic derives the raw material directly from the hand of nature, but the form, character and properties of the final product are determined by human contrivance, sometimes relying upon the plastic power of the hand, and at other times aided by natural forces, which man has learned to guide and control. Out of a mass of iron, the artisan can forge at his pleasure, a sword or a ploughshare. He can fashion from a block of wood a spinning wheel or a heathen idol. With a flask of mercury, he can silver a mirror, supply a hospital with a month's stock of calomel, or extract from ore and dross an ingot of gold. He can coin that same gold into Republican eagles, or royal sovereigns, gild with it the dome of a mosque or a capitol, or draw it out into a wire and twist it into a lady's necklace. He can convert a bale of cotton into muslin that shall rival the fineness of the spider's web, canvass and cordage for a ship of the line, or an explosive substance, an ounce of which shall rend that ship into ten thousand fragments. He can hew from the dead marble a chimney piece for a palace or a cottage, the mausoleum of a Napoleon, a baptismal font, or the speaking statue of a Washington. The whole art of the agriculturist on the other hand is exhausted in the multiplication of certain natural products all substantially like the original. Here the seed from the storehouse of the sower becomes as it were the raw material, or rather the model, and nature is the artificer, whom man compels to repeat and reproduce the works of her own mysterious cunning. The labors of the agriculturist are confined to the production and slight improvement of the comparatively few natural forms which he has learned to make subservient to his own uses; the toils and objects of the mechanic are as diversified as the wants and the inventive capacity of man.

But between these two great branches of productive industry, diverse as are their objects and their processes, there is neither interference nor competition; each depends upon and is in turn helpful to the other, and prosperity has crowned no country, which has adapted its legislation exclusively to the encouragement of either. As a general rule, it may be said, that mechanical operations absorb a much larger amount of agricultural products than mere agriculture consumes of the results of mechanical labor, and therefore the husbandman is directly interested in the prosperity of the mechanic, who is his best customer. A single operative in almost any branch of mechanical art works up a vastly greater amount of raw material than one agricultural laborer can grow, and produces a much larger quantity of the manufactured article than one laborer can consume. Three thousand spinners, weavers, dyers and finishers, will convert into dressed cloth all the wool grown in Vermont, and the cloth they will produce would furnish two full suits a year to every male inhabitant of the State. In the case of cottons, the manufacture of which is simple, and requires less manipulation, the disproportion between the quantity of raw material produced by one field laborer and consumed by one manufacturer is even greater; and a similar rule holds true, in general, of all the mechanic arts. The obvious reason of this disproportion between the results of labor is, that the mechanic performs the heaviest portion of his work by mechanical contrivances, which press into his service the inexhaustible inorganic forces of nature, and enable a single individual to wield more than the strength of a thousand, while the agriculturist accomplishes his task by mere bone and sinew, the unaided force of man and beast, and looks to the comparatively feeble and uncertain powers of organized nature to bring about the wished for result. There is another reason why the mechanic arts are of great and perpetually increasing value to the agriculturist. They are constantly discovering new uses, and thereby extending the demand for the raw material. Who can compute the increased value that the invention of paper has given to vegetable fibrous substances, or that of explosive properties to cotton? How many acres has the use of starch in manufacturing added to the culture of the potato, and how many field laborers find employment in growing madder and the teazle? The farmer is interested in the prosperity of the workshop, because it offers a market for his raw material and his surplus food, furnishes occupation in mechanical employments, and thereby reduces the number of rival laborers in agriculture, and cheaply supplies him with wares and implements, which must otherwise be, in the words of the homely proverb, "far fetched and dear bought."

But besides these considerations, there are other reasons of a higher character, why the farmer, in common with all wise and good citizens, should esteem the mechanic arts in an eminent degree worthy of patronage and encouragement, especially in a country whose yet undeveloped natural resources are so boundless and diversified. It is by means of these arts alone, that those internal improvements can be effected, which bind our wide confederacy together, unite our inland seas with the ocean—the common boundary and highway of nations, bring every producer within reach of a market, and tend to equalize the value of lands in all parts of our wide domain. On these topics I am sure I speak to neither ignorant nor uninterested ears; and the efforts which the people of Vermont are now making to secure themselves the advantages of the improvements to which I allude, are as creditable to the intelligence, as they are honorable to their public spirit. In the distribution of the bounty of the national government, the power of the larger states will never allow to our smaller commonwealths their just share, and whatever millions may be lavished on the favored West, we must be content with such improvements as the means accumulated by our own industry, aided by the enlightened liberality of our city capitalists, shall enable us to make. It is therefore a highly encouraging cause of hope and satisfaction, that Vermont has at length put her own shoulder to the wheel, without waiting for Hercules, and we have every reason to expect that another year or two will place our territory on as favorable a ground as localities more blessed by nature, or more pampered by the partial bounty of the general government. What great advantage over you in fact has the farmer who lives on a turnpike road ten miles from Boston, if the grain which was removed from your storehouse at sunset is, while he sleeps, hurried to market on the wings of steam, and delivered at the city depot before he has time to transport thither the corn he had measured and loaded up before he retired to rest? And why will not our water-power be as available as that of Massachusetts, when the cottons turned out by our factories today shall be afloat on the Atlantic tomorrow?

It is through the mechanic arts alone, that we can become truly independent of foreign nations, and establish an interchange between the producer, the manufacturer, and the consumer, which shall increase the wealth and lighten the burdens of each, by retaining among ourselves the net profits of labor, and thus avoiding the drains of the precious metals for supplies. The mechanic arts are worthy of patronage from their progressive character, and the promise they hold out to us of acquiring a complete mastery over inanimate nature. The progress of agriculture, within the last half century though great in itself and full of future promise, has been but a tardy movement, in comparison with the swift advancement of the mechanic arts. The steamboat, the locomotive, the power loom, and the power press, have all been brought into use since the beginning of the present century, and what a revolution have they wrought upon the face of the globe! How they have brought together and linked different states and countries! What millions have they clothed, and what millions enlightened! Suppose we were at once to be deprived of these great gifts of mechanic art, and suddenly cut off from the cheap and abundant supply of the means of knowledge, our necessary clothing doubled in cost, and our products reduced to half their value for want of speedy and economical means of transport to their market, our intelligence from the seat of government, from our distant friends, and from the old world, as well as our personal communication with other parts of our country, retarded and delayed for want of our accustomed means of transport and locomotion, what value should we not attach to these now almost unnoticed blessings, and what efforts and sacrifices should we not be ready to encounter to regain them? Yet we may well judge of the future from the past, and the progress of natural knowledge, upon which all mechanical art is founded, authorizes us to expect that the remaining half of the nineteenth century will be as fertile in improvements as the portion of it which has already elapsed. The mechanic arts are eminently democratic in their tendency. They popularize knowledge, they cheapen and diffuse the comforts and elegancies as well as the necessaries of life, they demand and develop intelligence in those who pursue them, they are at once the most profitable customers of the agriculturist, and the most munificent patrons of the investigator of nature's laws.

Thus, then, the several branches of productive industry, for the promotion of which you are associated, mutually cherish and depend upon each other. The herdsman, the ploughman and the mechanic are fellow laborers, not indeed competitors, but co-workers in a common cause, and every measure that tends to elevate any one of them at the expense of another, must in the end infallibly prove detrimental to the best interests of them all.

WILSON FLAGG

Thomas Wilson Flagg—he would drop the Thomas when he began to write—was born in Beverly, Massachusetts, in 1805, and lived in that state throughout his life. After studies at the Phillips Academy, he entered Harvard, but left soon thereafter and never pursued a career in his field of study, medicine. Instead he became a writer, first on politics and then much more substantially on the natural world. He walked the countryside around New England's villages, eschewing more heroic or adventurous journeys, and wrote about the plants, birds, and landscapes he observed. His published works include Studies in the Field and Forest *(1857),* The Woods and By-ways of New England *(1872), and* Birds and Seasons in New England *(1875).*

Mingled with his interest in nature was Flagg's attention to agriculture, its practitioners, and the rural countryside. Observing New England's farm communities during the middle years of the nineteenth century, Flagg saw many great changes that distressed him. In a note to readers at the beginning of The Woods and By-ways of New England, *he offered an explanation of his efforts as well as a warning. "My object is to inspire readers with a love of nature and simplicity of life, confident that the great fallacy of the present is that of mistaking the increase of our national wealth for the advancement of civilization. Our peril lies in the speed with which every work goes forward, rendering us liable, in our frantic efforts to grasp certain objects of immediate value, to leave ruin and desolation in our track which will render worthless all the desirable objects we have attained."*

"Agricultural Progress" (1859)

Dr. Franklin, on seeing a fly make his escape from a bottle, in which for a long period of years it had been corked up in a torpid state, expressed a wish that he could sleep half a century or more, and then awake, like the fly, to witness the progress which had been made in his beloved country. But if steam-power had been carried into operation to its present extent in Franklin's day, I do not believe he would have expressed any such wish. When I consider the inevitable tendency of this great invention to concentrate all wealth and power into the hands of capitalists, I feel as if I should be reluctant to wake up some ages hence, to view my country when the world is finished. Though it will be admitted that steam, in its application to travelling and to manufactures, has conferred great *apparent* benefits upon mankind, we still have reason to ponder seriously upon the ultimate consequences to small independent farmers, of the introduction of steam power into the operations of agriculture.

I read in the journals of the day, some weeks since, that a company had been formed in the western part of the State of New York, for agricultural purposes, and that they had purchased a "mammoth farm," on which they designed to operate by steam, in connection with the several magnificent inventions which have lately attracted the attention of our agricultural societies. However expedient this system of associated capital may be for the growth of manufactures, it would very soon be found destructive to the prosperity of individual farmers. These corporations, executing almost all their heavy labor by steam power and mammoth implements, would crowd out of the ranks of agriculture all those whose farms were of such small extent, that steam could not be profitably used by them. In competing with the companies, the small farmer would find himself in the situation of the hand-spinner and the hand-weaver, who should undertake to compete with the manufactories of Lowell and Lawrence.

Last year, the Illinois State Board of Agriculture offered a premium of $5000 for the best steam-plow—thus encouraging an invention calculated to make the business of farming profitable exclusively to great corporations or capitalists; to destroy the value of the present mode of farming, and to extirpate the whole class of small farmers from the State! All such inventions tend to make it necessary that agriculture should be carried on by large employments of capital, and on a magnificent scale of operations. All agricultural implements which are moved by steam must be profitable in a certain ratio to the extent of even and uninterrupted surface which is to be tilled. On small fields it would be impossible to use them with success. Hence follows the necessity of farming by associated capital, of greatly increasing the size of farms by combining many into one; and under such *improved* circumstances, the present system of farm labor could not stand in competition with steam-farming. The agricultural steam-company, with their implements carried by steam-power, would cultivate ten acres with about the same expense of labor which is now employed in cultivating one acre. If the moral education and physical improvement of laboring men were to be the effects of this new system of farming, there would be reason for rejoicing over the prospect of the change. But no such happy results would spring from it; laboring men, instead of being elevated into lords, would be degraded into mere machines.

Men are too prone to base their theories of human progress on the assumption that labor is a curse, and not, as it is undoubtedly, when it is free and justly rewarded—a blessing. But labor ceases to be free, in the highest sense, when the laborers are under the control and in the power of mammoth associations. Labor then becomes *servitude,* which is closely allied to *slavery.* No one would say, that under the present circumstances of the country, the operatives in our manufactories, however well paid, are as free as our farmers, masons, and carpenters. It should be remarked, also, that when labor is performed by powerful machines, man becomes a slave to the machinery; when, on the other hand, the implements in use are small, the machinery is the servant of man. The production may be greater in the former case; but the health and freedom of the masses are sacrificed to obtain it. The object of the statesman and the philanthropist should be to make the people free, virtuous and happy; and any increase of the wealth of the nation which must be obtained at the expense of the moral and physical welfare of the people, is not to be desired.

But it may be asked by some jealous friend of "progress," if it is right to refuse to agriculture those *aids* which have built up our manufactures? I would answer that we should refuse to agriculture any aid which is not beneficial to the agriculturist—for the farmer is of more importance than his crops. Let us not improve agriculture by any such means as will degrade man. If we could double the agricultural produce of the whole country at the present cost, by a system which would destroy the independence of our farmers, we should turn all our forces against it, as against the invasion of a foreign army.

In order to illustrate the consequences of this sort of "progress," we will apply it to an imagined case. We will suppose, for example, that in some indefinite period of the future, when steam-farming by associated capital has become nearly universal, there remains, in a certain part of the country, one of those farming villages which are now so common in our happy land. The farmers in this place are intelligent working-men, and small land-proprietors, who have but little capital except their lands and stock, and support themselves by industry and honest trade. After steam-plows, steam-rakes, steam mowing-machines, and other magnificent improvements connected with them, have swept over the country, they have arrived at last, at this antiquated village, where labor is free, and where the farmers are so old-fashioned and behind the times, as to own the lands they till, and carry on farming as we carry it on in the present barbarous age of political and social equality.

These industrious farmers have ascertained now by bitter experience, that by the use of hand implements and horse and cattle power, in the operations of the farm, they cannot compete with the great agricultural corporations, which by means of steam-power can produce at an expense of ten dollars, results which they could not produce at an expense of less than one hundred. The agent of a new company, chartered with ten millions of capital, offers to these unhappy men a price for their farms, which, though exceedingly low, is such as under their present circumstances they feel obliged to accept, especially as a promise accompanies the offer, to employ them as laborers on the soil, under the direction of the officers of the company. The majority consent to the sale, and the remainder are obliged to consent by a law of the legislature placing it in the power of corporations *"established for the public good,"* as it is now in the power of railroad corporations, to seize upon a refractory individual's land and estate, after paying him what a body of commissioners deem an equivalent for the property seized. These mammoth agricultural corporations, by means of bribery and political manoeuvreing, would easily obtain sufficient influence over legislative bodies to cause the enactment of such a law. This any one will believe who has had any political experience, and who knows how easily the worst measures may be carried by making them party tests.

Let us now examine the consequences in detail, after this little village of happy and independent laborers has been converted into a mammoth farm, owned by a company, and carried on by steam-power. At the commencement all the pleasant old farm-houses are removed, because they stand in the way of tillage, which is performed as much as possible in large, undivided lots. All fences and boundaries, except those by the roadside, are for the same reason taken down, to open many small fields into one. It has been ascertained, by experience, that no single field can be worked with the best advantage, unless it contains at least five hundred acres. If it contain a thousand, it is still better, since the larger the field, the more conveniently can it be worked by steam. Hence the preliminaries for steam-farming are necessarily a work of devastation. Many delightful groups of trees and shrubbery, some that skirted a winding brook, others that bordered the walls and fences, including many standard oaks and maples, are swept to the ground, rooted up by some giant infernal machine, as easily as a farmer pulls up weeds. All abruptly swelling ridges and other eminences—the charm of many a landscape—some of them beautifully crowned with trees and shrubs, and others velveted with green herbage, and forming numerous little valleys, now smiling in sunshine, and then sweetly sleeping under the summer shadows of trees, where the flocks found a comfortable resort in all weathers, are now graded into one vast level.

The brooks are conducted into canals, and carried along in straight courses for the convenience of labor and the purposes of irrigation; for it is necessary that their circuities should not interfere with the progress of the steam-plow. In fine, that pleasing variety of surface which beautified the landscape, when it was in possession of the original inhabitants; those quiet rustic lanes fringed with wild roses, hawthorns and viburnums, conducting from the dwelling-houses to the adjoining fields and woods; the comfortable enclosures that resounded with the lowing of cattle and the cheerful noise of poultry, and worst fate of all, the old farm-house, where the patriarch of a small estate presided over a happy family, happy, because they were free and healthfully employed—all, all are swept away by this besom of improvement.

And where are the inhabitants? The sturdy yeoman, who, though doomed to hard labor, found this labor sweet, because it was voluntary; the happy and independent swain who called no man master, and who was really a king in his own acres, is now a hired servant of the corporation. The farmers, their wives and their children, have all been reduced to servitude in this grand manufactory of corn and vegetables. The tiller of the soil has become a slave to his crops. Each thousand acres devoted to a single crop is managed by an agent imported from the city, who understands book-keeping, but was never accustomed to labor. He receives a large salary, and pays out their weekly pittance to the farm laborers. In order to facilitate operations, there is a minute division of labor, as in the cotton and woollen factories. Some of the farmers are employed exclusively as shovellers; some are used as drivers of cattle; some ride on the engine; others are employed continually to follow after the cattle and pick up their droppings, which are all nicely economized, and never allowed to lie and waste one minute upon the ground.

The several families, with the exception of those who emigrated to some other place, are tenants of wooden boxes, put up close to the ground, for the economizing of land. All these are in exact uniformity, and are owned by the corporation. I ought to add that the majority of the farmers, flattered with the hope of sudden wealth, invested all their capital—the proceeds of the sales of their estates—in the corporation stock, which they were soon obliged to sell, at an immense sacrifice, because the extravagance and dishonesty of the company's agents, absorbed all the profits, and cut down their dividends. In less than ten years, almost every one of these independent farmers was a poor man; and the village children who lived as free as the birds of the air in their humble rural homes, now work in platoons upon such parts of farm labor as they are able to perform. Before the village was sold, you might see these little children, with their satchels, going regularly to the district schools, clad in neat and various attire, skipping and playing on the route, full of gladness and freedom. Now they are called up in the morning by the ringing of a bell. They rise, they work, they eat, they go to bed and they sleep to the sound of a bell, that tolls dismally in their weary ears, the knell of all their former joys.

In the story of this once happy village and its inhabitants, we may read the fate of the whole country, should the steam-engine ever be introduced into the business of agriculture: and this would inevitably follow, if farming were to be carried on by corporations, involving large amounts of associated capital. Such a class as that of independent laboring farmers—the only *undegenerated* class in any civilized community—would cease to exist. If it be "progress" or "improvement" to convert all these valuable men into hirelings, under the agents of mammoth corporations—then we must admit the utility of the change. But I am not yet ready to admit any measures

to be progressive, which lessen the happiness and liberty of men, how much soever
they may increase the productiveness of the arts.

> Ill fares the land, to lurking ills a prey,
> Where wealth accumulates and men decay.
> Princes and lords may flourish and may fade;
> A breath can make them, as a breath has made;
> But a bold peasantry—their country's pride—
> When once destroyed, can never be supplied.

George Henry Evans and the Working Men's Movement

*"The mobs of great cities add just so much to the support of pure government," wrote
Thomas Jefferson, "as sores do to the strength of the human body." Agrarians who followed Jefferson feared that the United States was headed for serious trouble in the first
half of the nineteenth century as New York and other cities swelled in size. By this line of
thought, urban workers lacked much of the independence of a freehold yeomanry as well
as the self-discipline life on the farm demanded. Among those trying to head off this urban
nightmare was George Henry Evans, who hoped that an agrarian program might alleviate
the problems caused by America's growing working class. Born in Herefordshire, England,
in 1805, Evans emigrated to the United States with his family in 1820. He apprenticed
with a printer in Ithaca, New York, and remained active in printing and publishing for
much of his life. After moving to New York City, he took a leading role in that city's nascent
workers' movement, editing the* Working Man's Advocate *starting in 1829.*

*Evans left New York during the depression years after 1837 to farm in rural New
Jersey. His experiences on the land cemented an inclination toward agrarian reform.
When Evans returned to New York in 1844, he began to campaign for easier access to
public lands for ordinary citizens and for more safeguards against public lands falling
into the hands of speculators. He submitted a petition to his friend Robert Dale Owen,
then a congressman, promoting the cause of land reform, which Owen then presented to
Congress. Later, Evans's new organization, the National Reform Association, adopted
the slogan "Vote Yourself a Farm" to press for a national Homestead Act. Evans wanted
to set his movement squarely in the mainstream of American thought. He claimed to
seek "a condition of society more in accordance with our national professions as set forth
by Jefferson's immortal pen." A more direct link might be traced to Thomas Paine's 1797
pamphlet* Agrarian Justice, *which Evans read as a young man. Evans's focus on the future
of public lands helped open the national debate on homestead policy.*

"A Memorial to Congress" (1844)

The undersigned Citizens of New York respectfully represent that, in their opinion, the system of Land Traffic imported to this country from Europe is wrong in principle; that it is fast debasing us to the condition of a nation of dependant tenants, of which condition a rapid increase of inequality, misery, pauperism, vice, and crime are the necessary consequences; and that, therefore now, in the infancy of the Republic, we should take effectual measures to eradicate the evil, and establish a principle more in accordance with our republican theory, as laid down in the Declaration of Independence; to which end we propose that the General Government shall no longer traffic, or permit traffic, in the Public Lands yet in its possession, and that they shall be laid out in Farms and Lots for the free use of such citizens (not possessed of other land) as will occupy them, allowing the settler the right to dispose of his possession to any one not possessed of other land; and that the jurisdiction of the Public Lands be transferred to States only on condition that such a disposition should be made of them.

Your memorialists offer the following reasons for such a disposition of the lands as they propose:

1. It would increase the number of freeholders and decrease the anti-republican dependence of those who might not become freeholders; exactly reversing the state of things now in progress.
2. As the drain of the population would gradually be to where the land was free, the price of all land held for traffic would gradually decrease, till, ultimately, the land-holders would see greater advantages in an Agrarian plan that would make every man a freeholder, than in the system of land-selling, under which their children might become dependant tenants.
3. City populations would diminish gradually till every inhabitant could be the owner of a comfortable habitation; and the country populations would be more compactly settled, making less roads and bridges necessary, and giving greater facilities of education.
4. There need be no Standing Army, for there would be soon a chain of Townships along the frontiers, settled by independent freemen, willing and able to protect the country.
5. The danger of Indian aggressions would be materially lessened if our people only took possession of land enough for their use.
6. The strongest motive to encroachments by Whites on the rights of the Indians would be done away with by prohibiting speculation in land.
7. The ambition, avarice, or enterprise that would, under the present system, add acre to acre, would be directed, more usefully, to the improvement of those to which each man's possession was limited.

8. There would be no Repudiation of State Debts, for, let people settle the land compactly, and they could, and would, make all desirable improvements without going into debt.

9. National prosperity and the prosperity of the masses would be coincident, here again reversing the present order of things, of which England is a notable example.

10. Great facilities would be afforded to test the various plans of Association, which now engage the attention of so large a proportion of our citizens, and which have been found to work so well, so far as the accumulation of wealth and the prevention of crime and pauperism are concerned, in the case of those longest established, for instance, the Zoarites, Rappites, and Shakers.

11. The now increasing evil of office-seeking would be diminished, both by doing away with the necessity of many offices now in existence, and by enabling men to obtain a comfortable existence without degrading themselves to become office beggars. Cincinnatus and Washington could with difficulty be prevailed upon to take office, because they knew there was more real enjoyment in the cultivation of their own homesteads.

12. It would, in a great measure, do away the now necessary evil of laws and lawyers, as there could be no disputes about rents, mortgages, or land titles, and morality would be promoted by the encouragement and protection of industry.

13. As the people of England are now fast turning their attention to the recovery of their long-lost right to the soil, it would give them encouragement in their object, and enable them the sooner to furnish happy homes for the thousands who otherwise would come among us as exiles from their native land.

14. The principle of an Equal Right to the Soil once established, would be the recognition of a truth that has been lost sight of by civilization, and which, in our opinion, would tend powerfully to realize the glorious aspirations of philanthropists, universal peace and universal freedom.

"Vote Yourself a Farm" (1846)

Are you an American citizen? Then you are a joint-owner of the public lands. Why not take enough of your property to provide yourself a home? Why not vote yourself a farm?

Remember poor Richard's saying: "Now I have a sheep and a cow, every one bids me 'good morrow.'" If a man have a house and a home of his own, though it be a thousand miles off, he is well received in other people's houses; while the homeless wretch is turned away. The bare right to a farm, though you should never go near it, would save you from many an insult. Therefore, Vote yourself a farm.

Are you a party follower? Then you have long enough employed your vote to benefit scheming office-seekers; use it for once to benefit yourself—Vote yourself a farm.

Are you tired of slavery—of drudging for others—of poverty and its attendant miseries? Then, Vote yourself a farm.

Are you endowed with reason? Then you must know that your right to life hereby includes the right to a place to live in—the right to a home. Assert this right, so long denied mankind by feudal robbers and their attorneys. Vote yourself a farm.

George Washington Julian

Born in 1817 in Wayne County, Indiana, George Julian came of age politically as the nation began its slide toward the Civil War. After practicing law for a time, from the mid-1840s Julian increasingly devoted his efforts to the fight against slavery. He was elected to the Indiana state legislature as a Whig, then joined the Free-Soil Party in 1848 and was elected to Congress, where he served for one term. When the Whig and Free-Soil parties faded, Julian found a home in the new Republican Party. In 1860 he was elected to Congress as a Republican, serving until 1871. A member of the Joint Committee on the Conduct of the War, he was an early advocate of using emancipation to further the war effort of the Union.

Julian is best known as an uncompromising abolitionist, but he was also an enthusiastic advocate of public land policy reform—although he was careful to note that he was no "Agrarian" (as the label was used at that time), by which he meant that he did not advocate legislation that would abridge the rights of those who already owned land. In 1851 he spoke before the House of Representatives proposing his Homestead Bill, which was designed to make public domain lands easily available to families of limited means and to bypass the land speculators who often profited enormously from the nation's public land policy. In the speech, he makes clear how the two great concerns of his career were linked in his mind.

In 1862 Julian helped lead the successful effort to pass the Homestead Act. After his reelection that year, Julian became chairman of the Committee on Public Lands.

"Speech before Congress on the Homestead Bill" (1851)

Mr. Speaker,—The anxiety I feel for the success of the measure now before us, and its great importance, as I conceive, to the whole country, have induced me to beg the indulgence of the House in a brief statement of the reasons which urge me to give it my support. I do this the more willingly, because there has been a manifest disposition here, during the whole of the session, to suppress entirely the discussion of this bill, and at the same time, by parliamentary expedients, to avoid any direct action upon it. It seems to be troublesome to gentlemen. Many who are opposed to its principles appear to be haunted by the suspicion that the people are for it, and hence they will not vote directly against it. They prefer not to face it in any way. The proceedings on yesterday prove this. The House then refused to lay the bill on the table; but immediately afterwards, its reference to the Committee of the Whole, which was substantially equivalent, was carried by a large majority. There was an opportunity of evading the responsibility of a direct vote, and of accomplishing, by indirection, what gentlemen did not dare do by their open and independent action. I refer to these facts because I wish them to go before the people. I desire the country to understand the action of this body, in reference to the question under discussion.

Our present land system was established by act of Congress as far back as the year 1785. From that time to the 30th of last September the government has sold one hundred and two millions four hundred and eight thousand six hundred and forty acres. Within the same period it has donated about fifty millions of acres for the purposes of education, for internal improvements, for the benefit of private individuals and companies, and for military services. This calculation does not include the land granted by the Mexican Bounty Law of 1847, which has not yet spent its force, and which will exhaust from twelve to fifteen millions of acres. The Bounty Law of 1850 will subtract from the public domain the further sum of probably about fifty millions of acres. Besides all this, there were very large grants of land made at the last session of Congress for internal improvements; and there are at this time not less than sixty bills before us asking donations of land, larger or smaller, for various public and private purposes. Should the government, however, pause at the point we have now reached in the prosecution of our land policy, there will still remain, after deducting the sales and grants I have mentioned, the enormous sum of about fourteen hundred millions of acres. The management of this vast fund is devolved by the Constitution upon Congress, and its just disposition presents one of the gravest questions ever brought before the national legislature. The bill under consideration contemplates a radical change in the policy pursued by the government from its foundation to the present time. It abandons the idea of holding the public domain as a source of revenue; it abandons, at the same time, the policy of frittering it away by grants to the States or to chartered companies for special and local objects; and it makes it free, in limited portions, to actual settlers, on condition of occupancy and improvement. This, in my judgment, is the wisest appropriation of the public lands within the power of Congress to make, whether viewed in the light of economy, or the brighter light of humanity and justice.

I advocate the freedom of our public domain, in the first place, on the broad ground of natural right. I go back to first principles; and holding it to be wrong for governments to make merchandise of the earth, I would have this fundamental truth recognized by Congress in devising measures for the settlement and improvement of our vacant territory. I am no believer in the doctrines of Agrarianism, or Socialism, as these terms are generally understood. The friends of land reform claim no right to interfere with the laws of property of the several States, or the vested rights of their citizens. They advocate no *leveling* policy, designed to strip the rich of their possessions by any sudden act of legislation. They simply demand that, in laying the foundations of empire in the yet unpeopled regions of the great West, Congress shall give its sanction to the natural right of the landless citizen of the country to a home upon its soil. The earth was designed by its Maker for the nourishment and support of man. The free and unbought occupancy of it belonged, originally, to the people, and the cultivation of it was the legitimate price of its fruits. This is the doctrine of nature, confirmed by the teachings of the Bible. In the first peopling of the earth, it was as free to all its inhabitants as the sunlight and the air; and every man has, by nature, as perfect a right to a reasonable portion of it, upon which to subsist, as he has to inflate his lungs with the atmosphere which surrounds it, or to drink the waters which pass over its surface. This right is as inalienable, as emphatically *God-given*, as the right to liberty or life; and government, when it deprives him of it, independent of his own act, is guilty of a wanton usurpation of power, a flagrant abuse of its trust. In founding States, and rearing the social fabric, these principles should always have been recognized. Every man, indeed, on entering into a state of society, and partaking of its advantages, must necessarily submit the natural right of which I speak (as he must every other) to such regulations as may be established for the general good; yet it can never be understood that he has renounced it altogether, save by his own alienation or forfeiture. It attaches to him, and inheres in him, in virtue of his *humanity*, and should be sacredly guarded as one of those fundamental rights to secure which "governments are instituted among men."

The justness of this reasoning must be manifest to any one who will give the subject his attention. Man, we say, has a natural right to life. What are we to understand by this? Surely, it will not be contended that it must be construed strictly, as a mere right to *breathe*, looking no farther, and keeping out of view the great purpose of existence. The right to life implies what the law books call a "right of way" to its enjoyment. It carries necessarily with it the right to the *means* of living, including not only the elements of light, air, fire, and water, but *land* also. Without this man could have no habitation to shelter him from the elements, nor raiment to cover and protect his body, nor food to sustain life. These means of living are not only necessary, but absolutely indispensable. Without them life is impossible; and yet without land they are unattainable, except through the charity of others. They are at the mercy of the landholder. Does government then fulfill its mission when it encourages or permits the monopoly of the soil, and thus puts millions in its power, shorn of every right except the right to beg? The right to life is an empty mockery, if man is to be denied a place on the earth on which to establish a home for the shelter and nurture of his family, and employ his hands in obtaining the food and clothing necessary to his comfort. To say that God has given him the right to life, and at the same time that government may rightfully withhold the means of its enjoyment, except by the permission of others, is not simply an absurdity, but a libel on his Providence. It is true there are multitudes of landless poor in this country, and in all countries, utterly without the power to acquire homes upon the soil, who, nevertheless, are not altogether destitute of the essential blessings I have named; but they are dependent for them upon the saving grace of the few who have the monopoly of the soil. They are helpless pensioners upon the calculating bounty of those by whom they have been disinherited of their birthright. Was it ever designed that men should become vagrants and beggars by reason of unjust legislation, stripped of their right to the soil, robbed of the joys of home, and of those virtues and affections which ripen only in the family circle? Reason and justice revolt at such a conclusion. The gift of life, I repeat, is inseparable from the resources by which alone it can be made a blessing, and fulfill its great end. And this truth is beginning to dawn upon the world. The sentiment is becoming rooted in the great heart of humanity, that the right to a *home* attaches of necessity to the right to live, inasmuch as the physical, moral, and intellectual well-being of each individual cannot be secured without it; and that government is bound to guarantee it to the fullest practicable extent. This is one of the most cheering signs of the times. "The grand doctrine, that every human being should have the means of self-culture, of progress in knowledge and virtue, of health, comfort, and happiness, of exercising the powers and affections of a man,—this is slowly taking its place as the highest social truth."

But quitting the ground of right, I proceed to some considerations of a different character. I take it to be the clear interest of this government to render every acre of its soil as productive as labor can make it. More than one half the land already sold at the different land-offices, if I am not mistaken, has fallen into the cold grasp of the speculator, who has held it in large quantities for years without improvement, thus excluding actual settlers who would have made it a source of wealth to themselves and to the public revenue. This is not only a legalized robbery of the landless, but an exceedingly short-sighted policy. It does not, as I shall presently show, give employment to labor, nor productiveness to the soil, nor add to the treasury by increased returns in the shape of taxation. It is legislative profligacy. The true interest of agriculture is to widen the field of its operations as far as practicable, and then, by a judicious tillage, to make it yield the very largest resources compatible with the population of the country. The measure now before us will secure this object by giving independent homesteads to the greatest number of cultivators, thus imparting dignity to labor, and stimulating its activity. It may be taken for granted as a general truth, that a nation will be powerful, prosperous, and happy, in proportion to the number of independent cultivators of its soil. All experience demonstrates that it is most favorable to agriculture to have every plantation cultivated by its proprietor; nor is it less conducive to the same object, or less important to the general welfare, that every citizen who desires it should be the owner of a plantation, and engaged in its cultivation. The disregard of these simple and just principles in the actual policy of nations, has been one of the great scourges of the world. We now have it in our power, without revolution or violence, to carry them into practice, and reap their beneficent fruits; and a nobler work cannot engage the thoughts or enlist the sympathies of the statesman. No governmental policy is so wise as that which keeps constantly before the mind of the citizen the promotion of the public good, by a scrupulous regard for his private interest. This principle should be stamped upon all our legislation, since it will establish the strongest of all ties between him and the State. A philosophic writer of the last century, in sketching a perfectly-organized commonwealth, has the following:—

> As every man ploughed his own field, cultivation was more active, provisions more abundant, and individual opulence constituted the public wealth.
>
> As the earth was free, and its possession easy and secure, every man was a proprietor, and the division of property, by rendering luxury impossible, preserved the purity of manners.
>
> Every man finding his own well-being in the constitution of his country, took a lively interest in its preservation; if a stranger attacked it, having his field, his house, to defend, he carried into the combat all the animosity of a personal quarrel, and, devoted to his own interests, he was devoted to his country.

Here, sir, are principles worthy to guide our rulers in the disposition of the public lands. Give homes to the landless multitudes in the country, and you snatch them from crime and starvation, from the prison and the almshouse, and place them in a situation at once the most conducive to virtue, to the prosperity of the country, and to loyalty to its government and laws. Instead of paupers and outcasts, they will become independent citizens and freeholders, pledged by their gratitude to the government, by self-interest, and by the affections of our nature, to consecrate to honest toil the spot on which the family altar is to be erected and the family circle kept unbroken. They will feel, as never before, the value of free institutions, and the obligations resting upon them as citizens. Should a foreign foe invade our shores, having their homes and their firesides to defend, they would rush to the field of deadly strife, carrying with them "all the animosity of a personal quarrel." "Independent farmers," said President Jackson, "are everywhere the basis of society, and true friends of liberty;" and an army of such men, however unpracticed in the art of war, would be invincible. Carry out this reform of multiplying independent cultivators, and thus rendering labor at once honorable and gainful, and I verily believe more will be done than could be accomplished by any other means to break down our military establishments, and divert the vast sums annually expended in maintaining them to the arts of peace. It is emphatically a peace movement, since it will curb the war spirit by subsidizing to the public interest the "raw material," of which our armies are generally composed. By giving homes to the poor, the idle, the vicious, it will attach them to the soil, and cause them to feel, as producers of the country ought to feel, that upon *them* rest the burdens of war. The policy of increasing the number and independence of those who till the ground, in whatever light considered, commends itself to the government. England, and the countries of Western Europe, have risen in prosperity, just in proportion as freedom has been communicated to the occupiers of the soil. The work of tillage was at first carried on by slaves, then by villains, then by metayers, and finally by farmers; the improvement of those countries keeping pace with these progressive changes in the condition of the cultivator. The same observations would doubtless apply to other countries and to different ages of the world. But I need not go abroad for illustrations of this principle. Look, for example, at slave labor in this country. Compare Virginia with Ohio. In the former the soil is tilled by the slave. He feels no interest in the government, because it allows him the exercise of no civil rights. It does not even give him the right to himself. He has of course no interest in the soil upon which he toils. His arm is not nerved, nor his labor lightened by the thought of home, for to him it has no value or sacredness. It is no defense against outrage. His own offspring are the property of another. He does not toil for his family, but for a stranger. His wife and children may be torn from him at any moment, sold like cattle to the trader, and separated from him forever. Labor brings no new comforts to himself or his family. The motive from which he toils is the lash. He is robbed of his humanity by the system which has made him its victim. Can the cultivation of the soil by such a population add wealth or prosperity to the commonwealth? The question answers itself. I need not point to Virginia, with her great natural advantages, her ample resources in all the elements of wealth and power, yet dwindling and dying under the curse of slave labor. But cross the River Ohio, and how changed the scene! Agriculture is in the most thriving condition. The whole land teems with abundance. The owners of the soil are in general its cultivators, and these constitute the best portion of the population. Labor, instead of being looked upon as degrading, is thus rendered honorable and independent. The ties of interest, as well as the stronger ties of affection, animate the toils of the husbandman, and strengthen his attachment to the government; for the man who loves his home will love his country. His own private emolument and the public good are linked together in his thoughts, and whilst he is rearing a virtuous family on his own homestead, he is contributing wealth and strength to the State. Population is rapidly on the increase, whilst new towns are springing up almost as by magic. Manufactures and the mechanic arts, in general, are in a flourishing condition, whilst the country is dotted over with churches, school-houses, and smiling habitations. The secret of all this is the distribution of landed property, and its cultivation by freemen. But even in the virgin State of Ohio, the curse of land monopoly, or *white* slavery, is beginning to exhibit its bitter fruits, as it will everywhere, if unchecked by wise legislation. Let Congress, therefore, see to it, *in the beginning,* by an organic law for the public domain yet remaining unsold, that this curse shall be excluded from it. The enactment of such a law should not be delayed a single hour. Now is the "golden moment" for action. The rapidity with which our public lands have been melting away for the past few years under the prodigal policy of the government renders all-important the speedy interposition of Congress.

Mr. Speaker, I have spoken, incidentally, of slavery. This, I am aware, may be considered a violation of the "final settlement," the remarkably sanative measures, ratified by Congress a few months since. I beg leave to say, however, that I think the adoption of the policy for which I am contending will be a much better "settlement" of the slavery question than the one to which I refer. Donate the land lying within our Territories, in limited plantations, to actual settlers whose interest and necessity it will be to cultivate the soil with their own hands, and it will be a far more formidable barrier against the introduction of slavery than Mr. Webster's "ordinance of nature," or even the celebrated ordinance of Jefferson. Slavery only thrives on extensive estates. In a country cut up into small farms, occupied by as many independent proprietors who live by their own toil, it would be impossible,—there would be no room for it. Should the bill now under discussion become a law, the poor white laborers of the South, as well as of the North, will flock to our Territories; labor will become common and respectable; our democratic theory of equality will be realized; closely associated communities will be established; whilst education, so impossible to the masses where slavery and land monopoly prevail, will be accessible to the people through their common schools; and thus physical and moral causes will combine in excluding slavery forever from the soil. The freedom of the public lands is therefore an anti-slavery measure. It will weaken the slave power by lending the official sanction of the government to the natural right of man, as *man,* to a home upon the soil, and of course to the fruits of his own labor. It will weaken the system of chattel slavery, by making war upon its kindred system of wages slavery, giving homes and employment to its victims, and equalizing the condition of the people. It will weaken it, by repudiating the vicious dogma of the slaveholder that the laborious occupations are dishonorable and degrading. And it will weaken it, as I have just shown, by confining it within its present limits, and thus forcing its supporters to seek some mode of deliverance from its evils. Pass this bill, therefore, and whilst the South can have no cause to complain of Northern aggression, it will shake her peculiar institution to its foundations. Her three millions of slaves, now toiling, not under the stars, but the *stripes* of our flag, robbed of their dearest rights, inventoried as goods and chattels, and plundered of their humanity by law, may look forward with new hope to their final exodus from bondage. A number of Southern gentlemen, I am aware, view the subject differently. I am entirely willing that they should. I am satisfied to find them on the right side of the question. I speak only for myself, and claim no right to express any opinion but my own. Had this policy been adopted by the government in 1832, when General Jackson first recommended it, it is highly probable that Texas, whether in or out of the Union, would never have been a slave country. She would have been compelled to exclude slavery by adopting the same landed policy in order to secure the settlement of her domain. The same cause would have prevented our Mexican War, and thus have saved to the country the millions of money and thousands of lives that were sacrificed in that unsanctified struggle for the extension of human bondage.

EDMUND RUFFIN

Born in 1794 in Prince George County, Virginia, Edmund Ruffin was educated at home except for a brief spell at William and Mary College when he was sixteen. In 1813 his father died and Ruffin took control of the family farm. At the time, the fortunes of Virginia's farmers had declined to desperate levels, due in large part to soil exhaustion. Like John Taylor, Edmund Ruffin began to search for the formula to turn Virginia's agriculture back to productivity and prosperity. In fact, Ruffin read Taylor's Arator *essays closely and experimented with his techniques. When he found those methods of limited use on his own lands, Ruffin turned his attention to the fundamentals of soil chemistry and the value of soil amendments, especially marl. In* An Essay on Calcareous Manures, *Ruffin published the results of his experiments with marls and his analysis of why calcareous amendments were so valuable. He became a lifelong advocate of these methods, expanding the essay in successive editions to nearly five hundred pages. He also founded and edited the* Farmer's Register, *another of the agricultural journals of the day, and promoted the marling of soil in its pages.*

Ruffin was also intensely concerned with political matters and was one of the South's most vocal secessionists. Finding Virginia too Unionist for his taste, Ruffin moved to South Carolina, where in 1861 he was chosen to fire one of the first shots on Fort Sumter from the batteries of Charleston's "Palmetto Guards." Later, well into his sixties, Ruffin

Portrait of Private Edmund Ruffin, c. 1865.
Courtesy National Archives and Records
Administration, Washington, DC.

served as a "temporary" private in the First Bull Run battle. When the South lost the war, Ruffin wrapped himself in the Confederate flag and committed suicide.

Ruffin saw deep connections between his agricultural and political concerns. As he shows in the following essay, he believed the South's fortunes were closely tied to the Southern farmer's treatment of the land. Destructive practices had led to a decline in the South's political and social strength in relation to the North; by extension, practices that restored the health of the land could only help the South to rise.

From "An Address on the Opposite Results of Exhausting and Fertilizing Systems of Agriculture" (1852)

The particular object of the address which will now be read, is to exhibit in full, and place in contrast, the opposite results on a country and people, of *exhausting and improving systems of Agriculture.*

In every feeling and opinion there is no more true and zealous Southerner than myself. I have long studied the domestic life and institutions, and social and moral condition of the people of the slave-holding States, and in every important respect, I may truly say, that I concur with, approve, and sympathize with yourselves on these subjects. Yet it is my present design and business not to treat of our many points of perfect agreement of opinion, but of the few of difference; not to speak of your laudable works, but your errors; and to apply to the planters of South Carolina, censure where deserved as readily as I would applaud them in other respects, which have no relation to my present general subject. Even in the general system of southern agriculture, in which there is so much to condemn, I cannot but admire the energy and intelligence exercised by the cultivators to attain the object usually sought—which is to draw from the land the greatest *immediate* production and profit. If their object were instead, as it ought to be, the greatest *continued* products and profits, and that object were pursued with as much ability, the people of South Carolina would soon stand in as exalted a position of agricultural success, as now and heretofore, for social and moral qualities, as men and citizens. Even for the few years which have passed since I investigated and reported upon your abundant resources for fertilization, and urged their use, if these means had been properly applied, already the agricultural production of half the arable lands of the State might have been increased full fifty per cent. I may dare to express this opinion, inasmuch as on a newly purchased farm, I have myself more than tripled that amount of increase by the means recommended, and within the same short time since uttering the precepts for the like improvement here.

The great error of southern agriculture is the general practice of exhausting culture—the almost universal deterioration of the productive power of the soil—which power is the main and essential foundation of all agricultural wealth. The merchant, or manufacturer, who was using (without replacing) any part of his capital to swell his yearly income—or the ship-owner, who used as profit all his receipts from freight, allowing nothing for repairs, or deterioration of capital—would be accounted by all as in the sure road to bankruptcy. The joint-stock company that should (in good faith, as many have done by designed fraud) annually pay out something of what ought to be its reserved fund, or of its actual capital, to add so much to the dividends, would soon reach the point of being obliged to reduce the dividends below the original fair rate, and, in enough time, all the capital would be so absorbed. Yet this unprofitable procedure, which would be deemed the most marvelous folly in regard to any other kind of capital invested, is precisely that which is still generally pursued by the cultivators of the soil in all the cotton producing States, and which prevailed as generally, and much longer in my own country, and which, even now, is more usual there than the opposite course of fertilizing culture. The recuperative powers of nature are indeed continually operating, and to great effect, to repair the waste of fertility caused by the destructive industry of man, and but for this natural and imperfect remedy, all these Southern States, and most of the Northern likewise, would be already barren deserts, in which agricultural labours would be hopeless of reward, and civilized men could not exist. . . .

I do not mean that it necessarily follows that the planter who exhausts his land also lessens his general wealth. Would that it were so. For, then, such certain and immediate retribution would speedily stop the whole course of wrong doing and prevent all the consequent evils. It may be rarely, and it might be never the case, that the exhauster of land becomes absolutely poorer during the operation. He will have helped to impoverish his country, and to ruin it finally, (by the same general policy being continued,) he will have destroyed as much of God's bounties as the wasted fertility, if remaining, would have supplied forever, and as many human beings as those supplies would have supported, will be prevented from existing. And yet the mighty destroyer may have increased his own wealth. Nevertheless, he does not escape his own, and even the largest share of the general loss he has caused. While thus destroying, say $20,000 worth of fertility, the planter, by the exercise of industry, economy and talent in other departments of his business, or from other resources, may have grown richer by $10,000. But if, as I believe is always true, it is as cheap and profitable to save as to waste fertility, in the whole term of culture, then the planter, in this case, might have gained in all $30,000 of capital, if he had saved, instead of wasting, the original productive power of his land.

Even if admitting the common fallacy which prevails in every newly settled country, that it is profitable to each individual cultivator to wear out his land, still, by his doing so, and all his fellow proprietors doing the like, while each one might be adding to his individual wealth, the joint labours of all would be exhausting common stock of wealth, and greatly impairing the common welfare and interest of all. The average life of a man is long enough to reduce the fertility of his cultivated land to one-half, or less. Thus, one generation of exhausting cultivators, if working together, would reduce their country to one-half of its former production, and, in proportion, would be reduced the general income, wealth and means of living, population and the products of taxation, and, in time, would as much decline the measure of moral, intellectual and social advantages, the political power and military strength of the commonwealth. The destructive operations of the exhausting cultivator have most important influence far beyond his own lands and his own personal interests. He reduces the wealth and population of his country and the world, and obstructs the progress and benefits of education, the social virtues, and even moral and religious culture. For upon the productions of the earth depends more or less the measure to be obtained, by the people of any country, of these and all other blessings which a community can enjoy. There is, however, one very numerous class of exceptions to this general rule, which is, when an agricultural people, or interest, is tributary to some other people or interest, whether foreign or at home. Such exceptions are presented in different modes, by the agriculture of Cuba being tributary to Spain, of many other countries to their own despotic and oppressive home governments; and of the southern states of this confederacy, to greater or less extent, to different pauper and plundering interests of the northern states, which, through legislative enactments, have been mainly fostered and supported by levying tribute upon southern agriculture and industry.

The reason why such woeful results of impoverishment of lands, as have been stated, are not seen to follow the causes, and speedily, is that the causes are not all in action at once and in equal progress. The labours of exhausting culture, also, are necessarily suspended, as each of the cultivators' fields is successively worn out. And when tillage so ceases, and any space is thus left at rest, Nature immediately goes to work to recruit and replace as much as possible of the wasted fertility, until another destroyer, after many years, shall return again to waste, and in much shorter time than before, the smaller stock of fertility so renewed. Thus, the whole territory so scourged, is not destroyed at one operation. But though these changes and partial recoveries are continually, to some extent, counteracting the labours for destruction, still the latter work is in general progress. It may require (as it did in my native region) more than two hundred years from the first settlement, to reach the lowest degradation. But that final result is not the less certainly to be produced by the continued action of the causes. I have witnessed at home, nearly the last stage of decline. But I have also witnessed, subsequently, and over large spaces, more than the complete resuscitation of the land, and great improvement in almost every respect, not only to individual, but to public interests; not only in regard to fertility and wealth, but also in mental, moral and social improvement.

Inasmuch as my remarks would seem to ascribe the most exhausting system of cultivation especially to the slave-holding states, the enemies of the institution of slavery might cite my opinions, if without the explanation which will now be offered, as indicating that slave labour and exhausting tillage were necessarily connected as cause and effect. I readily admit that our slave labour has served greatly to facilitate our exhausting cultivation; but only because it is a great facility—far superior to any found in the non-slave-holding states—for all agricultural operations. Of course, if our operations are exhausting of fertility, then certainly our command of cheaper and more abundant labour enables us to do the work of exhaustion, as well as all other work, more rapidly and effectually. But if directed to improving, instead of destroying fertility, then this great and valuable aid of slave-labour will as much more advance improvement, as it has generally heretofore advanced exhaustion. The enunciation of this proposition is perhaps enough. But if any, from prejudice, should deny or doubt its truth, they may see the practical proofs on all the most improved and profitable farms of Lower and Middle Virginia. On the lands of our best improvers and farmers, such as Richard Sampson, Hill Carter, John A. Selden, William B. Harrison, Willoughby Newton, and many others, slave-labour is used not only exclusively and in larger than usual proportion, (because more required on very productive land,) but is deemed indispensable to the greatest profits, and operating to produce more increase of fertility, and more agricultural profit, than can be exhibited from any purely agricultural labours and capital north of Mason and Dixon's line.

There is another stronger reason for the greater exhausting effects of Southern agriculture, and, therefore, of tillage by slave-labour. The great crops of all the slave-holding States, and especially of the more Southern—corn, tobacco, and cotton—are all tilled crops. The frequent turning and loosening of the earth by the plow and hoe—and far more, when continued without intermission year after year—advance the decomposition and waste of all organic matter, and expose the soil of all but the most level surfaces to destructive washing by rains—and rains the more heavy and destructive in power, in proportion as approaching the South. The Northern farmer is guarded from the worst of these results, not because he uses free-labour, but because his labour is so scarce and dear that he uses as little as possible for his purposes. Besides this consideration, his climate is more suitable to grass than to grain, and his other large crops are much more generally broad-cast than tilled. These are sufficient causes why, in general, the culture of land in the Northern States should be less exhausting than in the Southern, without detracting anything from the superior advantages which we of the South enjoy, in the use of African slave-labour. . . .

There is not one of the industrial classes of mankind, more estimable for private worth and social virtues, than the landholders and cultivators of the Southern States. With them, unbounded hospitality is so universal, that it is not a distinguishing virtue—and, in truth, this virtue has been carried to such excess, as to become a vicious tendency. Honourable, high-minded, kindly in feeling and action, both to neighbours and to strangers—ready to sacrifice self-interest for the public weal— such are ordinary qualities and characteristics of southern planters. Many of the most intelligent men of this generally intelligent class, are ready enough to accept and to apply to themselves and their fellow-planters, the name of "land-killers." But while thus admitting, or even assuming this term of jocose reproach, they have not deemed as censurable or injurious, their conduct on which this reproach was predicated. They have regarded their "land-killing" policy and practice merely as affecting their own personal and individual interests—and if judged by their continued action, they must believe that their interests are thereby best promoted. Their error, in regard to their own interests, great as may be, is incomparably less than the mistake as to other and general interests not being thus affected. As I have already admitted, individuals may acquire wealth by this system of impoverishing culture, though the amount of accumulation is still much abated by the attendant waste of fertility. But with the impoverishment of its soil, a country, a people, must necessarily and equally be impoverished. Individual planters may desert the fields they have exhausted in South Carolina, and find new and fertile lands to exhaust in Alabama. And when the like work of waste and desolation is completed in Alabama, the spoilers, (whether with or without retaining a portion of the spoils,) may still proceed to Texas or to California. But South Carolina and Alabama, must, nevertheless, suffer and pay the full penalty of all the impoverishment so produced. The people who remain to constitute these States respectively, as communities, are not spared one tittle of the enormous evils produced—not only those of their own destructive labours, but of all the like and previous labours of their fellow citizens and predecessors who had fled from the ruin which they had helped to produce. And these evils to the community and to posterity, greater than could be effected by the most powerful and malignant foreign enemies of any country, are the regular and deliberate work of benevolent and intelligent men, of worthy citizens, and true lovers of their country.

I will not pursue this uninviting theme to its end—that lowest depression which surely awaits every country and people subjected to the effects of the "land-killing" policy. The actual extent of the progress toward that end, throughout the Southern States, ought to be sufficiently appalling, to induce a thorough change of procedure and reformation of the agricultural system of the South.

In addition to all increase of the other benefits of agricultural improvement which have been cited—pecuniary, social, intellectual and moral—there would be an equal increase of political power, both at home and abroad, which at this and the near approaching time, would be especially important to the well being and the defence of the Southern States, and the preservation of their yet remaining rights, and always vital interests. If Virginia, South Carolina, and the other older slave-holding States, had never been reduced in productiveness, but, on the contrary, had been improved according to their capacity, they would have retained nearly all the population they have lost by emigration, and that retained population, with its increase, would have given them more than a doubled number of representatives in the Congress of the United States. This greater strength would have afforded abundant legislative safeguards against the plunderings and oppressions of tariffs to protect Northern interests—compromises (so-called) to swell Northern power—pension and bounty laws for the same purposes—and all such acts to the injury of the South, effected by the greater legislative strength of the now more powerful, and to us, the hostile and predatory States of the confederacy. Even after Virginia, with more than Esau-like fatuity, had sacrificed her magnificent north-western territory, which now constitutes five great and fertile States, (and a surplus to make, by legislative fraud, a large part of the sixth State [Minnesota],) and all of which are now among the most hostile to the rights of the people of the South—if Virginia had merely retained and improved the fertility of her present reduced surface, her people would not have removed. Their descendants would now be south of the Ohio, ready and able to maintain the rights of the Southern States, instead of a large proportion, as now, serving to swell the numbers, and give efficient power to our most malignant enemies. The loss of both political and military strength, to Virginia and South Carolina, are not less than all other losses, the certain consequences of the impoverishment of their soil.

Thomas Cole, *View from Mount Holyoke, Northampton, Massachusetts, after a Thunderstorm—*
The Oxbow, 1836. Image copyright © The Metropolitan Museum of Art/Art Resource, New York.

3. The Machine in the Garden:
The Rise of American Romanticism

There is not much hard evidence that farmers have no love of nature. Those (including some farmers) who make fun of Romantics often seem to consider it self-evident that only people who are free of the grim reality of wresting a living from the soil can enjoy the luxury of finding beauty in the fields, woods, and wild things. But the relationship between farmers and nature has rarely been so starkly adversarial. Certainly no farmer would be likely to mistake nature's cruel side, but then, neither would any other human being who has ever been sick or watched a loved one die. But to conclude that enduring harsh weather, hard work, dirt, sweat, stink, and daily contact with the dung and blood of animals makes farmers callous people without time or taste to appreciate the beauty of nature or celebrate its fecundity is just as patronizing as to portray them as simple peasants who live happily in a snug, green, peaceable kingdom. In the agrarian worldview, love of nature has generally been subordinate to the struggle to gain an honorable and independent living from nature—but it has not been absent.

Anyone who has farmed knows that the natural world is a confounding mixture of benevolence and treachery, life and death in equal and endless measure. Human beings throughout the ages have been faced with the contradiction between the rich bounty of nature and the terror of its malevolence. The typical agrarian response, perhaps, is to accept this world with stoicism and wry humor, even as one gets on with the work at hand. But besides taking satisfaction in that work, rural people have always enjoyed natural pastimes such as hunting and fishing, nutting and berrying, swimming and skating. In the early nineteenth century, Lowell mill girls wrote longingly of missing their outdoor labor and leisure back home on the farm and often devoted their Sundays to walking in the countryside, gathering wild roses. When Henry Thoreau's mother and father (who were of plain New England stock) first brought their young children to the pond for a picnic, it is safe to say that they had not yet read *Walden.*

But for all of that, the Romantic movement of the early nineteenth century was something new. It changed how Americans thought about nature and about the place of farming in nature. Some Romantic ideas were gradually incorporated into the view of nature held by farmers, but others were less compatible and have remained troubling and divisive to this day. As a new agrarianism—one that sprang as much from urban and industrial alienation as from traditional rural culture—developed in the twentieth century, the Romantic strands of the movement gained in influence. This section is devoted to the roots of this more sympathetic view of the ideal relationship between farming and nature.

It is no surprise that, for all their down-to-earth practical knowledge, most farmers have looked to the heavens for help in understanding nature and in giving meaning to their lives within nature. Nature is a formidable and fickle master that for most of human history was regarded as a power superior to man, demanding propitiation. European cultivators came to America still holding to many of these ancient, pre-Christian beliefs about how to work with the powers of nature—planting by the moon, observing astrological signs and other omens, performing small rituals for luck and protection. But by that time such beliefs had mostly been subsumed within a Christian worldview that emphasized the proper dominion of man over nature: the world was given by God to man for his use. Like all of God's gifts this implied a covenant, and so in return for it God expected man to care for his creation. Improving upon wild nature was a Christian duty, but so was stewardship of nature. The agrarian view of nature was primarily utilitarian, but not completely unadorned. The beauty of nature reflected God's glory, while the power and cruelty of nature posed a severe trial. The reward for meeting this test on earth would be to regain a truly benign garden paradise in heaven.

For American agrarians, the earthly ideal was to be sought in a "middle landscape" lying between the unimproved wilderness and the corrupt cities. Here, in fields, pastures, and meadows, nature was cultivated like a garden. This pastoral geography mirrored an ideal moral landscape—again, midway between the frontier regions where men tended to revert to savagery, and the overcrowding of Europe, where the rich ate the poor. In the middle landscape men could be free of moral and political corruption, thanks to healthy, honest toil subduing nature and because they were economically independent. This worldview was not, at first, particularly anti-industrial—there wasn't yet much in the way of industry to oppose. Agrarians were profoundly suspicious of commerce and of government—of any way that men with influence might gain luxurious wealth while producing little or nothing of value themselves. But agrarians did not have the evils of mechanization in mind when they celebrated the virtues of farming.

The Romantics did. Romanticism arose among European intellectuals such as Johann Goethe and Thomas Carlyle about the beginning of the nineteenth century, and at heart it *was* a reaction against the industrial revolution and the rapid urbanization of that period, and against the coldly mechanical economic and scientific reasoning that accompanied it. Romantics were repelled by the world they saw rising around them, and in response they moved toward a sharp critique of industrial society and an embrace of wild nature that remain deeply embedded in environmental thinking to this day. Considering the abysmal working and living conditions of the laboring masses in the booming industrial cities of Europe, the Romantics weren't just imagining the forces that ruined human health and crushed the human soul—though in truth conditions were hardly better throughout much of the impoverished European countryside. But in turning from the ugliness of industrial civilization toward an idealized wild nature, Romantics left the pastoral "middle landscape" in an uncertain position. Perhaps the countryside could achieve a harmonious balance between the forces of wilderness and civilization, requiring only a slight modification of the agrarian ideal of stewardship. Certainly, much Romantic attraction encompasses the pastoral as well as the wild. But as the nineteenth century progressed, it appeared that the same forces of natural exploitation and social alienation driving industrialization were also invading the countryside, despoiling that as well. For Romantics, not just the quality of stewardship but the very idea of man's dominion over nature was open to question, on the farm as much as in the factory. This proved more difficult to reconcile with traditional agrarianism.

Romantic thinking was brought to America by the Transcendentalists, and the central figure was Ralph Waldo Emerson. Emerson himself was enthusiastic about technological progress and man's ability to manipulate nature for his own good—to "work it up," as he put it. However, to Emerson this utilitarian approach was man applying only "half his force" to nature. The better half of man's force was to also appreciate the beauty in nature and, with a kind of intuitive leap, attempt to grasp the ideal that transcended and informed nature. This was not just an abstract philosophical notion—Emerson hoped that in America such a balance between industrial progress and natural beauty could achieve physical reality. There was so much nature available in America—so much wide open space that was (compared to conditions in Europe) just beginning to be settled and farmed. It seemed possible that industry—factories, mill villages, railroads, cities—could be introduced into this landscape in ways that maintained harmony with nature. Nestled in the countryside, factories could appear pastoral and factory workers could find easy access to nature for their recreation. The machine and the garden could flourish together.

When pristine new red-brick factory buildings and neat boardinghouses for mill girls were erected in places like Waltham and Lowell, Massachusetts, in the 1810s and 1820s, still surrounded by farmland, it looked to some like this vision was actually coming to pass. Even Congressman Davy Crockett of Kentucky left us an admiring portrait following his visit to Lowell in 1834. But as the decades passed and the cities expanded into the countryside, as destitute immigrants streamed into America from Ireland and Germany, and as laissez-faire capitalist imperatives undermined the Christian paternalism of early American industrialists, the bloom was off the rose. Trenchant observers such as Albert Brisbane and Henry Thoreau saw clearly that the condition of mill workers in America was looking more and more like that of Europe, and had only one logical way to go. If left unopposed, the same principles would inevitably infect the countryside as well. "So is your pastoral life whirled past and away," concluded Thoreau after pondering the implications of the new railroad that first passed by Walden Pond in 1844. Some, such as Brisbane and the participants in utopian experiments like Brook Farm, began to espouse a more radical reorganization of labor, capital, and land in order to deliberately balance and harmonize industrial and agrarian work, and the uses of nature. Others, such as Andrew Jackson Downing, sought to defend the harmony of the pastoral landscape from the second homes of the industrial elite beginning to invade the countryside.

As they grew increasingly disenchanted with the state of industrial civilization, Romantics looked beyond the pastoral world to wilder, purer forms of nature. They began to celebrate the sublime in waterfalls and mountains, to visit wild places, to rough it by camping and hunting for recreation. As those wild places melted away before the continued expansion of American industrial society, this impulse culminated in the rise of the wilderness movement in the late nineteenth and early twentieth centuries. But Romantics also struggled to define the place of, and ultimately to protect, smaller pockets of wildness within the pastoral world. This urge appeared in the work of Susan Fenimore Cooper mainly as nostalgia for lingering elements of the pioneer landscape in Cooperstown, New York, her home—the towering, ancient white pines, the wildflowers along the stream—once the wilderness had been safely conquered. Here, as in her father's popular Leatherstocking novels, the call of the wild had a wistful tone—a bit of Romantic poignancy added to a fundamentally agrarian celebration of the improved, cultivated landscape. Although Cooper did include some call for restraint in clearing, it was muted—perhaps partly owing to her acceptance of the conventions of female writing in her day.

The defense of the wild took on a sharper edge in the work of Henry Thoreau. Concord had been settled a century and a half before Cooperstown, and farming there was not in its second or third generation but seventh or eighth. By the nineteenth century Concord's pastoral landscape was long established, and until Thoreau's lifetime had retained a quarter to a third of its cover in woodlands, swamps, and other relatively wild places. Yet before Thoreau's eyes the commercial expansion of agriculture was transforming even this deeply settled countryside, driving forest down to the bottom 10 percent. The last vestiges of wildness seemed to be disappearing. In response to a local explosion of not only cotton mills but milk cows, Thoreau made his case for the retention of wild nature, symbolized by the weeds, birds, and woodchucks with which he shared his beans.

In pleading for the "tonic of wildness" Thoreau was not proposing that the entire town revert to wilderness or that most farmers practice haphazard "half-cultivation" within their fields. If his bean field was never completely hoed, it was because he was trying to cultivate too much land—a metaphor for what he saw happening with agriculture in Concord generally. He was arguing for the preservation of at least that last tenth of the landscape for wild things, interwoven with an agrarian landscape. But in arguing for the preservation of wildness Thoreau was not content that wild places should be set aside, undisturbed, for passive enjoyment. For Thoreau, farming and civilization itself were renewed by contact, by vigorous engagement, with wild nature. Hence his admiration for the improving farmer who had to literally swim in his cold, boggy property in the course of draining and redeeming it, or his envy for his colonial ancestors who lived early enough in Concord's history to strip tan-bark from ancient oaks. This is much closer to the old agrarian spirit of dominion, and it poses a serious dilemma for complex pastoralists like Thoreau: How can you have your wildness and conquer it too? Near the end of his life, Thoreau did call for the formal protection of unmanaged wild areas within towns like Concord. But his larger question was, can the tonic of wildness be kept alive within a working agrarian landscape? The point was not simply the preservation of wilderness: it was the *employment* of wildness to preserve the world.

To this day, there is deep ambivalence toward farming in the minds of Romantics. If wild, unsullied nature is the best nature, is farming good or bad? Is cultivation healthy engagement allowing one to live closer to nature, or is it taking part in the despoiling of nature? This was the issue engaged by Thoreau. According to Thoreau, "Husbandry was once a sacred art," and perhaps could be again. But in Concord, as far as he was concerned, the line had been crossed: too much land had been cleared, and farmers had become devoted to commerce rather than to the nobility of farm work itself.

This brings us to a second difference between Romantics and traditional agrarians, which had to do with the nature of work and the object of work, which was "independence." To the agrarian, the goal of independence was not to free the individual but to make the household economically independent and able to stand on its own—the idea of a "competency." But to the Romantic, independence was very much about freeing oneself from the yoke of economic activity that served material desires so as to devote precious time to higher aims: reading, writing, art. Even as the American mainstream—farmers included—was moving away from the old agrarian ideal of a comfortable subsistence toward a new economic ideal of unlimited improvement, Romantics were pointing in the opposite direction: toward simplifying life, limiting material consumption, and keeping work to a minimum.

The agrarian had no doubt about the necessity and virtue of work: it went with the territory of farming. The yeoman's point of pride was to recognize whatever work needed to be done and to do it, rain or shine, without needing anybody's direction or asking anybody's leave. There was a simple expectation that there would always be more than enough work, and the farmer ought to have an appetite for all of it. But Romantics had their doubts about the value of work—especially repetitive physical labor. Many considered such work drudgery, and they sought to minimize it. The

question that seemed important to them was how to share work equitably and reward it fairly—quite unlike the agrarian's instinct to just do it and to despise anyone
who wasn't able to. Henry Thoreau was of Yankee stock and seems to have relished
the physical labor of hoeing his beans, but he was anxious (only partly tongue in
cheek) that it not become a "dissipation"—that is, that it not take up too much of
his time. He advised his readers to try to get all of their work done "while the dew
is on," and so he put in only a half day with his hoe—a mere seven hours from five
o'clock until noon! That freed the afternoon and evening for other pursuits such as
walking, reading, and writing. While a workday like that may not strike many of us
today as rank idleness, no agrarian would consider it a respectable way to run a farm.

Many Romantic utopians of the time were vegetarians who hoped that by eliminating the care and feeding of livestock and by eating a simple plant diet, they
could greatly reduce the labor of farming. But it requires a great deal of strength
and vigor to work the land by hand, and a lot of commitment to stick with simple
living. Many people—especially those without farming experience—find such a
lifestyle difficult to sustain long enough to acquire the necessary skills and habits.
In the process they often expose themselves to ridicule for flagrant incompetence
and hypocrisy, as exhibited in Louisa May Alcott's acerbic take on her father's communal experiment at Fruitlands. In the end it appears that the discipline to make
such deliberate simplicity succeed, and to deal with the social dynamics of group
living among idealists, is usually to be found only in the religious foundation of
groups such as the Shakers or various monastic orders. Such an austere life may
provide an admirable model of agrarian virtue, but is surely not for everyone—least
of all most Romantics.

One of the most striking features of the Romantic view of farming was that at
bottom these utopians were as interested in the *educational* value of farming as in
the farming itself. What many of them seem to have cared about most was how to
raise children so that they could flower as individuals, and a pastoral environment
seemed perfect for this task. Education was the true vocation of many of the Romantics featured in this section, and in the end they left their most lasting mark
on American society as teachers. It is significant that some of the most successful
features of today's neoagrarian movement revolve around education as well.

The antebellum Romantics and utopians brought a new strand to American
agrarianism. Often not born to farming or part of the existing yeoman tradition, the
Romantics tended to be intellectual elites, reacting against the alienation of urban,
industrial life. They did not come from the land; they wanted to go back to the land.
Of course, they often lacked the mental toughness and the practical knowledge to
succeed at farming and, at least at first, they did tend to "romanticize" it. They hoped
to enjoy farm work but also hoped not to have to do too much of it. They saw farming as a basis for enjoying a simplified, wholesome rural lifestyle close to nature,
one that facilitated social, intellectual, and artistic pursuits. The Romantics were
perhaps more interested than most agrarians in living peacefully with wild nature,

admiring wildlife and untamed forest. This can be difficult to harmonize with cultivation in practice—the hawks have a habit of eating your hens, and the crows your corn. A balance can be struck, though never perfectly—and successful agrarian cultures have struck it in various ways. But as American market and industrial culture drove toward more complete domination and exploitation of nature during the nineteenth century, Romantic agrarians began searching in the other direction —for an agriculture grounded in nature, leaving a wilder margin to rural life.

ALBERT BRISBANE

A new vision for rural production was put into effect a short distance from Emerson and Thoreau's Concord, in the industrial centers of northern Massachusetts. Capitalist Francis Cabot Lowell and his partners designed their great factories at Waltham and Lowell in hopes that the power of the industrial revolution could be harnessed without incurring the human costs they had witnessed in Britain's manufacturing centers. Their plans resulted, for instance, in the supervised housing provided to young female workers recruited from New England farms and the provisions made for the mill girls' wholesome recreation during their spare time. These early industrialists were uneasy about the reception their crowded mills would receive in agrarian, democratic America.

As late as 1830, the experiment seemed to be working out as planned, at least in the eyes of Edward Everett, Whig politician and orator. In an Independence Day speech at Lowell, Everett paid tribute to the "holy alliance" of labor and capital in the city's mills, which "established a mutually beneficial connection between those who have nothing but their muscular power and those who are able to bring into the partnership the masses of property requisite to carry on an extensive concern. . . . This I regard as one of the greatest triumphs of humanity, morals, and, I will add, religion."

Still, the mills of New England had their detractors, who became more numerous as time passed. Among them was Albert Brisbane, born in Batavia, New York, in 1809. As a young man, Brisbane traveled to Europe to study with Georg Wilhelm Friedrich Hegel in Germany and with Charles Fourier in France. Brisbane was especially drawn to Fourier's belief that the work that society needed to have done was naturally matched by the innate inclinations of individuals. Work need not be a bane; it just needed to be organized in a way that allowed people to find the jobs—and each individual might want varied labors—to which they were attracted.

Brisbane became the leading exponent of Fourier's ideas in the United States. He abhorred the Waltham-Lowell model of organizing labor for the mills, with its monotonous work routine and submission of the worker to the endless needs of the production system. Brisbane also decried the separation of workers from the natural world. He gave vent to his feelings about factory life in this 1846 essay from the Harbinger, *which was published at the Brook Farm after its leaders had aligned the community with Fourier-ism in the mid-1840s.*

"False Association, Established by the Capitalists, Contrasted with True Association" (1846)

The doctrine of INDUSTRIAL ASSOCIATION, as we advocate it, alone can save the laboring classes from one of the most heartless and degrading despotisms which ever existed,—from an *Industrial Feudalism,* or a gigantic system of industrial monopoly, in which Capital, with a rapacity and selfishness which have no term of comparison, will reign supreme, with a monied aristocracy for sovereign, and the laboring classes as its miserable serfs and hirelings. The *germs* of this system are already planted in society and have begun to grow; let us examine the result so far, and see what promise for the future. The germ which we refer to, is our joint-stock manufacturing system, as it is now being established by the capitalists of our land.

Attention should be called to these false and oppressive industrial Associations, in which the sweat and blood of the producing classes are slowly transmuted into gold, a sacrifice to satiate the lust of mammon, the main spring of action of this age; and a remedy should be proposed before the system becomes universal and all powerful.

We have lately visited the cities of Lowell and Manchester, and have had an opportunity of examining the factory system more closely than before. We had distrusted the accounts, which we had heard from persons engaged in the Labor Reform, now beginning to agitate New England; we could scarcely credit the statements made in relation to the exhausting nature of the labor in the mills, and to the manner in which the young women, the operatives, lived in their boarding-houses, six sleeping in a room, poorly ventilated.

We went through many of the mills, talked particularly to a large number of the operatives, and ate at their boarding-houses, on purpose to ascertain by personal inspection the facts of the case. We assure our readers that very little information is possessed, and no correct judgments formed, by the public at large, of our factory system, which is the first germ of the Industrial or Commercial Feudalism, that is to spread over our land.

The commercial press, and literary men who are in general the humble servants of the great capitalists and merchants, the successful speculators and stock-jobbers, give such a *couleur de rose* to the subject whenever they write upon it, that no true information can be obtained. Let us state very briefly the leading features of the factory system.

It is to be borne in mind that these large manufactories are *Associations;* for they are established by joint-stock companies, and worked by large numbers of people, so that they combine two characteristics of associations, union of laborers and union of capitalists; they are, in addition, *industrial* associations, for they are engaged in one great branch of industry—manufactures. We state this particularly, because we wish to contrast them with the industrial associations which we aim at establishing.

Now let us examine the *kind* of industrial Associations which are established in this Christian age, by the wealthiest and most skilful business men of the community, and with the sanction and approbation of Christian and democratic editors, who denounce our plan of Association as false, infidel and oppressive. Let us look into their arrangements, and see the condition of the people, whose labors and lives are spent in these Associations, formed and controlled solely by Capital.

In Lowell live between seven and eight thousand young women, who are generally daughters of farmers of the different States of New England; some of them are members of families that were rich the generation before. What a sad prognostic for the grand-daughters of many of the wealthy of the present day, and of some of those men who have built these dens of toil for children of the poor!

The operatives work *thirteen hours* a day in the summer time, and *from daylight to dark* in the winter. At half past four in the morning the factory bell rings, and at five the girls must be in the mills. A clerk, placed as a watch, observes those who are a few minutes behind the time, and effectual means are taken to stimulate to punctuality. This is the morning commencement of the industrial discipline—(should we not rather say industrial tyranny?) which is established in these Associations of this moral and Christian community. At seven the girls are allowed thirty minutes for breakfast, and at noon thirty minutes more for dinner, except during the first quarter of the year, when the time is extended to forty-five minutes. But within this time they must hurry to their boarding-houses and return to the factory, and that through the hot sun, or the rain and cold. A meal eaten under such circumstances must be quite unfavorable to digestion and health, as any medical man will inform us. At seven o'clock in the evening the factory bell sounds the close of the day's work.

Thus thirteen hours per day of close attention and monotonous labor are exacted from the young women in these manufactories. What remains to a being when he or she has given to toil so many hours? Nothing. Strength of body and mind, the desire for any intellectual pursuits or improvement, even the desire for amusements is gone. The latter effect would no doubt please many of our austere religious Journals. They would call it, probably, a "very wholesome system of restraint," checking the desires of "the flesh," and the "promptings of the devil." So fatigued,—we should say, exhausted and worn out, but we wish to speak of the system in the simplest language,—are numbers of the girls, that they go to bed soon after their evening meal, and endeavor by a comparatively long sleep to resuscitate their weakened frames for the toils of the coming day. When Capital has got thirteen hours of labor daily out of a being, it can get nothing more. It would be a poor speculation in an industrial point of view to *own* the operative; for the trouble and expense of providing for times of sickness and old age would more than counterbalance the difference between the price of wages and the expense of board and clothing. The far greater number of fortunes, accumulated by the North in comparison with the South, shows that hireling labor is more profitable for Capital than slave labor.

Now let us examine the nature of the labor itself, and the conditions under which it is performed. Enter with us into the large rooms, when the looms are at work. The largest that we saw is in the Amoskeag Mills at Manchester. It is four hundred feet long, and about seventy broad; there are five hundred looms, and twenty-one thousand spindles in it. The din and clatter of these five hundred looms under full operation, struck us on first entering as something frightful and infernal, for it seemed such an atrocious violation of one of the faculties of the human soul, the sense of hearing. After a while we became somewhat inured to it, and by speaking quite close to the ear of an operative and quite loud, we could hold a conversation, and make the inquiries we wished.

The girls attend upon an average three looms; many attend four, but this requires a very active person, and the most unremitting care. However, a great many do it. Attention to two is as much as should be demanded of an operative. This gives us some idea of the application required during the thirteen hours of daily labor. The atmosphere of such a room cannot of course be pure; on the contrary it is charged with cotton filaments and dust, which, we were told, are very injurious to the lungs. On entering the room, although the day was warm, we remarked that the windows were down; we asked the reason, and a young woman answered very *naively*, and without seeming to be in the least aware that this privation of fresh air was anything else than perfectly natural, that "when the wind blew, the threads did not work so well." After we had been in the room for fifteen or twenty minutes, we found ourselves, as did the persons who accompanied us, in quite a perspiration, produced by a certain moisture which we observed in the air, as well as by the heat.

Such is the atmosphere, such the din and clatter, in which the young women pass thirteen hours per day, for six days in the week. It struck us with amazement when we called to mind that persons had the courage and perseverance to go through with such efforts. It seemed to us as though a hundred dollars a day would be no compensation for passing the best hours of life in these industrial gallies, these infernal dens of labor, (to use a term expressive of the fact,) built by the most selfish passion, by the unlimited and insatiate lust of wealth.

We do not blame individuals for all this; we blame the whole spirit of our People, the tendencies of our Nation; we have scarcely any means of distinguishing ourselves except by fortune; art and science are not avenues to wealth and consideration; a high political standard can only be achieved by great talent; war is greatly lessened, so that the only thing we can do is to get rich, or be nobody. This has fanned the passion for wealth into a perfect mania, and made us the most money-making, grasping and rapacious people on earth, except, perhaps, the Hollanders and Jews.

If we follow the young girls from the manufactories to their boarding-houses, we find their domestic life as uncongenial and anti-social, as their labor is severe. Prolonged and absorbing application deadens the social sympathies, or rather exhausts the whole force of the mind; then a want of the union and mingling of the two sexes is, as we remarked, extremely pernicious; the continual presence and monotonous society of women alone, is most unfavorable to the development of the social affections. It is very "moral" however, it must be admitted; and this will excuse it, of course, in the eyes of the Saints of the Press. Yet not the less is it a sacrilege, a blighting of the sympathies of the heart, and it should not be tolerated.

The young women sleep upon an average six in a room; three beds to a room. There is no privacy, no retirement here; it is almost impossible to read or write alone, as the parlor is full and so many sleep in the same chamber. A young woman remarked to us, that if she had a letter to write, she did it on the head of a band-box, sitting on a trunk, as there was not space for a table. So live and toil the young women of our country in the boarding-houses and manufactories, which the rich and influential of our land have built for them.

The Editor of the Courier and Enquirer [Colonel Webb] has often accused the Associationists of wishing to reduce men "to herd together like beasts of the field." We would ask him whether he does not find as much of what may be called "herding together" in these modern industrial Associations, established by men of his own kidney, as he thinks would exist in one of the Industrial Phalanxes, which we propose.

We would put another question to Colonel Webb, while we speak of him; we would ask him, and in all candor, whether, if by some unforeseen accident, his daughters or any beings whom he loves as tenderly as his children, were forced to work in these manufactories and live in these boarding-houses, he would not prefer that they should be laid peacefully in their graves?

It will be asked how these young women can be induced voluntarily to work in the manufactories. We answer: Poverty, the want of a home, or an uncomfortable home, the desire of aiding parents who are involved, are among the reasons. Another is, that the manufacturing companies keep recruiters traversing the country, who obtain a dollar "a head" for every girl that they can secure for the mills. They make exaggerated representations as to the amount of money which the girls can earn, and excite hopes which lead them to abandon their homes for the manufacturing towns.

As regards the effect of the factory labor upon the health, we found it very deleterious. From numerous inquiries among the young women at the looms, the following is the general result of the information which we obtained: namely, that it requires a strong and healthy woman to work steadily for one year in a mill; that all must go into the country and recruit during a portion of the year; some require but six weeks, others two months, and many three, four, and even a greater length of time. A very intelligent operative informed us that she doubted whether the girls, if a period of years were taken, could make out much more than half of the full time; she said that she herself had only been able to work eight months in two years. We are perfectly certain from personal observation, that these long hours of labor in confined rooms, are very injurious to health, and we doubt whether it would be using too harsh terms to say, that the whole system is one of slow and legal assassination.

Such is the system of false and tyrannical Industrial Association which Capital is building up amongst us. It is rapidly monopolizing the different branches of manufactures, and it will be extended to agriculture, as soon as agricultural machinery is invented. Large joint-stock farms with vast and combined agricultural arrangements will be established, and as the little mechanics, the hand-loom weavers, and so forth, have fast been disappearing, and been forced to enter the large joint-stock manufactories, so the little farmers of our country will then disappear, and be brought into the large joint-stock farms, as hirelings of the feudal monopoly. A couple of generations more will accomplish this work, and see a commercial or industrial feudalism arise and be established, which will govern the world by the power of Capital, as did the military feudalism, or feudalism of the nobles in the past, by the power of the Sword. The great bankers and merchants will be the rulers, like the barons of old; the hireling masses, the serfs. Civilization commenced with a feudalism, and if there are not devotion and intelligence enough in the people to prevent such a catastrophe, it will, according to the law of contact of extremes, terminate in a feudalism. The age by its commercial and industrial excesses and its anarchical license, called free competition, is plunging headlong into this abyss, and a general monopoly of commerce and industry must inevitably be the result of the present universal conflict and incoherence, if a true system of Association is not established.

But without wandering so far into the future, let us keep in view the manufacturing system as now established, which is the first development of this industrial feudalism.

The leading characteristics of the false system of Industrial Association, which Capital is building up so rapidly, may be summed up, as follows:

1. Subjection of Labor to Capital, and of the Laboring Classes to the Capitalists.
2. No just division of profits; all the surplus is taken by commerce and capital.
3. No association of the laborer and the capitalist, but permanent conflict of interests between the two.
4. Prolonged and excessive toil.
5. Monotony of occupations, which is deadening to the intellect and ruinous to the body.
6. Strict system of industrial discipline enforced upon the mass. This goes so far even as to say where the operatives shall live—namely, in the boarding-houses of the companies—at what hours they shall be at home, regulations as regards attending church, and so forth. In the next generation perhaps, as the system spreads, it will determine their mode of education, and fit them properly for their position.
7. War of machinery upon the laboring classes, or machinery working against instead of for the mass.

8. Anarchical competition between the operatives for work; strife for the labor which capitalists have to give; decrease of wages and increase of the hours of toil.
9. Monotonous mode of life; extreme restrictions of social ties; deadening of the affections, particularly of the family sentiments, and of love, which woman most demands.
10. Radical selfishness, or the absolute power which is possessed by capital, wielded by capital for its interests alone, and without any regard whatever to the interests of the producing classes.

Such are a few of the beauties of the Industrial Associations which the rich and great of our land are establishing.

Now the Associationists wish to establish a system of Industrial Associations of their own; so far they have the same aim in view as the capitalists. But the false and tyrannical Associations of the latter are the very opposite, are an inverted image of the true Associations, based upon justice and liberty, which we wish to organize. Let us glance at a few of the features of our plan and contrast them with the foregoing.

1. Union of Agriculture and Manufactures; or a joint-prosecution on a large scale of a great variety of branches of Agriculture and the Mechanic Arts.
2. Equitable division of profits, securing to every person, man, woman, and child, the fruits of his or her labor, capital and talent.
3. Varied occupations in agriculture, manufactures, the arts and sciences, open to the free choice of all tastes and adapted to the capacities of both sexes and all ages. (From three to four hundred branches of industry should be pursued in a large Association of eighteen hundred persons.)
4. Industry dignified and rendered *Attractive,* by a proper organization.
5. Real liberty and independence in labor; the industrial classes will lay down all laws and regulations for the government of labor and their own affairs.
6. The land and machinery represented by stock, and owned by the members, that is, by those who cultivate and work them.
7. A thorough system of industrial and intellectual education, extended to all the children.
8. True and harmonious development of the faculties of the soul, and their legitimate satisfaction in a system of society adapted to them.
9. Equal opportunities in all the spheres of life, in intellectual development, in the choice of pursuits, and in social advancement and encouragement.
10. Unity of interests, combined action and general accord of all the elements of society.

Industrial reform, or a reform in our present false systems of trade, credit, labor for wages, division of profits, and the relation of capital to labor, must take place, and a new organization of labor, based upon principles of justice and right, must be devised and established. They form the grand problem which this age must solve. Without its solution, no further social progress is possible; we have achieved about all that can be achieved by political liberty and a just political organization. The next great step is a true organization of industry, which will form the *material* basis of the prosperity, the real liberty and intelligence of the people. We go further, and say that if this problem be not properly solved, the mass of the people of this country will, in a century more, be brought under an Industrial Despotism,—a vast system of commercial and industrial Monopoly, more rapacious, more vile and more oppressive than the political despotisms of the past. The manufacturing system in England, where the industrial feudalism is ripening rapidly, gives us a foretaste of what this system is to be when fully developed and universalized.

A true system of Industrial Association must be established, or a false system will prevail. Association is the point to which nature wishes to bring man, for it is her universal law, (except in the infancy or early growth of society,) and her ends will be attained. If the people possess the requisite intelligence and devotion, and can withstand the influence of their false guides and leaders—the commercial press and party politicians, and their own selfishness, they can organize rapidly and peacefully a true system, for the world is ready for it, and the science is discovered: if not, they will be brought into it by constraint and violence,—by the tyrannical power of capital, after passing through a period of false association; for such is the lot of the ignorant, the selfish, and the besotted; and they will wander through some generations of discipline, oppression and suffering, seeking blindly the end, before they attain it.

And now we ask opponents, which is the best system of Association, that which we propose, or that which your capitalists are establishing? Or, if this alternative of a true or false system of Industrial Association be denied, then, restricting the question to its narrowest limits:—Which is the most just and human organization, a joint-stock manufactory with its boarding-house arrangements such as our "wise and humane rich" are establishing, or an Association such as our "visionaries" and "infidels" propose?

ELIZABETH PALMER PEABODY

Elizabeth Palmer Peabody, educator, writer, publisher, and all-round enthusiast for social reform, was a leading figure in the Transcendentalist movement that took shape in and around Boston during the 1830s and 1840s. She was born in Billerica, Massachusetts, in 1804 into a family prone to financial trouble but marked by literary cultivation, especially in the person of Elizabeth's mother, Eliza. The Peabody family lived in Salem, Massachusetts, for much of Elizabeth's childhood before moving to the town of Lancaster in that state. After arriving in Lancaster, Elizabeth, then sixteen, opened a school, the first of many such educational ventures she would undertake. The years she spent in the bucolic town were among the happiest in Elizabeth's life, in part for the beauty of its setting; she and her sisters explored the local woods, meadows, and river.

Elizabeth was well prepared to teach. Her mother, also a teacher, had taught Elizabeth alongside her other pupils, and Elizabeth had learned Latin from her father at home. Moreover, she had in high degree the old Puritan virtues of self-examination and self-improvement. Elizabeth set out to learn Hebrew at age twelve in order to better understand the Old Testament. At thirteen she read the New Testament thirty times through in a three-month period, each time with a different point of doctrine in mind to untangle. As a girl in Salem, she discussed natural sciences with Nathaniel Bowditch and theology with a local minister. Nor did her efforts at self-education end when she became a teacher. Within two years she was studying Greek with Ralph Waldo Emerson and had become a confidante of William Ellery Channing.

Peabody took part in many of the encounters that made up the Transcendentalist movement. She was one of the few women in the Transcendental Club and the first to host one of its gatherings. She served under Bronson Alcott at his experimental Temple School in Boston, and her notes from that effort provided the basis for her influential book Record of a School. *In 1839 the bookshop Peabody opened became an important meeting place for Boston's freethinkers and reformers.*

Given her place among the Transcendentalists, it was inevitable that Peabody would be among the earliest to hear about the Brook Farm experiment—even if her sister Sophia and brother-in-law Nathaniel Hawthorne hadn't been a part of the founding generation at the farm. Peabody's interest in social reform and her love of nature drew her irresistibly to support the farm. She published the following article about the Brook Farm colony in the January 1842 issue of the Dial.

From "Plan of the West Roxbury Community" (1842)

In the last number of the Dial were some remarks, under the perhaps ambitious title, of "A Glimpse of Christ's Ideal Society;" in a note to which, it was intimated, that in this number, would be given an account of an attempt to realize in some degree this great Ideal, by a little company in the midst of us, as yet without name or visible existence. The attempt is made on a very small scale. A few individuals, who, unknown to each other, under different disciplines of life, reacting from different social evils, but aiming at the same object,—of being wholly true to their natures as men and women; have been made acquainted with one another, and have determined to become the Faculty of the Embryo University.

In order to live a religious and moral life worthy of the name, they feel it is necessary to come out in some degree from the world, and to form themselves into a community of property, so far as to exclude competition and the ordinary rules of trade;—while they reserve sufficient private property, or the means of obtaining it, for all purposes of independence, and isolation at will. They have bought a farm, in order to make agriculture the basis of their life, it being the most direct and simple in relation to nature.

A true life, although it aims beyond the highest star, is redolent of the healthy earth. The perfume of clover lingers about it. The lowing of cattle is the natural bass to the melody of human voices.

On the other hand, what absurdity can be imagined greater than the institution of cities? They originated not in love, but in war. It was war that drove men together in multitudes, and compelled them to stand so close, and build walls around them. This crowded condition produces wants of an unnatural character, which resulted in occupations that regenerated the evil, by creating artificial wants. Even when that thought of grief,

> I know, where'er I go
> That there hath passed away a glory from the Earth,

came to our first parents, as they saw the angel, with the flaming sword of self-consciousness, standing between them and the recovery of spontaneous Life and Joy, we cannot believe they could have anticipated a time would come, when the sensuous apprehension of Creation—the great symbol of God—would be taken away from their unfortunate children,—crowded together in such a manner as to shut out the free breath and the Universal Dome of Heaven, some opening their eyes in the dark cellars of the narrow, crowded streets of walled cities. How could they have believed in such a conspiracy against the soul, as to deprive it of the sun and sky, and glorious apparelled Earth!—The growth of cities, which were the embryo of nations hostile to each other, is a subject worthy the thoughts and pen of the philosophic historian. Perhaps nothing would stimulate the courage to seek, and hope to attain social good, so much, as a profound history of the origin, in the mixed nature of man, and the exasperation by society, of the various organized Evils under which humanity groans. Is there anything, which exists in social or political life, contrary to the soul's Ideal? That thing is not eternal, but finite, saith the Pure Reason. It has a beginning, and so a history. What man has done, man may *undo*. "By man came death; by man also cometh the resurrection from the dead."

The plan of the Community, as an Economy, is in brief this; for all who have property to take stock, and receive a fixed interest thereon; then to keep house or board in commons, as they shall severally desire, at the cost of provisions purchased at wholesale, or raised on the farm; and for all to labor in community, and be paid at a certain rate an hour, choosing their own number of hours, and their own kind of work. With the results of this labor, and their interest, they are to pay their board, and also purchase whatever else they require at cost, at the warehouses of the Community, which are to be filled by the Community as such. To perfect this economy, in the course of time they must have all trades, and all modes of business carried on among themselves, from the lowest mechanical trade, which contributes to the health and comfort of life, to the finest art which adorns it with food or drapery for the mind.

All labor, whether bodily or intellectual, is to be paid at the same rate of wages; on the principle, that as the labor become merely bodily, it is a greater sacrifice to the individual laborer, to give his time to it; because time is desirable for the cultivation of the intellect, in exact proportion to ignorance. Besides, intellectual labor involves in itself higher pleasures, and is more its own reward, than bodily labor.

Another reason, for setting the same pecuniary value on every kind of labor, is, to give outward expression to the great truth, that all labor is sacred, when done for a common interest. Saints and philosophers already know this, but the childish world does not; and very decided measures must be taken to equalize labors, in the eyes of the young of the community, who are not beyond the moral influences of the world without them. The community will have nothing done within its precincts, but what is done by its own members, who stand all in social equality;—that the children may not "learn to expect one kind of service from Love and Goodwill, and another from the obligation of others to render it,"—a grievance of the common society stated, by one of the associated mothers, as destructive of the soul's simplicity. Consequently, as the Universal Education will involve all kinds of operation, necessary to the comforts and elegances of life, every associate, even if he be the digger of a ditch as his highest accomplishment, will be an instructor in that to the young members. Nor will this elevation of bodily labor be liable to lower the tone of manners and refinement in the community. The "children of light" are not altogether unwise in their generation. They have an invisible but all-powerful guard of principles. Minds incapable of refinement, will not be attracted into this association. It is an Ideal community, and only to the ideally inclined will it be attractive; but these are to be found in every rank of life, under every shadow of circumstance. Even among the diggers in the ditch are to be found some, who through religious cultivation, can look down, in meek superiority, upon the outwardly refined, and the book-learned.

Besides, after becoming members of this community, none will be engaged merely in bodily labor. The hours of labor for the Association will be limited by a general law, and can be curtailed at the will of the individual still more; and means will be given to all for intellectual improvement and for social intercourse, calculated to refine and expand. The hours redeemed from labor by community, will not be reapplied to the acquisition of wealth, but to the production of intellectual goods. This community aims to be rich, not in the metallic representative of wealth, but in the wealth itself, which money should represent; namely, LEISURE TO LIVE IN ALL THE FACULTIES OF THE SOUL. As a community, it will traffic with the world at large, in the products of Agricultural labor; and it will sell education to as many young persons as can be domesticated in the families, and enter into the common life with their own children. In the end, it hopes to be enabled to provide—not only all the necessaries, but all the elegances desirable for bodily and for spiritual health; books, apparatus, collections for science, works of art, means of beautiful amusement. These things are to be common to all; and thus that object, which alone gilds and refines the passion for individual accumulation, will no longer exist for desire, and whenever the Sordid passion appears, it will be seen in its naked selfishness. In its ultimate success, the community will realize all the ends which selfishness seeks, but involved in spiritual blessings, which only greatness of soul can aspire after.

And the requisitions on the individuals, it is believed, will make this the order forever. The spiritual good will always be the condition of the temporal. Every one must labor for the community in a reasonable degree, or not taste its benefits. The principles of the organization therefore, and not its probable results in future time, will determine its members. These principles are coöperation in social matters, instead of competition or balance of interests; and individual self-unfolding, in the faith that the whole soul of humanity is in each man and woman. The former is the application of the love of man; the latter of the love of God, to life. Whoever is satisfied with society, as it is; whose sense of justice is not wounded by its common action, institutions, spirit of commerce, has no business with this community; neither has any one who is willing to have other men (needing more time for intellectual cultivation than himself) give their best hours and strength to bodily labor, to secure himself immunity therefrom. And whoever does not measure what society owes to its members of cherishing and instruction, by the needs of the individuals that compose it, has no lot in this new society. Whoever is willing to receive from his fellow men that, for which he gives no equivalent, will stay away from its precincts forever.

But whoever shall surrender himself to its principles, shall find that its yoke is easy and its burden light. Everything can be said of it, in a degree, which Christ said of his kingdom, and therefore it is believed that in some measure it does embody his Idea. For its Gate of entrance is straight and narrow. It is literally a pearl *hidden in a field.* Those only who are willing to lose their life for its sake shall find it. Its voice is that which sent the young man sorrowing away. "Go sell all thy goods and give to the poor, and then come and follow me." "Seek first the kingdom of Heaven, and its righteousness, and all other things shall be added to you."

This principle, with regard to labor, lies at the root of moral and religious life; for it is not more true that "money is the root of all evil," than that *labor is the germ of all good.* . . .

It seems impossible that the little organization can be looked on with any unkindness by the world without it. Those, who have not the faith that the principles of Christ's kingdom are applicable to real life in the world, will smile at it, as a visionary attempt. But even they must acknowledge it can do no harm, in any event. If it realizes the hopes of its founders, it will immediately become a manifold blessing. Its moral *aura* must be salutary. As long as it lasts, it will be an example of the beauty of brotherly love. If it succeeds in uniting successful labor with improvement in mind and manners, it will teach a noble lesson to the agricultural population, and do something to check that rush from the country to the city, which is now stimulated by ambition, and by something better, even a desire for learning. Many a young man leaves the farmer's life, because only by so doing can he have intellectual companionship and opportunity; and yet, did he but know it, professional life is ordinarily more unfavorable to the perfection of the mind, than the farmer's life; if the latter is lived with wisdom and moderation, and the labor mingled as it might be with study. This community will be a school for young agriculturalists, who may learn within its precincts, not only the skilful practice, but the scientific reasons of their work, and be enabled afterwards to improve their art continuously. It will also prove the best of normal schools, and as such, may claim the interest of those, who mourn over the inefficiency of our common school system, with its present ill-instructed teachers. . . .

There may be some persons, at a distance, who will ask, to what degree has this community gone into operation? We cannot answer this with precision, for we do not write as organs of this association, and have reason to feel, that if we applied to them for information, they would refuse it, out of their dislike to appear in public. We desire this to be distinctly understood. But we can see, and think we have a right to say, that it has purchased the Farm, which some of its members cultivated for a year with success, by way of trying their love and skill for agricultural labor;—that in the only house they are as yet rich enough to own, is collected a large family, including several boarding scholars, and that all work and study together. They seem to be glad to know of all, who desire to join them in the spirit, that at any moment, when they are able to enlarge their habitations, they may call together those that belong to them.

LOUISA MAY ALCOTT

Louisa May Alcott was born on November 29, 1832, the second child in the large, close-knit family raised by Abigail May Alcott and Louisa's famously eccentric father, Bronson—"the most transcendental of all the Transcendentalists." Bronson Alcott was a onetime peddler, educator, and freelance philosopher. He was a devoted husband and father, but impractical in the extreme and endlessly impecunious.

After the failure of various teaching ventures, including his well-known school at the Masonic Temple in Boston, Bronson Alcott decided to attempt communal living on a farm in Harvard, Massachusetts. A paltry handful of apple trees on the site of the farm earned it the name Fruitlands. Joining the Alcotts were several friends and admirers, including the Englishman Charles Lane, who provided most of the start-up money for the farm. Another who joined was Joseph Palmer, the only member who had any real farming experience.

The Alcotts settled at Fruitlands in June of 1843, with others arriving later that summer. The society's principles included a strict vegetarian ethic, which extended to a reluctance to use horses or oxen as draft animals. The Alcotts' daily regimen included cold-water baths even as the months passed into autumn. In planting, Alcott and his disciples favored "aspiring vegetables," those that grew upward, over those like potatoes that grew down in the soil. Though the laborers' efforts were erratic, the farm did attain a meager harvest in the autumn. Unfortunately, harvesttime coincided with a reform meeting to which Bronson and several other men set forth, leaving Abigail and the children to gather what they could for the winter.

By the end of the year many of the members had left, and the Alcotts were threatened with starvation. In early winter they abandoned the effort, making it one of the least successful of the many experiments in communal living conducted during the 1840s and 1850s. The Alcott family's poverty wouldn't end until after 1868 when Louisa May first published Little Women, *which enjoyed immediate and lasting success. In the meantime, the family scraped by largely thanks to the efforts of Abigail and Louisa May, who took in sewing and entered into domestic service. Louisa May's practicality is reflected in her parody of the Fruitlands episode, "Transcendental Wild Oats," a barely fictionalized retelling of the Alcotts' experience in communal farming.*

From "Transcendental Wild Oats" (1873)

On the first day of June, 184–, a large wagon, drawn by a small horse and containing a motley load, went lumbering over certain New England hills, with the pleasing accompaniments of wind, rain, and hail. A serene man with a serene child upon his knee was driving, or rather being driven, for the small horse had it all his own way. A brown boy with William Penn style of countenance sat beside him, firmly embracing a bust of Socrates. Behind them was an energetic-looking woman, with a benevolent brow, satirical mouth, and eyes brimful of hope and courage. A baby reposed upon her lap, a mirror leaned against her knee, and a basket of provisions danced about at her feet, as she struggled with a large, unruly umbrella. Two blue-eyed little girls, with hands full of childish treasures, sat under one old shawl, chatting happily together.

In front of this lively party stalked a tall, sharp-featured man, in a long blue cloak; and a fourth small girl trudged along beside him through the mud as if she rather enjoyed it.

The wind whistled over the bleak hills; the rain fell in a despondent drizzle, and twilight began to fall. But the calm man gazed as tranquilly into the fog as if he beheld a radiant bow of promise spanning the gray sky. The cheery woman tried to cover every one but herself with the big umbrella. The brown boy pillowed his head on the bald pate of Socrates and slumbered peacefully. The little girls sang lullabies to their dolls in soft, maternal murmurs. The sharp-nosed pedestrian marched steadily on, with the blue cloak streaming out behind him like a banner; and the lively infant splashed through the puddles with a duck-like satisfaction pleasant to behold.

Thus these modern pilgrims journeyed hopefully out of the old world, to found a new one in the wilderness.

The editors of *The Transcendental Tripod* had received from Messrs. Lion & Lamb (two of the aforesaid pilgrims) a communication from which the following statement is an extract:

"We have made arrangements with the proprietor of an estate of about a hundred acres which liberates this tract from human ownership. Here we shall prosecute our effort to initiate a Family in harmony with the primitive instincts of man.

"Ordinary secular farming is not our object. Fruit, grain, pulse, herbs, flax, and other vegetable products, receiving assiduous attention, will afford ample manual occupation, and chaste supplies for the bodily needs. It is intended to adorn the pastures with orchards, and to supersede the labor of cattle by the spade and the pruning-knife.

"Consecrated to human freedom, the land awaits the sober culture of devoted men. Beginning with small pecuniary means, this enterprise must be rooted in a reliance on the succors of an ever-bounteous Providence, whose vital affinities being secured by this union with uncorrupted field and unworldly persons, the cares and injuries of a life of gain are avoided.

"The inner nature of each member of the Family is at no time neglected. Our plan contemplates all such disciplines, cultures, and habits as evidently conduce to the purifying of the inmates.

"Pledged to the spirit alone, the founders anticipate no hasty or numerous addition to their numbers. The kingdom of peace is entered only through the gates of self-denial; and felicity is the test and the reward of loyalty to the unswerving law of Love."

This prospective Eden at present consisted of an old red farm-house, a dilapidated barn, many acres of meadow-land, and a grove. Ten ancient apple trees were all the "chaste supply" which the place offered as yet; but, in the firm belief that plenteous orchards were soon to be evoked from their inner consciousness, these sanguine founders had christened their domain Fruitlands.

Here Timon Lion intended to found a colony of Latter Day Saints, who, under his patriarchal sway, should regenerate the world and glorify his name for ever. Here Abel Lamb, with the devoutest faith in the high ideal which was to him a living truth, desired to plant a Paradise, where Beauty, Virtue, Justice, and Love might live happily together, without the possibility of a serpent entering in. And here his wife, unconverted but faithful to the end, hoped, after many wanderings over the face of the earth, to find rest for herself and a home for her children.

"There is our new abode," announced the enthusiast, smiling with a satisfaction quite undamped by the drops dripping from his hatbrim, as they turned at length into a cart-path that wound along a steep hillside into a barren-looking valley.

"A little difficult of access," observed his practical wife, as she endeavored to keep her various household goods from going overboard with every lurch of the laden ark.

"Like all good things. But those who earnestly desire and patiently seek will soon find us," placidly responded the philosopher from the mud, through which he was now endeavoring to pilot the much-enduring horse.

"Truth lies at the bottom of a well, Sister Hope," said Brother Timon, pausing to detach his small comrade from a gate, whereon she was perched for a clearer gaze into futurity.

"That's the reason we so seldom get at it, I suppose," replied Mrs. Hope, making a vain clutch at the mirror, which a sudden jolt sent flying out of her hands.

"We want no false reflections here," said Timon, with a grim smile, as he crunched the fragments under foot in his onward march.

Sister Hope held her peace, and looked wistfully through the mist at her promised home. The old red house with a hospitable glimmer at its windows cheered her eyes; and considering the weather, was a fitter refuge than the sylvan bowers some of the more ardent souls might have preferred.

The newcomers were welcomed by one of the elect precious—a regenerate farmer, whose idea of reform consisted chiefly in wearing white cotton raiment and shoes of untanned leather. This costume, with a snowy beard, gave him a venerable, and at the same time a somewhat bridal appearance.

The goods and chattels of the Society not having arrived, the weary family reposed before the fire on blocks of wood, while Brother Moses White regaled them with roasted potatoes, brown bread and water, in two plates, a tin pan, and one mug—his table service being limited. But, having cast the forms and vanities of a depraved world behind them, the elders welcomed hardship with the enthusiasm of new pioneers, and the children heartily enjoyed this foretaste of what they believed was to be a sort of perpetual picnic.

During the progress of this frugal meal, two more brothers appeared. One was a dark, melancholy man, clad in homespun, whose peculiar mission was to turn his name hind part before and use as few words as possible. The other was a bland, bearded Englishman, who expected to be saved by eating uncooked food and going without clothes. He had not yet adopted the primitive costume, however; but contented himself with meditatively chewing dry beans out of a basket.

"Every meal should be a sacrament, and the vessels used beautiful and symbolical," observed Brother Lamb, mildly, righting the tin pan slipping about on his knees. "I priced a silver service when in town, but it was too costly; so I got some graceful cups and vases of Britannia ware."

"Hardest things in the world to keep bright. Will whiting be allowed in the community?" inquired Sister Hope, with a housewife's interest in labor-saving institutions.

"Such trivial questions will be discussed at a more fitting time," answered Brother

Timon, sharply, as he burnt his fingers with a very hot potato. "Neither sugar, molasses, milk, butter, cheese, nor flesh are to be used among us, for nothing is to be admitted which has caused wrong or death to man or beast."

"Our garments will be linen till we learn to raise our own cotton or some substitute for woollen fabrics," added Brother Abel, blissfully basking in an imaginary future as warm and brilliant as the generous fire before him.

"Haou abaout shoes?" asked Brother Moses, surveying his own with interest.

"We must yield on that point till we can manufacture an innocent substitute for leather. Bark, wood, or some durable fabric will be invented in time. Meanwhile, those who desire to carry out our idea to the fullest extent can go barefooted," said Lion, who liked extreme measures. . . .

Such farming probably was never seen before since Adam delved. The band of brothers began by spading garden and field; but a few days of it lessened their ardor amazingly. Blistered hands and aching backs suggested the expediency of permitting the use of cattle till the workers were better fitted for noble toil by a summer of the new life.

Brother Moses brought a yoke of oxen from his farm—at least, the philosophers thought so till it was discovered that one of the animals was a cow; and Moses confessed that he "must be let down easy, for he couldn't live on garden sarse entirely."

Great was Dictator Lion's indignation at this lapse from virtue. But time pressed, the work must be done; so the meek cow was permitted to wear the yoke and the recreant brother continued to enjoy forbidden draughts in the barn, which dark proceeding caused the children to regard him as one set apart for destruction.

The sowing was equally peculiar, for, owing to some mistake, the three brethren, who devoted themselves to this graceful task, found when about half through the job that each had been sowing a different sort of grain in the same field; a mistake which caused much perplexity, as it could not be remedied; but, after a long consultation and a good deal of laughter, it was decided to say nothing and see what would come of it.

The garden was planted with a generous supply of useful roots and herbs; but, as manure was not allowed to profane the virgin soil, few of these vegetable treasures ever came up. Purslane reigned supreme, and the disappointed planters ate it philosophically, deciding that Nature knew what was best for them, and would generously supply their needs, if they could only learn to digest her "sallets" and wild roots. . . .

They preached vegetarianism everywhere and resisted all temptations of the flesh, contentedly eating apples and bread at well-spread tables, and much afflicting hospitable hostesses by denouncing their food and taking away their appetites, discussing the "horrors of shambles," the "incorporation of the brute in man," and "on elegant abstinence the sign of a pure soul." But, when the perplexed or offended ladies asked what they should eat, they got in reply a bill of fare consisting of "bowls of sunrise for breakfast," "solar seeds of the sphere," "dishes from Plutarch's chaste table," and other viands equally hard to find in any modern market.

Reform conventions of all sorts were haunted by these brethren, who said many wise things and did many foolish ones. Unfortunately, these wanderings interfered with their harvest at home; but the rule was to do what the spirit moved, so they left their crops to Providence and went a-reaping in wider and, let us hope, more fruitful fields than their own.

Luckily, the earthly providence who watched over Abel Lamb was at hand to glean the scanty crop yielded by the "uncorrupted land," which, "consecrated to human freedom," had received "the sober culture of devout men."

About the time the grain was ready to house, some call of the Oversoul wafted all the men away. An easterly storm was coming up and the yellow stacks were sure to be ruined. Then Sister Hope gathered her forces. Three little girls, one boy (Timon's son), and herself, harnessed to clothes-baskets and Russia-linen sheets, were the only teams she could command; but with these poor appliances the indomitable woman got in the grain and saved food for her young, with the instinct and energy of a mother-bird with a brood of hungry nestlings to feed.

This attempt at regeneration had its tragic as well as comic side, though the world only saw the former.

With the first frosts, the butterflies, who had sunned themselves in the new light through the summer, took flight, leaving the few bees to see what honey they had stored for winter use. Precious little appeared beyond the satisfaction of a few months of holy living.

At first it seemed as if a chance to try holy dying also was to be offered them. Timon, much disgusted with the failure of the scheme, decided to retire to the Shakers, who seemed to be the only successful community going.

"What is to become of us!" asked Mrs. Hope, for Abel was heartbroken at the bursting of his lovely bubble.

"You can stay here, if you like, till a tenant is found. No more wood must be cut, however, and no more corn ground. All I have must be sold to pay the debts of the concern, as the responsibility is mine," was the cheering reply.

"Who is to pay us for what we have lost? I gave all I had—furniture, time, strength, six months of my children's lives—and all are wasted. Abel gave himself body and soul, and is almost wrecked by hard work and disappointment. Are we to have no return for this, but leave to starve and freeze in an old house, with winter at hand, no money, and hardly a friend left, for this wild scheme has alienated nearly all we had. You talk much about justice. Let us have a little, since there is nothing else left."

But the woman's appeal met with no reply but the old one: "It was an experiment. We all risked something, and must bear our losses as we can."

With this cold comfort, Timon departed with his son, and was absorbed into the Shaker brotherhood, where he soon found that the order of things was reversed, and it was all work and no play.

ANDREW JACKSON DOWNING

Andrew Jackson Downing, born in 1815, was raised in Newburgh, New York, along the Hudson River. His father owned a nursery in which Downing worked and which he eventually took over. Through this work and his intense interest in local plant and animal life, Downing struck up a series of fortuitous friendships, including one with the Austrian Baron de Liderer, a botanist and mineralogist. These relationships fostered Downing's interest in landscape design and channeled many of the latest European notions about design his way.

Turning these ideas over in his mind and seeking to apply them to the American land-scape, Downing wrote A Treatise on the Theory and Practice of Landscape Gardening, Adapted to North America, *published in 1841. The book was printed in many editions over the course of the next eighty years. Downing also took a strong interest in architecture. He designed his own home in Newburgh and published several works on building, including the 1850 book* The Architecture of Country Houses. *Downing's writings on architecture and landscaping are full of suggestions and exhortations for improved taste. Historian Steven Stoll had this in mind when he described Downing as a sort of Martha Stewart for his time. But Downing's deeper interest should be taken into account: culti-vating an aesthetic fully integrated into the economic and social life of the American countryside and consistent with the traditions of rural virtue. Downing's wide readership suggests he struck a responsive chord in writing about these concerns.*

"Cockneyism in the Country" (1849)

When a farmer, who visits the metropolis once a year, stares into the shop win-dows in Broadway, and stops now and then with an indefinite curiosity at the corners of the streets, the citizens smile, with the satisfaction of superior knowledge, at the awkward airs of the countryman in town.

But how shall we describe the conduct of the true *cockneys* in the country? How shall we find words to express our horror and pity at the cockneyisms with which they deform the landscape? How shall we paint, without the aid of Hogarth and Cruikshanks, the ridiculous insults which they often try to put upon nature and truth in their cottages and country-seats?

The countryman in town is at least modest. He has, perhaps, a mysterious though mistaken respect for men who live in such prodigiously fine houses, who drive in coaches with liveried servants, and pay thousands for the transfer of little scraps of paper, which they call stocks.

But the true cit is brazen and impertinent in the country. Conscious that his clothes are designed, his hat fabricated, his tilbury built, by the only *artists* of their several professions on this side of the Atlantic, he pities and despises all who do not bear the outward stamp of the same coinage. He comes in the country to rusticate, (that is, to recruit his purse and his digestion,) very much as he turns his horse out to grass; as a means of gaining strength sufficient to go back again to the only arena in which it is worth while to exhibit his powers. He wonders how people can live in the country from choice, and asks a solemn question, now and then, about passing the winter there, as he would about a passage through Behring's Straits, or a pic-nic on the borders of the Dead Sea.

But this is all very harmless. On their own ground, country folks have the advantage of the cockneys. The scale is turned then; and knowing perfectly well how to mow, cradle, build stone walls and drive oxen,—undeniably useful and substantial kinds of knowledge,—they are scarcely less amused at the fine airs and droll ignorances of the cockney in the country, who does not know a bullrush from a butternut, than the citizens are in town at their ignorance of an air of the new opera, or the step of the last redowa.

But if the cockney visitor is harmless, the cockney resident is not. When the downright citizen retires to the country,—not because he has any taste for it, but because it is the fashion to have a country house,—he often becomes, perhaps for the first time in his life, a dangerous member of society. There is always a certain influence about the mere possessor of wealth, that dazzles us, and makes us see things in a false light; and the cockney has wealth. As he builds a house which costs five times as much as that of any of his country neighbors, some of them, who take it for granted that wealth and taste go together, fancy the cockney house puts their simple, modest cottages to the blush. Hence, they directly go to imitating it in their moderate way; and so, a quiet country neighborhood is as certainly tainted with the malaria of cockneyism, as it would be by a ship-fever, or the air of the Pontine marshes.

The cockneyisms which are fatal to the peace of mind, and more especially to the right feeling of persons of good sense and propriety in the country, are those which have perhaps a real meaning and value in town; which are associated with excellent houses and people there; and which are only absurd and foolish when transplanted, without the least reflection or adaptation, into the wholly different and distinct condition of things in country life.

It would be too long and troublesome a task to give a catalogue of these sins against good sense and good taste, which we every day see perpetrated by people who come from town, and who, we are bound to say, are far from always being cockneys; but who, nevertheless, unthinkingly perpetrate these ever to be condemned cockneyisms. Among them, we may enumerate, as illustrations,—building large houses, only to shut up the best rooms and live in the basement; placing the first story so high as to demand a long flight of steps to get into the front door; placing the dining-room below stairs, when there is abundant space on the first floor; using the iron railings of street doors in town to porches and piazzas in the country; arranging suites of parlors with folding doors, precisely like a town house, where other and far more convenient arrangements could be made; introducing plate glass windows, and ornate stucco cornices in cottages of moderate size and cost; building large parlors for display, and small bed-rooms for daily use; placing the house so near the street (with acres of land in the rear) as to destroy all seclusion, and secure all possible dust; and all the hundred like expedients, for producing the utmost *effect* in a small space in town, which are wholly unnecessary and uncalled for in the country.

We remember few things more unpleasant than to enter a cockney house in the country. As the highest ideal of beauty in the mind of its owner is to reproduce, as nearly as possible, a fac-simile of a certain kind of town house, one is distressed with the entire want of fitness and appropriateness in every thing it contains. The furniture is all made for display, not for use; and between a profusion of gilt orna-ments, embroidered white satin chairs, and other like finery, one feels that one has no rest for the sole of his foot.

We do not mean, by these remarks, to have it understood that we do not admire really beautiful, rich and tasteful furniture, or ornaments and decorations belonging to the interior and exterior of houses in the country. But we only admire them when they are introduced in the right manner and the right place. In a country house of large size—a mansion of the first class—where there are rooms in abundance for all purposes, and where a feeling of comfort, luxury, and wealth, reigns throughout, there is no reason why the most beautiful and highly finished decorations should not be seen in its drawing-room or saloon,—always supposing them to be tasteful and appropriate; though we confess our feeling is, that a certain *soberness* should distinguish the richness of the finest mansion in the country from that in town. Still, in a villa or mansion, where all the details are carefully elaborated, where there is no neglect of essentials in order to give effect to what first meets the eye, where every thing is substantial and genuine, and not trick and tinsel,—there one expects to see more or less of the luxury of art in its best apartments.

But all this pleasure vanishes in tawdry and tinsel *imitation* of costly and ex-pensive furniture, to be found in cockney country houses. Instead of a befitting harmony through the whole house, one sees many minor comforts visibly sacrificed to produce a little extra show in the parlor; mock "fashionable" furniture, which, instead of being really fine, has only the look of finery, usurps in the principal room the place of the becoming, unpretending and modest fittings that belong there; and one is constantly struck with the *effort* which the cottage is continually making to look like the town house, rather than to wear its own more appropriate and becom-ing modesty of expression.

The pith of all that should be said on this subject, lies in a few words, viz., that *true taste lies in the union of the beautiful and the significant.* Hence, as a house in the country is quite distinct in character and uses, in many respects, from a house in town, it should always be built and furnished upon a widely different principle. It is far better, in a country house, to have an abundance of space, as many rooms as possible on a floor, the utmost convenience of arrangement, and a thorough realization of comfort throughout, than a couple of very fine apartments, loaded with showy furniture, "in the latest style," at the expense of the useful and convenient every where else.

And we may add to this, that the superior charm of significance or appropriateness is felt instantly by every one, when it is attained—though *display* only imposes on vulgar minds. We have seen a cottage where the finest furniture was of oak in simple forms, where every thing like display was unknown, where every thing costly was eschewed, but where you felt, at a glance, that there was a prevailing taste and fitness, that gave a meaning to all, and brought all into harmony; the furniture with the house, the house with the grounds, and all with the life of its inmates. This cottage, we need scarcely say, struck all who entered it with a pleasure more real and enduring than that of any costly mansion in the land. The pleasure arose from the feeling that all was significant; that the cottage, its arrangement, its furniture, and its surroundings, were all in keeping with the country, with each other and with their uses; and that no cockneyisms, no imitations of city splendor, had violated the simplicity and modesty of the country.

There must with us be progress in all things; and an American cannot but be proud of the progress of taste in this country. But as a great portion of the improvements, newly made in the country, are made by citizens, and not unfrequently by citizens whose time has been so closely occupied with business, that they have had no opportunity to cultivate a taste for rural matters, it is not surprising that we should continually see transplanted, as unexceptionable things, the ideas in houses, furniture, and even in gardens, which have been familiar to them in cities.

As, however, it is an indisputable axiom, that there are laws of taste which belong to the country and country life, quite distinct from those which belong to town, the citizen always runs into cockneyisms when he neglects these laws. And what we would gladly insist upon, therefore, is that it is only what is appropriate and significant in the country, (or what is equally so in town and country,) that can be adopted, without insulting the natural grace and freedom of umbrageous trees and green lawns.

He who comes from a city, and wishes to build himself a country-seat, would do well to *forget* all that he considers the standard of excellence in houses and furniture in town, (and which are, perhaps, really excellent there,) and make a pilgrimage of inspection to the best country houses, villas and cottages, with their grounds, before he lays a stone in his foundation walls, or marks a curve of his walks. If he does this, he will be certain to open his eyes to the fact, that, though there are good models in town, for town life, there are far better models in the country, for country life.

AN ORNAMENTAL FARM HOUSE.

FIG. 31.

An Ornamental Farm House. From Andrew Jackson Downing, *Cottage Residences* (1842). Courtesy University of Kansas Special Collections, Lawrence.

SUSAN FENIMORE COOPER

Susan Fenimore Cooper, born in 1813, was the daughter of novelist James Fenimore Cooper and granddaughter of Judge William Cooper, the founder of Cooperstown, New York. Her father sought out a good education for Susan and her sisters; when he had exhausted the options in New York, he took his family to Paris in 1826. Susan learned several languages while there and also settled into her role (in which she continued until his death in 1851) as her father's copyist before returning to the United States in 1833.

While young, Susan absorbed a strong interest in the natural world from her family. Her maternal grandfather took her on drives through his farmlands, quizzing her along the way, asking her to identify trees, for instance, by their bark or leaves. Her father, too, enjoyed his land and garden, sharing his delight with Susan.

Cooper lived most of her life in Cooperstown, and her long experience and close observation gave her a rich perspective on the village's transformation from frontier town to mature agricultural community. It also gave her an attachment to the land itself, which fed her concern for its integrity and the dangers posed by unrestrained agricultural development. This experience, and her constant attention to Cooperstown's flora, fauna, and weather, provide the basis for Rural Hours *(1850), from which the following excerpt is taken.*

From *Rural Hours* (1850)

Wednesday, [June] 27th.—Charming day; thermometer 80. Toward sunset strolled in the lane.

The fields which border this quiet bit of road are among the oldest in our neighborhood, belonging to one of the first farms cleared near the village; they are in fine order, and to look at them, one might readily believe these lands had been under cultivation for ages. But such is already very much the character of the whole valley; a stranger moving along the highway looks in vain for any striking signs of a new country; as he passes from farm to farm in unbroken succession, the aspect of the whole region is smiling and fruitful. Probably there is no part of the earth, within the limits of a temperate climate, which has taken the aspect of an old country so soon as our native land; very much is due, in this respect, to the advanced state of civilization in the present age, much to the active, intelligent character of the people, and something, also, to the natural features of the country itself. There are no barren tracts in our midst, no deserts which defy cultivation; even our mountains are easily tilled—arable, many of them, to their very summits—while the most sterile among them are more or less clothed with vegetation in their natural state. Altogether, circumstances have been very much in our favor.

While observing, this afternoon, the smooth fields about us, it was easy, within the few miles of country in sight at the moment, to pick out parcels of land in widely different conditions, and we amused ourselves by following upon the hillsides the steps of the husbandsman, from the first rude clearing, through every successive stage of tillage, all within range of the eye at the same instant. Yonder, for instance, appeared an opening in the forest, marking a new clearing still in the rudest state, black with charred stumps and rubbish; it was only last winter that the timber was felled on that spot, and the soil was first opened to the sunshine, after having been shaded by the old woods for more ages than one can tell. Here, again, on a nearer ridge, lay a spot not only cleared, but fenced, preparatory to being tilled; the decayed trunks and scattered rubbish having been collected in heaps and burnt. Probably that spot will soon be ploughed, but it frequently happens that land is cleared of the wood, and then left in a rude state, as wild pasture ground; an indifferent sort of husbandry this, in which neither the soil nor the wood receives any attention; but there is more land about us in this condition than one would suppose. The broad hillside facing the lane in which we were walking, though cleared perhaps thirty years since, has continued untilled to the present hour. In another direction, again, lies a field of new land, ploughed and seeded for the first time within the last few weeks; the young maize plants, just shooting out their glossy leaves, are the first crop ever raised there, and when harvested, the grain will prove the first fruits the earth has ever yielded to man from that soil, after lying fallow for thousands of seasons. Many other fields in sight have just gone through the usual rotation of crops, showing what the soil can do in various ways; while the farm before us has been under cultivation from the earliest history of the village, yielding every season, for the last half century, its share of grass and grain. To one familiar with the country, there is a certain pleasure in thus beholding the agricultural history of the neighborhood unfolding before one, following upon the farms in sight these progressive steps in cultivation.

The pine stumps are probably the only mark of a new country which would be observed by a stranger. With us, they take the place of rocks, which are not common; they keep possession of the ground a long while; some of those about us are known to have stood more than sixty years, or from the first settlement of the country, and how much longer they will last, time alone can tell. In the first years of cultivation, they are a very great blemish, but after a while, when most of them have been burnt or uprooted, a gray stump here and there, among the grass of a smooth field, does not look so very much amiss, reminding one, as it does, of the brief history of the country. Possibly there may be something of partiality in this opinion, just as some lovers have found to admire a freckled face, because the rosy cheek of their sweetheart was mottled with brown freckles; people generally may not take the same view of the matter. When uprooted, the stumps are drawn together in heaps and burnt, or frequently they are turned to account as fences, being placed on end, side by side, their roots interlocking, and a more wild and formidable barrier about a quiet field cannot well be imagined. These rude fences are quite common in our neighborhood, and being peculiar one rather likes them; it is said that they last much longer than other wooden fences, remaining in good condition for sixty years.

But there are softer touches also, telling the same story of recent cultivation. It frequently happens, that walking about our farms, among rich fields, smooth and well worked, one comes to a low bank, or some little nook, a strip of land never yet cultivated, though surrounded on all sides by ripening crops of eastern grains and grasses. One always knows such places by the pretty native plants growing there. It was but the other day we paused to observe a spot of this kind in a fine meadow, near the village, neat and smooth, as though worked from the days of Adam. A path made by the workmen and cattle crosses the field, and one treads at every step upon plantain, that regular path-weed of the Old World; following this track, we come to a little runnel, which is dry and grassy now, though doubtless at one time the bed of a considerable spring; the banks are several feet high, and it is filled with native plants; on one side stands a thorn-tree, whose morning shadow falls upon grasses and clovers brought from beyond the seas, while in the afternoon, it lies on gyromias and moose-flowers, sarsaparillas and cahoshes, which bloomed here for ages, when the eye of the red man alone beheld them. Even within the limits of the village spots may still be found on the bank of the river, which are yet unbroken by the plough, where the trailing arbutus, and squirrel-cups, and May-wings tell us so every spring; in older regions, these children of the forest would long since have vanished from all the meadows and villages, for the plough would have passed a thousand times over every rood of such ground. . . .

Tuesday, [July] 3d.—[Upon a visit to a neighbor]

. . . How pleasant things look about a farm-house! There is always much that is interesting and respectable connected with every better labor, every useful or harmless occupation of man. We esteem some trades for their usefulness, we admire others for their ingenuity, but it seems natural to like a farm, or a garden, beyond most workshops.

From the window of the room in which we were sitting, we looked over the whole of Mr. B——'s farm; the wheat-field, corn-field, orchard, potato-patch, and buckwheat-field. The farmer himself, with his wagon and horses, a boy and a man, were busy in a hay-field, just below the house; several cows were feeding in the meadow, and about fifty sheep were nibbling on the hillside. A piece of woodland was pointed out on the height above, which supplied the house with fuel. We saw no evergreens there; the trees were chiefly maple, birch, oak, and chestnut; with us, about the lake, every wood contains hemlock and pine.

Finding we were interested in rural matters, our good friend offered to show us whatever we wished to see, answering all our many questions with the sweet, old smile peculiar to herself. She took us to the little garden; it contained potatoes, cabbages, onions, cucumbers, and beans; and a row of currant-bushes was the only fruit; a patch of catnip, and another of mint, grew in one corner. Our farmers, as a general rule, are proverbially indifferent about their gardens. There was no fruit on the place besides the apple-trees of the orchard; one is surprised that cherries, and pears, and plums, all suited to our hilly climate in this country, should not receive more attention; they yield a desirable return for the cost and labor required to plant and look after them.

Passing the barn, we looked in there also; a load of sweet hay had just been thrown into the loft, and another was coming up the road at the moment. Mr. B—— worked his farm with a pair of horses only, keeping no oxen. Half a dozen hens and some geese were the only poultry in the yard; the eggs and feathers were carried, in the fall, to the store at B—— Green, or sometimes as far as our own village.

They kept four cows; formerly they had had a much larger dairy; but our hostess had counted her threescore and ten, and being the only woman in the house the dairy-work of four cows, she said, was as much as she could well attend to. One would think so; for she also did all the cooking, baking, washing, ironing, and cleaning for the family, consisting of three persons; besides a share of the sewing, knitting, and spinning. We went into her little buttery; here the bright tin pans were standing full of rich milk; everything was thoroughly scoured, beautifully fresh, and neat. A stone jar of fine yellow butter, whose flavor we knew of old, stood on one side, and several cheeses were in press. The wood-work was all painted red. . . .

A great spinning-wheel, with a basket of carded wool, stood in a corner, where it had been set aside when we arrived. There was a good deal of spinning done in the family; all the yarn for stockings, for flannels, for the cloth worn by the men, for the colored woolen dresses of the women, and all the thread for their coarse toweling, was spun in the house by our hostess, or her grand-daughter, or some neighbor hired for the purpose. Formerly, there had been six step-daughters in the family, and then, not only all the spinning, but the weaving and dyeing also, were done at home. They must have been notable women, those six step-daughters; we heard some great accounts of day's spinning and weaving done by them. The presses and cupboards of the house were still full to overflowing with blankets, white and colored flannels, colored twilled coverlets for bedding, besides sheets, tablecloths, and patched bed-quilts, all their own work. In fact, almost all the clothing of the family, for both men and women, and everything in the shape of bedding and toweling used by the household, was home-made. Very few dry-goods were purchased by them; hats and shoes, some light materials for caps and collars, a little ribbon, and a printed calico now and then, seemed to be all they bought. Nor was this considered at all remarkable; such is the common way of living in many farmers' families. It has been calculated that a young woman who knows how to spin and weave can dress herself with ease and comfort, as regards everything necessary, for twelve dollars a year, including the cost of the raw materials; the actual allowance for clothing made by the authorities of this county, to farmers' daughters, while the property remained undivided, has been fifteen dollars, and the estimate is said to have included everything necessary for comfort, both winter and summer clothing. . . .

The food of the family, as well as their clothing, was almost wholly the produce of their own farm; they dealt but little with either grocer or butcher. In the spring, a calf was killed; in the fall, a sheep and a couple of hogs; once in a while, at other seasons, they got a piece of fresh meat from some neighbor who had killed a beef or a mutton. They rarely ate their poultry; the hens were kept chiefly for eggs, and their geese for feathers. The common piece of meat, day after day, was corned pork from their pork-barrel; they usually kept, also, some corned beef in brine, either from their own herd, or a piece procured by some bargain with a neighbor. The bread was made from their own wheat, and so were the hoe-cakes and griddle-cakes from the Indian meal and buckwheat of their growth. Butter and cheese from their dairy were on table at every meal, three times a day. Pies were eaten very frequently, either of apples, pumpkins, dried fruits, or coarse minced-meat; occasionally they had pie without any meat for their dinner; puddings were rare; Yankee farmers generally eating much more pastry than pudding. Mush and milk was a common dish. They ate but few eggs, reserving them for sale. Their vegetables were almost wholly potatoes, cabbage, and onions, with fresh corn and beans, when in season, and baked beans with pork in winter. Pickles were put on table at every meal. Their sugar and molasses were made from maple, only keeping a little white sugar for company or sickness. They drank cider from their own orchard. The chief luxuries of the household were tea and coffee, both procured from the "stores," although it may be doubted if the tea ever saw China; if like much of that drunk about the country, it was probably of farm growth also. . . .

Monday, [July] 23d.—Just at the point where the village street becomes a road and turns to climb the hillside, there stands a group of pines, a remnant of the old forest. There are many trees like these among the woods; far and near such may be seen rising from the hills, now tossing their arms in the stormy winds, now drawn in still and dark relief against the glowing evening sky. Their gaunt, upright forms standing about the hill-tops, and the ragged gray stumps of those which have fallen, dotting the smooth fields, make up the sterner touches in a scene whose general aspect is smiling. But although these old trees are common upon the wooded heights, yet the group on the skirts of the village stands alone among the fields of the valley; their nearer brethren have all been swept away, and these are left in isolated company, differing in character from all about them, a monument of the past.

It is upon a narrow belt of land, a highway and a corn-field on one side, a brook and an orchard on the other, that these trees are rooted; a strip of woodland connected with the forest on the hills above, and suddenly cut off where it approaches the first buildings of the village. There they stand, silent spectators of the wonderful changes that have come over the valley. Hundreds of winters have passed since the cones which contained the seed of that grove fell from the parent tree; centuries have elapsed since their heads emerged from the topmost wave of the sea of verdure to meet the sunshine, and yet it is but yesterday that their shadows first fell, in full length, upon the sod at their feet.

Sixty years since, those trees belonged to a wilderness; the bear, the wolf, and the panther brushed their trunks, the ungainly moose and the agile deer browsed at their feet; the savage hunter crept stealthily about their roots, and painted braves passed noiselessly on their warpath beneath their shade. How many successive generations of the red man have trod the soil they overshadowed, and then sat down in their narrow graves—how many herds of wild creatures have chased each other through that wood, and left their bones to bleach among the fern and moss, there is no human voice can tell. We only know that the summer winds, when they filled the canvas of Columbus and Cabot, three hundred years ago, came sweeping over these forest pines, murmuring then as we hear them murmur to-day. . . .

At length, nearly three long centuries after the Genoese had crossed the ocean, the white man came to plant a home on this spot, and it was then the great change began; the axe and the saw, the forge and the wheel, were busy from dawn to dusk, cows and swine fed in thickets whence the wild beasts had fled, while the ox and the horse drew away in chains the fallen trunks of the forest. The tenants of the wilderness shrunk deeper within its bounds with every changing moon; the wild creatures fled away within the receding shades of the forest, and the red man followed on their track; his day of power was gone, his hour of pitiless revenge had passed, and the last echoes of the war-whoop were dying away forever among these hills, when the pale-faces laid their hearth-stones by the lake shore. The red man, who for thousands of years had been lord of the land, no longer treads the soil; he exists here only in uncertain memories, and in forgotten graves.

Such has been the change of the last half century. Those who from childhood have known the cheerful dwellings of the village, the broad and fertile farms, the well beaten roads, such as they are to-day, can hardly credit that this has all been done so recently by a band of men, some of whom, white-headed and leaning on their staves, are still among us. Yet such is the simple truth. This village lies just on the borders of the tract of country which was opened and peopled immediately after the Revolution; it was among the earliest of those little colonies from the sea-board which struck into the wilderness at that favorable moment, and whose rapid growth and progress in civilization have become a by-word. Other places, indeed, have far surpassed this quiet borough; Rochester, Buffalo, and others of a later date, have become great cities, while this remains a rural village; still, whenever we pause to recall what has been done in this secluded valley during the lifetime of one generation, we must needs be struck with new astonishment. And throughout every act of the work, those old pines were there. Unchanged themselves, they stand surrounded by objects over all of which a great change has passed. The open valley, the half-shorn hills, the paths, the flocks, the buildings, the woods in their second growth, even the waters in the different images they reflect on their bosom, the very race of men who come and go, all are different from what they were; and those calm old trees seem to heave the sigh of companionless age, as their coned heads rock slowly in the winds. . . .

It needs but a few short minutes to bring one of these trees to the ground; the rudest boor passing along the highway may easily do the deed; but how many years must pass ere its equal stand on the same spot! Let us pause to count the days, the months, the years; let us number the generations that must come and go, the centuries that must roll onward, ere the seed sown from this year's cones shall produce a wood like that before us. The stout arm so ready to raise the axe to-day, must grow weak with age, it must drop into the grave; its bone and sinew must crumple into dust long before another tree, tall and great as those, shall have grown from the cone in our hand. Nay, more, all the united strength of sinew, added to all the powers of mind, and all the force of will, of millions of men, can do no more toward the work than the poor ability of a single arm; these are of the deeds which time alone can perform. But allowing even that hundreds of years hence other trees were at length to succeed these with the same dignity of height and age, no other younger

Louis Rémy Mignot, *Cooperstown from Three Mile Point*, c. 1855. Reproduced by permission of the Fenimore Art Museum, New York State Historical Association, Cooperstown, NY.

wood can ever claim the same connection as this, with a state of things now passed away forever; they cannot have that wild, stern character of the aged forest pines. This little town itself must fall to decay and ruin; its streets must become choked with bushes and brambles; the farms of the valley must be anew buried within the shades of a wilderness; the wild deer and the wolf and the bear must return from beyond the great lakes; the bones of the savage men buried under our feet must rise and move again in the chase, ere trees like those, with the spirit of the forest in every line, can stand on the same ground in wild dignity of form like those old pines now looking down upon our homes.

Henry David Thoreau

Henry David Thoreau needs no introduction. Or maybe he does. The Thoreau of fame was a misanthropic recluse who withdrew to the wilderness—and an easy man to mock when the reader discovers that Walden Pond is only about a mile from Concord village, and that young Henry often went home for dinner, and sometimes brought his dirty laundry, too. As a lover of wildness he was known to despise farmers, without whom, of course, he would have starved.

But the real Henry Thoreau was a more complicated fellow than that straw man. Born in Concord in 1817, Thoreau was very much a son of the village, and closely devoted to his family. Before his eyes Concord was going through a transition from a small agrarian community to an aspiring commercial and industrial town. Thoreau's family was of that entrepreneurial bent, struggling to succeed in the new middle class. His father started a small pencil manufacturing business that later showed some success, thanks largely to Henry's innovations, which allowed the family to ship not just pencils but an improved graphite mixture far and wide for use in typesetting. Henry Thoreau was an able gardener, carpenter, surveyor, and engineer. An exacting workman, he had all the makings of the kind of Yankee mechanic who drove the New England industrial revolution.

Thoreau's family stretched to send their promising son to Harvard, but after teaching school with his older brother John for a few years while in college and after graduation, his life took a turn away from the beaten track. When John died suddenly of lockjaw, Henry Thoreau closed the school and determined to become a writer, inspired by his Concord neighbor Ralph Waldo Emerson. He worked in the Emerson household as a gardener and tutor until in 1845 Emerson allowed the younger man to live in his newly acquired woodlot at Walden Pond for two years—essentially a writer's retreat. (After the sojourn was over Emerson retained the solid little house that Thoreau had built, so the arrangement was mutually advantageous.)

For the rest of his life Thoreau lived with his family and worked from time to time at the family business, especially after his father died. But as far as possible he limited his work to a few surveying jobs each year to pay his expenses so that he could devote his time to exploring Concord's rivers and woods and to writing. Along with his mother, sister, and aunt, Thoreau was a fierce abolitionist, and it was for his fiery writing on that subject that he was perhaps best known in his lifetime.

Thoreau knew Concord's farmers well and admired many of them. But his deep ambivalence toward them is reflected in *Walden*, especially in the bean field chapter of that book. Thoreau watched the rapid decline of Concord's remaining forests as pastures and hayfields for commercial dairy farming spread. He also saw farmers moving from their traditional agrarian aspiration for "comfort and independence" to the risky business of farming for the market. His experiment in growing two and a half acres of beans, not for consumption but for sale, on what he knew to be some of the most sterile, sandy soil in Concord, was a small-scale financial bust. It can be read as a satirical parable aimed at the often even less profitable extensive commercial farming that was then sweeping the last wild places from Concord.

From *Walden* (1854)

I see young men, my townsmen, whose misfortune it is to have inherited farms, houses, barns, cattle, and farming tools; for these are more easily acquired than got rid of. Better if they had been born in the open pasture and suckled by a wolf, that they might have seen with clearer eyes what field they were called to labor in. Who made them serfs of the soil? Why should they eat their sixty acres, when man is condemned to eat only his peck of dirt? Why should they begin digging their graves as soon as they are born? They have got to live a man's life, pushing all these things before them, and get on as well as they can. How many a poor immortal soul have I met well-nigh crushed and smothered under its load, creeping down the road of life, pushing before it a barn seventy-five feet by forty, its Augean stables never cleansed, and one hundred acres of land, tillage, mowing, pasture, and wood-lot! The portionless, who struggle with no such unnecessary inherited encumbrances, find it labor enough to subdue and cultivate a few cubic feet of flesh.

But men labor under a mistake. The better part of the man is soon plowed into the soil for compost. By a seeming fate, commonly called necessity, they are employed, as it says in an old book, laying up treasures which moth and rust will corrupt and thieves break through and steal. It is a fool's life, as they will find when they get to the end of it, if not before. . . .

When I consider my neighbors, the farmers of Concord, who are at least as well off as the other classes, I find that for the most part they have been toiling twenty, thirty, or forty years, that they may become the real owners of their farms, which commonly they have inherited with encumbrances, or else bought with hired money—and we may regard one third of that toil as the cost of their houses—but commonly they have not paid for them yet. It is true, the encumbrances sometimes outweigh the value of the farm, so that the farm itself becomes one great encumbrance, and still a man is found to inherit it, being well acquainted with it, as he says. On applying to the assessors, I am surprised to learn that they cannot at once name a dozen in the town who own their farms free and clear. If you would know the history of these homesteads, inquire at the bank where they are mortgaged. The man who has actually paid for his farm with labor on it is so rare that every neighbor can point to him. I doubt if there are three such men in Concord. What has been said of the merchants, that a very large majority, even ninety-seven in a hundred, are sure to fail, is equally true of the farmers. With regard to the merchants, however, one of them says pertinently that a great part of their failures are not genuine pecuniary failures, but merely failures to fulfil their engagements, because it is inconvenient; that is, it is the moral character that breaks down. But this puts an infinitely worse face on the matter, and suggests, beside, that probably not even the other three succeed in saving their souls, but are perchance bankrupt in a worse sense than they who fail honestly. Bankruptcy and repudiation are the springboards from which much of our civilization vaults and turns its somersets, but the savage stands on

Woodblock print from *Walden* (1854). Courtesy
Thoreau Institute at
Walden Woods, Lincoln, MA.

the unelastic plank of famine. Yet the Middlesex Cattle Show goes off here with *éclat* annually, as if all the joints of the agricultural machinery were suent.

The farmer is endeavoring to solve the problem of a livelihood by a formula more complicated than the problem itself. To get his shoestrings he speculates in herds of cattle. With consummate skill he has set his trap with a hair spring to catch comfort and independence, and then, as he turned away, got his own leg into it. This is the reason he is poor; and for a similar reason we are all poor in respect to a thousand savage comforts, though surrounded by luxuries. . . .

Before I finished my house, wishing to earn ten or twelve dollars by some honest and agreeable method, in order to meet my unusual expenses, I planted about two acres and a half of light and sandy soil near it chiefly with beans, but also a small part with potatoes, corn, peas, and turnips. The whole lot contains eleven acres, mostly growing up to pines and hickories, and was sold the preceding season for eight dollars and eight cents an acre. One farmer said that it was "good for nothing but to raise cheeping squirrels on." I put no manure on this land, not being the owner, but merely a squatter, and not expecting to cultivate so much again, and I did not quite hoe it all once. I got out several cords of stumps in plowing, which supplied me with fuel for a long time, and left small circles of virgin mould, easily distinguishable through the summer by the greater luxuriance of the beans there. The dead and for the most part unmerchantable wood behind my house, and the driftwood from the pond, have supplied the remainder of my fuel. I was obliged to hire a team and a man for plowing, though I held the plow myself. My farm outgoes for the first season were, for implements, seed, work, etc., $14.72½. The seed corn was given me. This never costs anything to speak of unless you plant more than enough. I got twelve bushels of beans, and eighteen bushels of potatoes, beside some peas and sweet corn. The yellow corn and turnips were too late to come to anything. My whole income from the farm was

	$ 23	44.
Deducting the outgoes, ...	14	72½
there are left, ...	$ 8	71½

beside produce consumed and on hand at the time this estimate was made of the value of $4.50,—the amount on hand much more than balancing a little grass which I did not raise. All things considered, that is, considering the importance of a man's soul and of to-day, notwithstanding the short time occupied by my experiment, nay, partly because of its transient character, I believe that that was doing better than any farmer in Concord did that year.

The next year I did better still, for I spaded up all the land which I required, about a third of an acre, and I learned from the experience of both years, not being in the least awed by many celebrated works on husbandry, Arthur Young among the rest, that if one would live simply and eat only the crop which he raised, and raise no more than he ate, and not exchange it for an insufficient quantity of more luxurious and expensive things, he would need to cultivate only a few rods of ground, and that it would be cheaper to spade up that than to use oxen to plough it, and to select a fresh spot from time to time than to manure the old, and he could do all his necessary farm work as it were with his left hand at odd hours in the summer; and thus he would not be tied to an ox, or horse, or cow, or pig, as at present. I desire to speak impartially on this point, and as one not interested in the success or failure of the present economical and social arrangements. I was more independent than any farmer in Concord, for I was not anchored to a house or farm, but could follow the bent of my genius, which is a very crooked one, every moment. Beside being better off than they already, if my house had been burned or my crops had failed, I should have been nearly as well off as before. . . .

Meanwhile my beans, the length of whose rows, added together, was seven miles already planted, were impatient to be hoed, for the earliest had grown considerably before the latest were in the ground; indeed they were not easily to be put off. What was the meaning of this so steady and self-respecting, this small Herculean labor, I knew not. I came to love my rows, my beans, though so many more than I wanted. They attached me to the earth, and so I got strength like Antaeus. But why should I raise them? Only Heaven knows. This was my curious labor all summer,—to make this portion of the earth's surface, which had yielded only cinquefoil, blackberries, johnswort, and the like, before, sweet wild fruits and pleasant flowers, produce instead this pulse. What should I learn of beans or beans of me? I cherish them, I hoe them, early and late I have an eye to them; and this is my day's work. It is a fine broad leaf to look on. My auxiliaries are the dews and rains which water this dry soil, and what fertility is in the soil itself, which for the most part is lean and effete. My enemies are worms, cool days, and most of all woodchucks. The last have nibbled for me a quarter of an acre clean. But what right had I to oust johnswort and the rest, and break up their ancient herb garden? Soon, however, the remaining beans will be too tough for them, and go forward to meet new foes.

When I was four years old, as I well remember, I was brought from Boston to this my native town, through these very woods and this field, to the pond. It is one of the oldest scenes stamped on my memory. And now to-night my flute has waked the echoes over that very water. The pines still stand here older than I; or, if some have fallen, I have cooked my supper with their stumps, and a new growth is rising all around, preparing another aspect for new infant eyes. Almost the same johnswort springs from the same perennial root in this pasture, and even I have at length helped to clothe that fabulous landscape of my infant dreams, and one of the results of my presence and influence is seen in these bean leaves, corn blades, and potato vines.

I planted about two acres and a half of upland; and as it was only about fifteen years since the land was cleared, and I myself had got out two or three cords of stumps, I did not give it any manure; but in the course of the summer it appeared by the arrowheads which I turned up in hoeing, that an extinct nation had anciently dwelt here and planted corn and beans ere white men came to clear the land, and so, to some extent, had exhausted the soil for this very crop.

Before yet any woodchuck or squirrel had run across the road, or the sun had got above the shrub oaks, while all the dew was on, though the farmers warned me against it,—I would advise you to do all your work if possible while the dew is on,—I began to level the ranks of haughty weeds in my bean-field and throw dust upon their heads. Early in the morning I worked barefooted, dabbling like a plastic artist in the dewy and crumbling sand, but later in the day the sun blistered my feet. There the sun lighted me to hoe beans, pacing slowly backward and forward over that yellow gravelly upland, between the long green rows, fifteen rods, the one end terminating in a shrub oak copse where I could rest in the shade, the other in a blackberry field where the green berries deepened their tints by the time I made another bout. Removing the weeds, putting fresh soil about the bean stems, and encouraging this weed which I had sown, making the yellow soil express its summer thought in bean leaves and blossoms rather than in wormwood and piper and millet grass, making the earth say beans instead of grass,—this was my daily work. As I had little aid from horses or cattle, or hired men or boys, or improved implements of husbandry, I was much slower, and became much more intimate with my beans than usual. But labor of the hands, even when pursued to the verge of drudgery, is perhaps never the worst form of idleness. It has a constant and imperishable moral, and to the scholar it yields a classic result. A very *agricola laboriosus* was I to travellers bound westward through Lincoln and Wayland to nobody knows where; they sitting at their ease in gigs, with elbows on knees, and reins loosely hanging in festoons; I the home-staying, laborious native of the soil. But soon my homestead was out of their sight and thought. It was the only open and cultivated field for a great distance on either side of the road, so they made the most of it; and sometimes the man in the field heard more of the travellers' gossip and comment than was meant for his ear: "Beans so late! peas so late!"—for I continued to plant when others had begun to hoe,—the ministerial husbandman had not suspected it. "Corn, my boy, for fodder; corn for fodder." "Does he *live* there?" asks the black bonnet of the gray coat; and the hard-featured farmer reins up his grateful dobbin to inquire what you are doing where he sees no manure in the furrow, and recommends a little chip dirt, or any little waste stuff, or it may be ashes or plaster. But here were two acres and a half of furrows, and only a hoe for cart and two hands to draw it,—there being an aversion to other carts and horses,—and chip dirt far away. Fellow-travellers as they rattled by compared it aloud with the fields which they had passed, so that I came to know how I stood in the agricultural world. This was one field not in Mr. Colman's report. And, by the way, who estimates the value of the crop which Nature yields in the still wilder fields unimproved by man? The crop of *English* hay is carefully weighed, the moisture calculated, the silicates and the potash; but in all dells and pond-holes in the woods and pastures and swamps grows a rich and various crop only unreaped by man. Mine was, as it were, the connecting link between wild and cultivated fields; as some states are civilized, and others half-civilized, and others savage or barbarous, so my field was, though not in a bad sense, a half-cultivated field. They were beans cheerfully returning to their wild and primitive state that I cultivated, and my hoe played the *Ranz des Vaches* for them. . . .

It was a singular experience that long acquaintance which I cultivated with beans, what with planting, and hoeing, and harvesting, and threshing, and picking over and selling them,—the last was the hardest of all,—I might add eating, for I did taste. I was determined to know beans. When they were growing, I used to hoe from five o'clock in the morning till noon, and commonly spent the rest of the day about other affairs. Consider the intimate and curious acquaintance one makes with various kinds of weeds,—it will bear some iteration in the account, for there was no little iteration in the labor,—disturbing their delicate organizations so ruthlessly, and making such invidious distinctions with his hoe, levelling whole ranks of one species, and sedulously cultivating another. That's Roman wormwood,—that's pigweed,—that's sorrel,—that's piper-grass,—have at him, chop him up, turn his roots upward to the sun, don't let him have a fibre in the shade, if you do he'll turn himself t'other side up and be green as a leek in two days. A long war, not with cranes, but with weeds, those Trojans who had sun and rain and dews on their side. Daily the beans saw me come to their rescue armed with a hoe, and thin the ranks of their enemies, filling up the trenches with weedy dead. Many a lusty crest-waving Hector, that towered a whole foot above his crowding comrades, fell before my weapon and rolled in the dust.

Those summer days which some of my contemporaries devoted to the fine arts in Boston or Rome, and others to contemplation in India, and others to trade in London or New York, I thus, with the other farmers of New England, devoted to husbandry. Not that I wanted beans to eat, for I am by nature a Pythagorean, so far as beans are concerned, whether they mean porridge or voting, and exchanged them for rice; but, perchance, as some must work in fields if only for the sake of tropes and expression, to serve a parable-maker one day. It was on the whole a rare amusement, which, continued too long, might have become a dissipation. Though I gave them no manure, and did not hoe them all once, I hoed them unusually well as far as I went, and was paid for it in the end. . . .

Ancient poetry and mythology suggest, at least, that husbandry was once a sacred art; but it is pursued with irreverent haste and heedlessness by us, our object being to have large farms and large crops merely. We have no festival, nor procession, nor ceremony, not excepting our cattle-shows and so called Thanksgivings, by which the farmer expresses a sense of the sacredness of his calling, or is reminded of its sacred origin. It is the premium and the feast which tempt him. He sacrifices not to Ceres and the Terrestrial Jove, but to the infernal Plutus rather. By avarice and self-ishness, and a grovelling habit, from which none of us is free, of regarding the soil as property, or the means of acquiring property chiefly, the landscape is deformed, husbandry is degraded with us, and the farmer leads the meanest of lives. He knows Nature but as a robber. Cato says that the profits of agriculture are particularly pious or just (*maximeque pius quaestus*), and according to Varro the old Romans "called the same earth Mother and Ceres, and thought that they who cultivated it led a pious and useful life, and they alone were left of the race of King Saturn."

We are wont to forget that the sun looks on our cultivated fields and on the prai-ries and forests without distinction. They all reflect and absorb his rays alike, and the former make but a small part of the glorious picture which he beholds in his daily course. In his view the earth is all equally cultivated like a garden. Therefore we should receive the benefit of his light and heat with a corresponding trust and magnanimity. What though I value the seed of these beans, and harvest that in the fall of the year? This broad field which I have looked at so long looks not to me as the principal cultivator, but away from me to influences more genial to it, which water and make it green. These beans have results which are not harvested by me. Do they not grow for woodchucks partly? The ear of wheat (in Latin *spica*, obsoletely *speca*, from *spe*, hope) should not be the only hope of the husbandman; its kernel or grain (*granum*, from *gerendo*, bearing) is not all that it bears. How, then, can our harvest fail? Shall I not rejoice also at the abundance of the weeds whose seeds are the granary of the birds? It matters little comparatively whether the fields fill the farmer's barns. The true husbandman will cease from anxiety, as the squirrels manifest no concern whether the woods will bear chestnuts this year or not, and finish his labor with every day, relinquishing all claim to the produce of his fields, and sacrificing in his mind not only his first but his last fruits also.

Gift for the Grangers, 1873. Lithograph by Strobridge & Co. Courtesy Library of Congress, Washington, DC.

4. Agriculture in an Industrializing Nation, 1860–1910

If there were an easy correlation between the destructive power expended in the Civil War and the nation's reserves of energy for facing its postbellum future, one might have expected the United States to subside into exhaustion. Yet something like the opposite happened. Despite occasional economic slumps, the nation entered a period of remarkable growth after 1865 that dramatically reshaped its geographic and social landscape. These conditions presented many farmers with seemingly insurmountable challenges and set the stage for a mass movement to protect their standing from the rapid rise of competing interests. By most estimates that movement failed, and the end of the century also saw the end of America as a primarily agrarian society.

The changes of the second half of the nineteenth century grew out of mutually reinforcing demographic, technological, economic, and political developments. The United States' population more than doubled, growing from just under 40 million in 1870 to nearly 92 million in 1910. The American economy also grew rapidly. Total economic output between 1865 and the beginning of the First World War increased by about 4 percent annually, for an eightfold increase during the period. Industrial growth was especially strong. In the mid-1890s the United States surpassed the United Kingdom as the world's leading manufacturer. By 1910 the country was producing twice the amount of manufactured goods as its nearest rival, Germany. The nation's growing industrial puissance seemed unstoppable.

One pillar of this growth was the exceptionally fast evolution of the technologies upon which industry and the broader economy depended. The Bessemer and open-hearth processes, for example, revolutionized steel production. Likewise, energy production underwent dramatic changes. As late as 1850, the bulk of the power—above three-quarters—used in basic economic pursuits was provided by animals, cordwood, and charcoal. By 1900, as steam technology improved and new transportation systems opened access to huge reserves, coal had become the leading energy source, while hydroelectric power and petroleum were getting their start. Less obvious than the changes in the steel and energy industries but also of revolutionary importance were improvements in the machine tool and metal-working industries. These innovations filtered throughout the economy, bringing better performance in all kinds of high-speed and precision tools, from sewing machines and typewriters to factory equipment, used in a breathtaking array of new applications that transformed life for Americans.

Technological breakthroughs dovetailed with economic specialization and economies of scale to drive the growth of the corporation. Improved machinery allowed mass production of goods by big businesses, which were able to outcompete smaller-scale producers and artisans such as local blacksmiths, tanners, and coopers. The new jobs provided by these large manufacturers drew workers to industrial towns, leading to a decline in rural population in older, settled parts of country. In many country villages it wasn't so much the farmers who began to disappear during this period; it was everybody else. The railroad was fundamental to these changes, especially in knitting the nation's regions into a single market of mass-produced goods and commodities. Such an undertaking could be accomplished only by large organizations. And the railroads were far larger than the businesses of the prewar years, employing tens of thousands of people versus the rough thousand or so who might have worked in the largest early corporations.

The evolution of the railroad industry in the nineteenth century also illuminates the intersection between economic and industrial growth and the nation's laws. Railroads, including both transcontinental and smaller regional lines, were often subsidized by federal, state, and municipal governments. Subsidies took a variety of forms, including direct funding, generous loans, and massive land grants. Further, the development of contract and corporate law kept the railroads and other large businesses mostly free of regulatory burdens for many years. The earliest transcontinental railroads were the direct offspring of state action, notably the Pacific Railway Act of 1862. The larger the railroads and other industries grew, the more important was their health to the rest of the economy, and the greater the voice they had in government. The rise of these special interests, while not new, gained irresistible momentum during the late nineteenth century.

Agriculture also saw dramatic growth in the decades after 1860. The railroads opened vast new areas of the continent, and their owners sold land to prospective settlers as a means of raising income for further track construction. Through these sales, the Homestead Act, and other acts that moved the public domain quickly into private hands, the nation added hundreds of thousands of new farms. Between 1870 and 1900, the total acreage of the nation's land under cultivation more than doubled, letting loose a flood of agricultural commodities to the world's markets. But how much good this dramatic expansion of production did farmers is a very different question.

These new farms were far from uniform across the country. Improved transportation fostered regional specialization like never before. The hog and corn culture of the Midwest was one example, as was the migration of wheat farming to the Great Plains. The iconic American ranching industry was born on the Texas plains and from there followed cattle trails north through the prairies and west into the Great Basin—a symbol of rugged individualism, perhaps, but in reality driven by a massive influx of Eastern and English capital. In the Northeast, agriculture shifted away from meat and grains toward dairy, hay, and market gardening. The slave plantations of the South were replaced by the "debt peonage" of sharecropping, and cotton production went right on expanding into the Southwest. In California agriculture was growing by leaps and bounds, aided by the same kinds of technological and biological innovations that were altering the economy all across the nation. While vast wheat fields dominated the Central Valley until late in the century, new specialties were beginning to appear. For example, the state's wine industry, led by the Hungarian émigré Agoston Haraszthy, was evolving from its roots in Spanish missions into a mass commodity. The navel orange, derived from a single sterile mutation in a Brazilian orchard, was first planted in southern California in 1870 and quickly became a major export from the region, while other fruits, nuts, and vegetables began to take hold in California's fertile inland valleys.

The technological changes that transformed the nation's economy also led to dramatic increases in labor productivity on the farm. Machinery such as John Deere's famous steel plow and Cyrus McCormick's reaper allowed farmers to cultivate more land. Nor were those the only inventions at the farmers' disposal. As advertisements in the agricultural press of the last decades of the nineteenth century show, all manner of other revolutionary new implements were being employed—windmills, barbed wire, steam threshers, corn silos, grain elevators. With these tools and the railroad to link them to national markets, farmers were able to move onto the forbidding prairies and to begin mining some of the richest soils in the world with unprecedented efficiency. Even without much improvement in per-acre yield, the nation saw a great boom in agricultural production.

Technologies that put increased productive power in the hands of farmers also made them more capable of playing havoc with the environment. Much of the expansion of agriculture took place onto unfamiliar lands that proved easily susceptible to degradation. Given the climate and soils of the grasslands, for example, wide-scale plowing often led to rapid erosion. Similar "resource mining" applied to the pine forests of the upper Midwest and South left burned, cutover wastelands in its wake. The degradation of land through shortsighted agriculture and logging helped spark the nascent conservation movement, whose leaders were more often critical of "uneducated" farmers than of the market forces driving them. The disconnect between the rhetoric of conservation leaders and the practices of many ordinary farmers who were trying to succeed in an increasingly competitive market would widen in the following decades.

New farming methods were capital intensive compared to earlier practices, and the need for capital in turn pushed farmers further toward commercial production and specialization. But farmers found themselves at the mercy of prices for their products that were difficult to predict and beyond their control, as they involved increased production for the world market: not only in America but in Canada, Australia, Argentina, and other countries. And farmers were now contending with much larger and more powerful players, such as the railroads, whose rate charges for freight often seemed at best illogical and at worst flagrantly predatory.

The situation in which farmers found themselves during the decades after the Civil War thoroughly undermined the promise of economic independence that Crèvecoeur had articulated a century earlier. Integration into a vast market set constraints on farmers that they hadn't been subjected to before. These structures curtailed farmers' ability to control their own destiny. For example, those who resisted highly capitalized methods but still wanted to sell their surplus were competing against farmers who made use of new machinery and could produce crops at a lower cost. And yet the reliance on capital-dependent materials meant that most farmers needed to integrate themselves even more fully into the economic system, often going into debt to do so, further undermining their independence. Likewise, with farmers able to till land more efficiently and all competing for greater acreage, land prices rose, driving many into tenancy or insupportable mortgages. These pressures blasted the hope that America would remain the place where ordinary men and women could thrive as independent farmer-citizens. The European oppressions of serfdom and tenancy, successfully escaped in the New World, were in danger of being replaced by a new subordinate, powerless status within the industrial economy. An unavoidable question arose: Could the market-driven agriculture of the United States still serve the country's agrarian ideal?

The years from 1870 to 1895 were particularly painful for many American farmers, especially when compared to the waxing fortunes of those in other walks of life. While prices of products across the entire economy were dropping as productivity improved, the prices for farm products were falling faster than the rest, so that "the farmer had to run faster just to avoid losing ground," as economic historians Gary Walton and Hugh Rockoff put it. Frustration among farmers, especially in the West and South, led to a prolonged period of protest and upheaval that climaxed with the Populist "People's Party" bid for national political power.

The agrarian uprising caught fire among people who held to a core vision that dated back to the nation's founding. This vision, according to historian Robert McMath, "was based on the simple idea that the producer deserves the fruits of his or her work. . . . Or expressed biblically (and believers in producerism *often* expressed it biblically), 'the laborer is worthy of his hire.'" William Peffer, a Populist leader and senator from Kansas, expressed this central belief together with the old agrarian inclination to blame corruption and influence peddling for the injustice: "[The farmers] toiled while others took the increase; they sowed, but others reaped the harvest. It is avarice that despoiled the farmer. Usury absorbed his substance. He sweat gold and the money changers coined it. And now, when misfortunes gather about and calamity overtakes him, he appeals to those he has enriched only to learn how poor and helpless he is alone."

Feeling cheated out of the just rewards of their labor, aggrieved farmers responded by organizing to seek redress. They had models to follow, including the labor groups that had sprung up in the 1830s and 1840s and the Knights of Labor, an organization that had spread across the country after its founding in 1869, with branches in hundreds of small to medium-sized cities. The Patrons of Husbandry, better known as the Grange, was the first major agricultural organization to take shape after the Civil War. Many others would follow, including local cooperatives designed to strengthen the power of farmers to make bulk purchases and to market their crops.

These organizations offered a way for farmers to gather and discuss mutual problems, to publish newsletters, and to provide platforms for rural orators to spread their message. The message could be angry and the delivery high-voltage. A Farmers' Declaration of Independence circulated in 1873 adapted the language of 1776 to lay out grievances:

> All experience hath shown that mankind are more disposed to suffer while evils are sufferable than to right themselves by abolishing the laws to which they are accustomed. But when a long train of abuses and usurpations, pursuing invariably the same object, evinces a desire to reduce a people under the absolute despotism of combinations that, under the fostering care of government and with wealth wrung from the people, have grown to such gigantic proportions as to overshadow all the land and wield an almost irresistible influence for their own selfish purposes in all its halls of legislation, it is their right—it is their duty—to throw off such tyranny and provide new guards for their future security.

For these farmers, the government was failing to defend the interests of freehold farmers, the backbone of the democratic Republic, against the rising power of the "combinations." Banks were leading offenders, but railroads were especially reviled, in part because of the complete dependence of the farmer on the rail system and practices such as charging exorbitant prices for transporting crops, even for short runs, in areas where there was little competition. "The history of the present railway monopoly," the Declaration continued, "is a history of repeated injuries and oppressions, all having in direct object the establishment of an absolute tyranny over the people of these states unequaled in any monarchy of the Old World."

The 1880s and 1890s brought the culmination of rural protest, the Populist movement. The program of the Southern Farmers' Alliance, the strongest of the farmers' organizations at the height of the revolt, indicates that thoughtful men and women in the movement grappled earnestly with intractable problems and arrived at reasonable, often nuanced solutions (whatever their practicability). Key elements of the Populist program included cooperative buying programs for farmers to increase their purchasing power as well as cooperative warehousing so that farmers would not be forced to sell crops when the market was glutted. The Farmers' Alliance called for currency reform and developed a "subtreasury" plan to give farmers access to otherwise unavailable credit. In a bid for increased economic and political democracy, the reformers called for a graduated income tax and direct election of senators—reforms that would be adopted in the decades to follow. More radically, the Farmers' Alliance pushed for regulation of railroads and in some cases for outright government control of them.

These proposals should be seen in the context of their time. Populists expressed a desire to protect the well-being of farmers but also to preserve the sovereignty of the people as a whole in the face of unprecedented economic and social challenges. As William O. Dawes, a leader of the Farmers' Alliance, put it, members had an "obligation to stand as a great conservative body against the encroachment of monopolies and in opposition to the growing corruption of wealth and power."

While the rhetoric of the agrarian Populists may have rallied the faithful, it led to easy caricature by those less sympathetic. William Allen White, a young journalist, vigorously opposed the Populists, believing that the interests of his home state, Kansas, were best served by economic modernization. In 1896 he famously asked, "What's the Matter with Kansas?" By way of answer he offered:

> We all know; yet here we are at it again. We have any old mossback Jacksonian who snorts and howls because there is a bathtub in the state house; we are running that old jay for governor. We have another shabby, wild-eyed, rattle-brained fanatic who has said openly in a dozen speeches that "the rights of the user are paramount to the rights of the owner;" we are running him for chief justice so that capital will come tumbling over itself to get into the state. . . . Then, for fear some hint that the state had become respectable might percolate through the civilized portions of the nation, we have decided to send three or four harpies out lecturing, telling the people that Kansas is raising hell and letting the corn go to weeds.

The Populists were unable to rally enough popular support in urbanizing America to break through the entrenched sectional divisions in national politics. They went down to defeat in 1892, and again in their alliance with the Democratic Party behind William Jennings Bryan in 1896. Their platform, mixing radical economic reforms with a vision of an older agrarian democracy that no longer resonated with many Progressive Americans such as William Allen White, proved unequal to the power of the industrialists.

Before long, both White's invective and the rhetorical flamboyance of the Populist orators began to recede, due in part to changing economic circumstances. On the horizon was a "golden age" for farmers, lasting from the late 1890s to the end of

World War I, during which the balance between agricultural production and demand for the crops shifted back in the farmers' favor. That brief return of farm prosperity masked the lasting outcome of this pivotal period in American agrarianism. As Charles and Mary Beard observed in *The Rise of American Civilization,* the late nineteenth century "marked the absorption of agriculture into the industrial vortex, endlessly sustained by capitalism, science, and machinery." The United States was no longer an agrarian nation.

THOMAS STARR KING

Thomas Starr King, born in New York City in 1824, was the son of an itinerate Universalist minister who died when Thomas was fifteen, whereupon the boy left school to support his family. Possessed of an immense intellectual energy, King pursued his studies informally, reading widely and mastering Latin, Greek, and several modern European languages. He studied theology under Hosea Ballou, star of the Universalist firmament, and flourished among Boston's progressive religious leadership. King embarked on his own ministerial career in 1846.

In 1860 King left the East, accepting the call to lead the First Unitarian Church of San Francisco. The city's mingling of boomtown wealth and frontier rawness presented an opportunity that King eagerly welcomed. His learned sermons pleased culture-hungry San Franciscans, and King, feeling freed from Boston's rigid social system, was energized by the way Californians embraced him. The result was a remarkable outpouring of civic engagement by the young minister. He traveled throughout the state offering lectures, visiting hospitals, throwing support to various clubs and causes. Most famously, he launched a speaking campaign to encourage wavering Californians to side with the Union in the Civil War. King's exertions weakened him and may have contributed to his death from diphtheria in 1864, at the age of only thirty-nine.

During his travels King marveled at California's landscapes. He was so taken by the Sierra Nevada and the Yosemite Valley that he planned to write a book about them. He was also stirred by California's agricultural abundance, the subject of the speech excerpted below. That rich endowment fueled King's argument that California was an unequaled gift, a new promised land for agrarian America. What would become of that gift over the next 150 years is one of the key stories of American agriculture.

"Over the Rich Orchards of the Santa Clara Valley from
Hamilton, California," 1906. International Stereograph Co.
Courtesy Library of Congress, Washington, DC.

From "Address before the San Joaquin Valley Agricultural Society" (1862)

Nearly four months ago I had the privilege of climbing to the top of Mount Diablo, which rises like an enchanted billow, from the plain. (From San Francisco we see only one mountain and one peak. Here you see two. Is this a sign that in Stockton you have had a double share of the power of Satan to contend with?) How glorious the view was from the highest peak in May! Sweeps and slopes of green, such as no artist's colors at the East could imitate, the San Joaquin plains beneath, emerged from their flood, embroidered with flowers, and bursting into the promise which this week fulfills—the San Joaquin itself so dingy that it looked as though it flowed molten from Copperopolis—the sea showing its unruffled azure far-off between the cliffs of the Golden Gate; and on the east the snowy guards of all your opulence, the mighty bulwark of the Sierra, visible for two hundred and fifty miles—its lower slopes as rich in gold as their crests at evening with the gold of sunset; its further slopes veined with silver only less white and pure than those great crests at noon!

One sees in a moment from that elevation, in early May, what a bounteous and wonderful district it is which your Society represents. But suppose that some one with a powerful telescope could be lifted to that eminence in early September, and make his first acquaintance then with your district and the State! What would his impression be? Would he not suppose that he was lifted over a boundless desert? Would he not believe that the six rainless months were a virtual curse of Providence sealed into the sallow landscape? Would he not imagine that if any inhabitants dwelt there, they were fed either by manna or by bacon from the East? Would he not behold, in the wide-spread desolation, and in the hot, thick air, the fulfillment of the doom of death upon Nature—"Dust thou art, and unto dust thou shalt return?"

What would our visitor on this height be likely to say, if told that the landscape, so brown and lugubrious beneath him, inclosed an agricultural opulence of which the figures seem almost miraculous; that its grain crops average double those of the Eastern States; that fruits were then ripening all around him in surpassing luxuriance and beauty; and that the growth of the grape in that blasted landscape, during the last three years, surpasses any thing known in the most favored districts of the Rhine lands—France or Italy? What would he say, if his telescope should bring within vision all that the District Fairs and the State Fair will collect during this month to attest the strength and richness of our soil? He would see, to his amazement, that the State which seems given over, in a general view, to the "abomination of desolation," is really the field of two immense "horns of plenty,"—one widening downward from the pinnacle of Mt. Shasta, the other widening upward from the mountains of San Bernardino, crammed with the riches of granaries and orchards, and overflowing all upon the metropolis in the center of the coast line by the Golden Gate! He would see that we are called upon by our copious blessings to be the most grateful and the most patriotic people on the globe.

Let me say, first, that the farmers here are to be congratulated on the intrinsic nobleness of their office and labor. All honest labor is noble. But in respect to physical toil, it is impossible to conquer the instinct of the race, which assigns greater dignity to the skillful industry expended directly upon mother earth.

If an aroma could always attend gold, telling you what ways it was gained, whether it was inherited or won by enterprise and skill,—and if earned, whether in ways useful or hurtful to the higher interests of society, there would be no danger of a mean worship of money. If a man's silver and gold told the story at once whether he earned it in making sugar or turning it into liquor—in raising wheat or in speculating on it—in weaving honest cloth or in weaving shoddy—in putting soles to shoes for soldiers or sham ones which prove that the makers hadn't any soul at all—in spinning cotton or in serving as one of the crowd of unnecessary agents in its distribution, money would carry its own judgment with it.

In any such system, the farmer need not fear to let the aroma of his money expend itself far and wide. It would sprinkle the wholesomeness of winds, the perfume of blossoms, the strengthening smell of the soil, the fragrance of noblest uses.

The farmer that pays his debts can't get rich dishonestly in the sight of heaven. There can't be too much wheat, too many noble cattle, too much wool, an excess of excellent peaches and pears, too many pumpkins, or even too great a crowd of cabbages, if they are not eaten so immoderately as to come to a head again on human shoulders.

The two noblest classes of labor are the extremes—those expended on the material soil, and upon the mental and spiritual regions—those that improve the earth and that make humanity more fertile—the men who give us beets and grapes, and the men who give us ideas; the productive thinkers who show the fields can double their products without waste, and those who improve the capacity of the human mind and hand; the men who labor wisely for the fulfillment of the world's prayer. "Give us this day our daily bread," and the men who, by their genius and service, prove to us the immense significance of that other passage of instruction, "Man shall not live by bread alone, but by every word that proceedeth out of the mouth of God."

The land is the noblest of the gifts of God to humanity. A full treatise on agriculture—its annals and vicissitudes—would be a history of human society from Eden to the staking out of the last "claim" in Iowa or Oregon. The first step from the nomadic state upward toward stable civilization is into the feeling of personal possession of the soil. The fence is the first rude boundary between savagism and civilization.

It requires 800 acres of land, we are told, to supply a hunter as much food as half an acre will furnish under cultivation. And on the 800 acre system of supply, society is impossible; education is impossible; trades and arts are out of the question; combinations of power and interchange of products and help are unattainable. Just in proportion as the land is better tilled, and a smaller quantity of it is made to yield rich returns, the progress of the race is aided, and becomes manifest.

In dealing with the land man is called to be a co-worker with the infinite mind. This is the foundation of the nobleness of the farmer's office.

The air is given to us. We can not alter its constitution, or change its currents. The sea is not placed under our dominion. We can not freshen it, or increase its saltness; we can not level or raise its billows. The rain is ordained for each latitude, and we can not hasten or vary the bounty of the clouds. Minerals are provided in a definite, unalterable measure by the creative force. But the soil we can make our own. We can increase and renew its richness. God does not make it to be a fixed or self-perpetuating blessing like the atmosphere and the ocean. It is a trust. So much He will do for it; but a very great deal is left for us to be faithful in. In the management of the soil, the Creator takes us into partnership. And on our fidelity, within bounds of our trust, the progress of success depends.

The greatness of the trust is seen in this—that agriculture requires the greatest amount and variety of knowledge, and is everywhere latent in its development. We are only now entering upon the study of it. Nation after nation has withered and shriveled because it could not manage its land—because it had not science enough, vigor enough, virtue enough to organize the State so that the soil could be thoroughly tilled and refreshed. As soon as the land begins to yield regularly decreasing stores, so that small farms are absorbed into larger ones, and poverty creeps toward the farmer's hearth, there is radical evil in the State. Its prosperity is not rightly based. Its roots are feeble. It has begun to die. It is not able to sustain the tremendous partnership with Providence in making the soil creative.

In fact, we shall not reach the right point for appreciating the eminence of agriculture as a duty, a profession, and a trust, until we see that the earth is not yet finished. The Creator has left part of the fashioning to man, or rather waits work [*sic*] through man in perfecting it. The air comes up to the divine idea. The sea also answers to the majesty of God's first conception of it. The clouds correspond in their charms of form and glory of color to the archetypes of them in the divine imagination. The highest mountain tops, of splintered crag or dazzling snow, can not be improved any more than they can be altered by the power and wit of man.

But the earth does not yet fulfill the divine intention. It was not made for nettles, nor for the Manzanita and chaparral. It was made for grain, for orchards, for the vine, for the comfort and luxuries of thrifty homes. It was made for these through the educated, organized, and moral labor of man. As plows run deeper, as irrigation is better understood and observed, as the capacity of different soils are comprehended, as types of vegetation are improved, as economy in the renewal of the vitality of the land is learned and practiced, the process of creation goes on; chaos subsides; the divine power and beauty appear in Nature. . . .

Still we must come back to the position that agriculture is a very serious trust. There is little cause for gratulation [*sic*] and complacency, if in all this work we have not been studying the conditions of long continued fertility—if we have been "skinning the land." The race which does this, and is content to do it, after the fact is clearly revealed and the consequences are foreshown, is simply barbarous. All its immense dividends are gained by paying out the capital. It is traveling the swifter to bankruptcy. It is mortgaging and spending the patrimony of its children. It is wasting wealth and energy at such a rate that it will not be able to renew the present lands when they shall be exhausted, and will not have the capital or the enterprise to begin to attach the richest soils which are the last to be approached. There is not probably in all the United States a tract of a hundred miles square which is cultivated in a way to get dividends and save capital both—not certainly a tract of that size where the capital is increasing in power of productiveness. Wherever you find any such districts, you will find them in the least-favored places—in states like Massachusetts and Rhode Island. There are very few districts yet where labor and industry are diversified enough within small compasses to furnish near markets, and so permit the land to be properly enriched, that its vigor may be sustained. Agriculture, as we have said before, is so important that its complete success is interwoven with a right distribution and order of occupations—the symmetry of the State. We have not sufficiently diversified industry on our coast, as yet, to make farming profitable to the land and to the tillers of it besides. So much the more reason, therefore, for notes of alarm and calls to the greater prudence, and economy, and science in what we do till. The agricultural societies are of inestimable importance in this regard. So our future and our civilization—and they should be preaching in our ears the principles of the indispensable gospel of economy—*smaller farms, more labor on them, and reverence for manure.* The man who, by putting the amount of labor on twenty acres which he spreads over fifty, could get the same product, is bound to retrench his limits and save the fertility of that extra thirty acres for a future emigrant. Not to do it is to live by marauding upon nature, not by cultivating the soil. . . .

Rejoice, all of you that are called to the dignity, and trusts, and delights of the farmer or the horticulturist! Rejoice that you belong to a class through whom God is finishing his creation, and who, in enlarging the Divine bounty, are adding to the beauty of the world! Whether an acre, a garden spot, or a section is under your charge, feel more deeply your commission, be glad in the responsible honor of your lot. Study your calling more. Resolve to add to the fertility of your domain. Remember that weeds, and all tares, and slovenly labor are of the devil, and tend backward to chaos. Remember that economy is the fountain of all agricultural opulence. Subdue the lust for immense ranches. If you have fifty acres, and burn to have fifty more, annex fifty that lie *beneath* those you now own, and gain your title by a subsoil plow. Own deep thus by agriculture, not wide by scratchiculture. Increase the beauty of your homestead, by taste which costs nothing; by the training of noble trees and lovely flowers, whose shade and grace will be a dividend of which you can't be cheated, and a gracious spring of good influence in the memory of your children.

California is sketched out by the Almighty as a vast canvas, such as no tribe of men ever received, for the genius and fidelity of colonists to fill with beauty. One of our own citizens has recently indulged an artist's dream of what the State might look like a hundred years hence. He sees in vision "long ribbons of fields stretching to Fort Tejon—each field a different color—green grapes, brown furrows, emerald vines, fringing hedges; grains growing, cream-colored grains, grains aureate and russet; houses dotted along like violets in flower-beds; houses dotted along like dew-drops in clover field; houses reaching forth like mosses in the crystal brook; houses clumped, houses grouped, hamlets modest, hamlets blooming and luxuriant like gorgeous creepers; villages with spires, towns with burnished domes goldened by the sun, and silvered by the moon; cities with minarets, cities with columns, cities with tall needle chimneys pouring up to God the frankincense of labor; terraced foot-hills laughing with generous villas, sloping forelands alive with herds; swelling mounds nestling with vines, oval knolls crowned with festoons of fruit blossoms, breathing sweet perfume to the sky; mountain gorges rolling out metals, mountain peaks staring at opposite peaks from bold-faced palaces, mountain rivulets murmuring to trellised rose-hidden cottages, mountain vales creeping away to love God in dreamy repose."

This gorgeous rhetoric from the pen of your gifted townsman, Rev. Mr. Anderson, may be the cool prose of 1962. Every wise farmer and gardener will help to make it so. It should make hearts swell with sacred pride to know that this generation can contribute to such a future and insure to our posterity a land in which the snow of the Sierra and of Shasta shall emboss and crown such magnificence. And then not only may every California farmer sit under his own vine and fig-tree, but every Californian may drink tea plucked and cured under his own sky; may grind coffee freely from an Arabia at his doors; may sweeten it with sugar landed from no ship that has ever ventured beyond the Golden Gate; may take rice with it raised in our tules; may see the cotton for his household baled in his own county—not by slave labor—and sped for weaving to California mills; may buy his linen stamped with the marks also of domestic produce and skill; may purchase silks on which no duty is paid to a custom-house; and may smoke, in gratitude for his luxuries, tobacco raised in the Virginia within our own bounds.

If we are faithful to our duties, in 1962 the millions that shall live here, can sing with new meaning the old passover song of Palestine:—

> Thou crownest the year with thy goodness;
> Thy footsteps drop fruitfulness;
> They drop it upon the pastures of the wilderness,
> And the hills are girded with gladness,
> The pastures are clothed with flocks,
> And the valleys are covered with corn;
> They shout—yea, they sing for joy.

ORSON HYDE

Some of the utopian (or intentional) communities of the early nineteenth century thrived into the latter portion of the century and beyond. The Mormons, or the Church of Jesus Christ of Latter-Day Saints, were one of the most successful of the new religious sects. The Mormons had their start in the "Burned-Over District" of western New York State, so-called after the many religious enthusiasms that swept through the region. Frequently persecuted for their beliefs, they settled briefly in Ohio, Missouri, Illinois, and Iowa before finding a permanent home in the valley of the Great Salt Lake in Utah and portions of the surrounding Great Basin.

Although not religiously enjoined to an agrarian way of life in the manner of the Amish, the Mormons have been widely admired for their careful, community-oriented approach to farming, and particularly to irrigation, in the dry country of the intermontane West. As J. H. Ward wrote in the Latter-Day Saints' Millennial Star *in 1884, "Here, in the solitude of the desert, with only the bleak mountains, the gilded morning, the painted sunset, and the glimmer of that tideless sea about them, they gave obedience to the com-*

mands of God, the fulfillment of which has compelled even the reluctant admiration of their enemies. The desert has been made to blossom as the rose, and they who were considered too vile to dwell among so-called Christian people have, in the providence of God, built up a civilization so far superior to all that has preceded it, that travelers from distant lands are struck with admiration on beholding it."

The discourse by Apostle Orson Hyde provides an insight into both the philosophy and the labor required to bring farms into the valley of the Great Salt Lake. In this excerpt, from 1867, Hyde intertwines the themes of religious and agricultural productivity, promoting a measured, careful approach to cultivation of both the land and the spirit. Hyde was one of the original members of the Mormon Church, born in 1805 in Oxford, Connecticut, and baptized into the church in 1831. A successful missionary, Hyde traveled to Europe and the Middle East to proselytize, and observed closely the cultures and agriculture of Palestine, Syria, and Lebanon while traveling in the Holy Land. Hyde brought his interest in dry-land agriculture to Utah in 1852, and he wrote and taught about the importance of adhering to the limits of the land on Utah farms, as the passage from his discourse recounts.

From "Instructions" (1867)

There is a good deal of ambition among our people to cultivate a great quantity of ground, the result of which is, that we cultivate our lands poorly in comparison to what we would if we were contented with a smaller area, and would confine our labors to it. We have found some difficulty with regard to water, and complaints have been made about a scarcity of water in many places, when, indeed, I suppose the Lord has apportioned the water to the amount of land he intended should be cultivated. I do not think that these things are passed over unnoticed by Him without some kind of arrangement or calculation. He understands perfectly well what the elements are capable of producing, and how many of His people may be established here or there with profit and with advantage. I have labored most industriously since I have acquired a little experience myself, to induce my brethren to direct their energies upon smaller tracts of land; for I have noticed where men would attempt to raise a crop off forty acres of land, that they could not get their crops in in season, and frequently the frost came early and destroyed a great portion of them. This is bestowing our labor for that which does not profit. Now, would it not be better to confine our energies to a small tract of land, put in our crops in due season, have ample time to do it, do it well, and then it would only require one-half or one-third the amount of water to mature them, and they would mature in advance of the frost?

I do not know how it is in other sections of the country, but I presume it is more or less with them like the circumstances I will relate. I have known men, single handed, attempt to raise twenty-five and thirty acres of grain, when it is more than

Thomas C. Larson, farmers working in a field, c. 1900–1920. Reproduced by permission of L. Tom Perry Special Collections, Harold B. Lee Library, Brigham Young University, Provo, UT.

any one man can well do; the result is, they find themselves troubled to get the water; they run from break of day until dark at night, wearing themselves out, and with all they can do they cannot bestow that attention upon their fields which they need, and they only get from eighteen to twenty bushels of wheat to the acre. When men have confined themselves to ten acres of land, having plowed it well the season before, all the foul weeds killed out and the soil left clean, the seed sown at an early day in the Spring, and put in in good order, I have known such fields to produce from forty to sixty bushels of good plump wheat to the acre. Besides, when fields are so cultivated, less water is used; the necessary labor can be performed without being hurried, and a plentiful harvest of golden sheaves reward the toil of the laborer.

This season, in all probability, our crops will fall short of other years some thirty thousand bushels of wheat, by reason of the early frosts. While I regret this loss, I am happy to say that there is plenty of good wheat in the granary, or in the Egypt of Utah; and I think the loss this year, through early frosts, will aid very much in enforcing the principles which I have endeavored to advance, namely, to confine our labors to smaller tracts of land and put in our crops in good time; that while they are growing luxuriantly and yielding bountifully, filling our bins with golden grain, we be not worn out with toil before the days allotted to us to live are expired; but we still have our strength, time to build comfortable houses for our families to live in, barns and sheds, and to prepare shelter for our stock.

I find the longer we live in these valleys that the range is becoming more and more destitute of grass; the grass is not only eaten up by the great amount of stock that feed upon it, but they tramp it out by the very roots; and where grass once grew luxuriantly, there is now nothing but the desert weed, and hardly a spear of grass is to be seen.

Between here and the mouth of Emigration kanyon [*sic*], when our brethren, the Pioneers, first landed here in '47, there was an abundance of grass over all those benches; they were covered with it like a meadow. There is now nothing but the desert weed, the sage, the rabbit-bush, and such like plants, that make very poor feed for stock. Being cut short of our range in the way we have been, and accumulating stock as we are, we have nothing to feed them with in the winter and they perish. There is no profit in this, neither is it pleasing in the sight of God our Heavenly Father that we should continue a course of life like unto this. Hence, in my labors I have exerted an influence, as far as I have been able, to cultivate less land in grain and secure to ourselves meadows that we might have our hay in the time and in the season thereof, shades for our stock, barns, and stables for our horses, and good houses for our families, where they may be made comfortable and happy, and that we may not be everlasting slaves, running, as it were, after an *ignus fatuus,* or jack in the lantern, falling a false light, but that we may confine ourselves to a proper and profitable course of life. I do say, that a man's life consisteth not in the abundance of the things that he possesses, nor upon the vast amount he extends his jurisdiction over, but it consists in a little well cared for, and everything in order. When we confine ourselves and our labors to small tracts of land, we shall then find time to do everything that is necessary to be done; but if we branch out so largely in plowing, sowing and reaping, we have no time to make necessary improvements around our homes and in our cities; in fact, we have so much to do that we can do nothing at all.

Now I speak of these things, my brethren, not because I think that they are the most edifying to you, but I speak of them because I consider that a temporal salvation is as important as a spiritual one. It is salvation in every respect that we are laboring to obtain, not only to make ourselves comfortable and happy, so far as the physical energies of the body are concerned, but, also, that the mind should not constantly be on the strain day and night. There should be a little time for relaxation and rest to both body and mind, that while our bodies are resting the mind may be fresh to plan and arrange for our personal comfort and how to make everything snug and tidy around us. How much more agreeable is life when everything is in order and good regulation is maintained in and around our homes and cities. This is what I have endeavored, in my weak way, to instill into the minds of the Saints. In some instances I have been successful, and where men have adopted the course I have suggested, they have invariably borne testimony in its favor. I would rather have half a dozen cows in the winter, and have them well taken care of, than to have twenty and have fourteen of them die for want of feed and proper attention, which would leave me only six. I would rather only have the six to begin with, then I would not have the mortification of seeing so many suffer and die. In the present condition of the ranges, we cannot indulge in the hope of raising such large herds of stock as we have done heretofore; but we have got to keep about what will serve us, and take care of them well; then we can enjoy ourselves, and we are not the authors of misery to any part of creation.

We are trying to get into this way; it is a slow operation, and it seems that men's inordinate desire for wealth and extensive possessions is hard to overcome. They hate to be limited; they think their fields are not large enough for their strength; but it is a good thing to have a little strength on hand all the time, and not let out the very last link, because there might be an emergency that would really require it. If we drive a pair of horses all the time at their utmost speed they are soon worn out; and if you want to make a trip very speedily, you cannot do it, your animals are run down, you have not husbanded their strength, and they are not capable of performing the journey you wish; whereas, if they are properly driven, judiciously fed, and their strength properly husbanded, when you want to make a sudden dash you have the power to do it. We are not unlike, in this respect, to other portions of the animal creation. Perhaps I have said enough upon this subject.

We have had our difficulties to encounter in the south; it has not been all sunshine and fair weather with us, but we have got along as well as we could. Perhaps that is saying too much, it is saying a good deal; I do not know that I dare say it. I look back frequently upon my past life and find many places that I think I could have bettered; but were I to live my life over again I do not know that I could do any differently. I will, however, let the past take care of itself, and for the future seek to do the will of God and keep myself in subjection to it.

I have no objections to men obtaining wisdom and learning from books, whether old or new; that is all right and good enough; but I consider it is better to have the Spirit of God in our hearts, that we may know the truth when we hear it; and not only know it when we hear it, but be capable by that Spirit of bringing forth things that we never heard. I feel that it is our privilege, brethren and sisters, to have this principle dwelling within us; and when I see men laboring through books, ancient and modern, to find but little that is good, I am reminded of those who run over forty acres of land in a superficial manner, and only reap a little, when a small quantity of land, well watered and well cultivated, would be sure to yield a rich harvest.

HAMLIN GARLAND

Born in 1860 near the town of West Salem, Wisconsin, Hamlin Garland was, as he put it, a son of the middle border. By this he meant the lands along the frontier, which pushed west across the middle of the country as he grew up. Like many Midwestern settlers, his father was a New Englander who moved across the country via the Erie Canal. After fighting in the Union army in the Civil War, the elder Garland returned to his fam-

ily and his Wisconsin farm, but before long got the urge to move west again. The family next settled near Osage, Iowa, where Hamlin spent much of his youth. Before his father's peregrinations ended, however, the Garlands were farming in the Dakota Territory, which was swelling with settlers in the 1880s.

Garland had fond memories of his early years in Wisconsin, surrounded by family on his maternal side in a close-knit farming community. As the family skipped toward the West, and he grew old enough to help with farm work, that contentment dissipated. And although he was strongly sympathetic to farmers and an active supporter of the Populist movement, he chided those who romanticized agriculture. They "omit the mud and the dust and the grime, they forget the army worm, the flies, the heat, as well as the smells and drudgery of the barns," he wrote. "Milking the cows is spoken of in the traditional fashion as a lovely pastoral recreation, when as a matter of fact it is a tedious job. . . . We hated it in summer when the mosquitoes bit and the cows slashed with their tails, and we hated it still more in the winter time when they stood in crowded malodorous stalls."

Garland's intelligence and acute aesthetic sense made schooling attractive, though it was difficult to obtain a formal education. Having heard his father's tales of the glories of New England—including Boston's theaters—young Hamlin was also attracted to the East. As soon as he could, he left for Massachusetts and set out upon the literary career for which he is remembered today.

The following selection is taken from his memoir A Son of the Middle Border, *first published in 1917. In the book, he recounts his coming of age in the Midwest, his ventures to the East, and the establishment of his career as a writer. The excerpted chapter, "A Visit to the West," tells of a return trip to Iowa in 1887, after a six-year sojourn in Boston.*

From *A Son of the Middle Border* (1917)

At twenty-seven years of age, and after having been six years absent from Osage, the little town in which I went to school, I found myself able to re-visit it. My earnings were still humiliatingly less than those of a hod-carrier, but by shameless economy I had saved a little over one hundred dollars and with this as a traveling fund, I set forth at the close of school, on a vacation tour which was planned to include the old home in the Coulee, the Iowa farm, and my father's house in Dakota. I took passage in a first class coach this time, but was still a long way from buying a berth in a sleeping car.

To find myself actually on the train and speeding westward was deeply and pleasurably exciting, but I did not realize how keen my hunger for familiar things had grown, till the next day when I reached the level lands of Indiana. Every field of wheat, every broad hat, every honest treatment of the letter "r" gave me assurance that I was approaching my native place. The reapers at work in the fields filled my mind with visions of the past. . . .

Once out of the city, I absorbed "atmosphere" like a sponge. It was with me no longer (as in New England) a question of warmed-over themes and appropriated characters. Whittier, Hawthorne, Holmes, had no connection with the rude life of these prairies. Each weedy field, each wire fence, the flat stretches of grass, the leaning Lombardy trees,—everything was significant rather than beautiful, familiar rather than picturesque.

Something deep and resonant vibrated within my brain as I looked out upon this monotonous commonplace landscape. I realized for the first time that the east had surfeited me with picturesqueness. It appeared that I had been living for six years amidst painted, neatly arranged pasteboard scenery. Now suddenly I dropped to the level of nature unadorned, down to the ugly unkempt lanes I knew so well, back to the pungent realities of the streamless plain.

Furthermore I acknowledged a certain responsibility for the conditions of the settlers. I felt related to them, an intolerant part of them. Once fairly out among the fields of northern Illinois everything became so homely, uttered itself so piercingly to me that nothing less than song could express my sense of joy, of power. This was my country—these my people.

It was the third of July, a beautiful day with a radiant sky, darkened now and again with sudden showers. Great clouds, trailing veils of rain, enveloped the engine as it roared straight into the west,—for an instant all was dark, then forth we burst into the brilliant sunshine careening over the green ridges as if drawn by run-away dragons with breath of flame.

It was sundown when I crossed the Mississippi river (at Dubuque) and the scene which I looked out upon will forever remain a splendid page in my memory. The coaches lay under the western bluffs, but away to the south the valley ran, walled with royal purple, and directly across the flood, a beach of sand flamed under the sunset light as if it were a bed of pure untarnished gold. Behind this an island rose, covered with noble trees which suggested all the romance of the immemorial river. The redman's canoe, the explorer's batteau, the hunter's lodge, the emigrant's cabin, all stood related to that inspiring vista. For the first time in my life I longed to put this noble stream into verse.

All that day I had studied the land, musing upon its distinctive qualities, and while I acknowledged the natural beauty of it, I revolted from the gracelessness of its human habitations. The lonely box-like farm-houses on the ridges suddenly appeared to me like the dens of wild animals. The lack of color, of charm in the lives of the people anguished me. I wondered why I had never before perceived the futility of woman's life on a farm.

I asked myself, "Why have these stern facts never been put into our literature as they have been used in Russia and in England? Why has this land no storytellers like those who have made Massachusetts and New Hampshire illustrious?"

These and many other speculations buzzed in my brain. Each moment was a revelation of new uglinesses as well as of remembered beauties.

At four o'clock of a wet morning I arrived at Charles City, from which I was to take "the spur" for Osage. Stiffened and depressed by my night's ride, I stepped out upon the platform and watched the train as it passed on, leaving me, with two or three other silent and sleepy passengers, to wait until seven o'clock in the morning for the "accommodation train." I was still busy with my problem, but the salient angles of my interpretation were economic rather than literary. . . .

At last the train came, and as it rattled away to the north and I drew closer to the scenes of my boyhood, my memory quickened. The Cedar rippling over its limestone ledges, the gray old mill and the pond where I used to swim, the farm-houses with their weedy lawns, all seemed not only familiar but friendly, and when at last I reached the station (the same grimy little den from which I had started forth six years before), I rose from my seat with the air of a world-traveller and descended upon the warped and splintered platform, among my one time friends and neighbors, with quickened pulse and seeking eye.

It was the fourth of July and a crowd was at the station, but though I recognized half the faces, not one of them lightened at sight of me. The 'bus driver, the ragged old dray-man (scandalously profane), the common loafers shuffling about, chewing and spitting, seemed absolutely unchanged. One or two elderly citizens eyed me closely as I slung my little Boston valise with a long strap over my shoulder and started up the billowing board sidewalk toward the center of the town, but I gave out no word of recognition. Indeed I took a boyish pride in the disguising effect of my beard. . . .

As I walked the street I met several neighbors from Dry Run as well as acquaintances from the Grove. Nearly all, even the young men, looked worn and weather-beaten and some appeared both silent and sad. Laughter was curiously infrequent and I wondered whether in my days on the farm they had all been as rude of dress, as misshapen of form and as wistful of voice as they now seemed to be. "Have times changed? Has a spirit of unrest and complaining developed in the American farmer?"

I perceived the town from the triple viewpoint of a former resident, a man from the city, and a reformer, and every minutest detail of dress, tone and gesture revealed new meaning to me. Fancher and Gammons were feebler certainly, and a little more querulous with age, and their faded beards and rough hands gave pathetic evidence of the hard wear of wind and toil. At the moment nothing glozed the essential tragic futility of their existence.

Then down the street came "The Ragamuffins," the little Fourth of July procession, which in the old days had seemed so funny, so exciting to me. I laughed no more. It filled me with bitterness to think that such a makeshift spectacle could amuse anyone. "How dull and eventless life must be to enable such a pitiful travesty to attract and hold the attention of girls like Ella and Flora," I thought as I saw them standing with their little sister to watch "the parade."

From the window of a law office, Emma and Matilda Leete were leaning and I decided to make myself known to them. Emma, who had been one of my high admirations, had developed into a handsome and interesting woman with very little of the village in her dress or expression, and when I stepped up to her and asked, "Do you know me?" her calm gray eyes and smiling lips denoted humor. "Of course I know you—in spite of the beard. Come in and sit with us and tell us about yourself."

As we talked, I found that they, at least, had kept in touch with the thought of the east, and Ella understood in some degree the dark mood which I voiced. She, too, occasionally doubted whether the life they were all living was worth while. "We make the best of it," she said, "but none of us are living up to our dreams."

Her musical voice, thoughtful eyes and quick intelligence, re-asserted their charm, and I spent an hour or more in her company talking of old friends. It was not necessary to talk down to her. She was essentially urban in tone while other of the girls who had once impressed me with their beauty had taken on the airs of village matrons and did not interest me. If they retained aspirations they concealed the fact. Their husbands and children entirely occupied their minds. . . .

Next day I rode forth among the farms of Dry Run, retracing familiar lanes, standing under the spreading branches of the maple trees I had planted fifteen years before. I entered the low stone cabin wherein Neighbor Button had lived for twenty years (always intending sometime to build a house and make a granary of this), and at the table with the family and the hired men, I ate again of Ann's "riz" biscuit and sweet melon pickles. It was not a pleasant meal, on the contrary it was depressing to me. The days of the border were over, and yet Arvilla his wife was ill and aging, still living in pioneer discomfort toiling like a slave.

At neighbor Gardner's home, I watched his bent complaining old wife housekeeping from dawn to dark, literally dying on her feet. William Knapp's home was somewhat improved but the men still came to the table in their shirt sleeves smelling of sweat and stinking of the stable, just as they used to do, and Mrs. Knapp grown more gouty, more unwieldy than ever (she spent twelve or fourteen hours each day on her swollen and aching feet), moved with a waddling motion because, as she explained, "I can't limp—I'm just as lame in one laig as I am in t'other. But 'tain't no use to complain, I've just so much work to do and I might as well go ahead and do it."

I slept that night in her "best room," yes, at last, after thirty years of pioneer life, she had a guest chamber and a new "bed-room soot." With open pride and joy she led Belle Garland's boy in to view this precious acquisition, pointing out the soap and towels, and carefully removing the counterpane! I understood her pride, for my mother had not yet acquired anything so luxurious as this. She was still on the border!

Next day, I called upon Andrew Ainsley and while the women cooked in a red-hot kitchen, Andy stubbed about the barnyard in his bare feet, showing me his hogs and horses. Notwithstanding his town-visitor and the fact that it was Sunday, he came to dinner in a dirty, sweaty, collarless shirt, and I, sitting at his oil-cloth covered table, slipped back, deeper, ever deeper among the stern realities of the life from which I had emerged. I recalled that while my father had never allowed his sons or the hired men to come to the table unwashed or uncombed, we usually ate while clothed in our sweaty garments, glad to get food into our mouths in any decent fashion, while the smell of the horse and the cow mingled with the savor of the soup. There is no escape even on a modern "model farm" from the odor of the barn.

Every house I visited had its individual message of sordid struggle and half-hidden despair. Agnes had married and moved away to Dakota, and Bess had taken upon her girlish shoulders the burdens of wifehood and motherhood almost before her girlhood had reached its first period of bloom. In addition to the work of being cook and scrub-woman, she was now a mother and nurse. As I looked around upon her worn chairs, faded rag carpets, and sagging sofas,—the bare walls of her pitiful little house seemed a prison. I thought of her as she was in the days of her radiant girlhood and my throat filled with rebellious pain.

All the gilding of farm life melted away. The hard and bitter realities came back upon me in a flood. Nature was as beautiful as ever. The soaring sky was filled with shining clouds, the tinkle of the bobolink's fairy bells rose from the meadow, a mystical sheen was on the odorous grass and waving grain, but no splendor of cloud, no grace of sunset could conceal the poverty of these people, on the contrary they brought out, with a more intolerable poignancy, the gracelessness of these homes, and the sordid quality of the mechanical daily routine of these lives.

I perceived beautiful youth becoming bowed and bent. I saw lovely girlhood wasting away into thin and hopeless age. Some of the women I had known had withered into querulous and complaining spinsterhood, and I heard ambitious youth cursing the bondage of the farm. "Of such pain and futility are the lives of the average man and woman of both city and country composed," I acknowledged to myself with savage candor, "Why lie about it?"

Some of my playmates opened their acrid hearts to me. My presence stimulated their discontent. I was one of them, one who having escaped had returned as from some far-off glorious land of achievement. My improved dress, my changed manner of speech, everything I said, roused in them a kind of rebellious rage and gave them unwonted power of expression. Their mood was no doubt transitory, but it was as real as my own.

Men who were growing bent in digging into the soil spoke to me of their desire to see something of the great eastern world before they died. Women whose eyes were faded and dim with tears, listened to me with almost breathless interest whilst I told them of the great cities I had seen, of wonderful buildings, of theaters, of the music of the sea. Young girls expressed to me their longing for a life which was better worth while, and lads, eager for adventure and excitement, confided to me their secret intention to leave the farm at the earliest moment. "I don't intend to wear out my life drudging on this old place," said Wesley Fancher with a bitter oath.

In those few days, I perceived life without its glamor. I no longer looked upon these toiling women with the thoughtless eyes of youth. I saw no humor in the bent forms and graying hair of the men. I began to understand that my own mother had trod a similar slavish round with never a full day of leisure, with scarcely an hour of escape from the tugging hands of children, and the need of mending and washing clothes. I recalled her as she passed from the churn to the stove, from the stove to the bedchamber, and from the bedchamber back to the kitchen, day after day, year after year, rising at daylight or before, and going to her bed only after the evening dishes were washed and the stockings and clothing mended for the night.

The essential tragedy and hopelessness of most human life under the conditions into which our society was swiftly hardening embittered me, called for expression, but even then I did not know that I had found my theme. I had no intention at the moment of putting it into fiction.

WILLA CATHER

Willa Cather was born in 1873 in Black Creek, near Winchester, Virginia, but her family left when she was nine for the high plains of Nebraska. The transition from the Shenandoah Valley to the Western prairies was a shock, and Cather later described her sense of dislocation, explaining that she almost died from homesickness after having been "thrown out" into a land "as bare as a piece of sheet iron." The themes of uprooting and transplantation came to dominate her literary oeuvre, as her fictional characters struggle to find meaning in strange landscapes while trying to make a home. Cather came of age in and around Red Cloud, Nebraska, a trading town along the Republican River. The immigrant farmers from Scandinavia and central Europe who populated this region became fodder for her later fiction.

Eventually, the sweeping landscapes of the prairie gave Cather a "love of great spaces, of rolling open country like the sea"; as she once observed, "It's the grand passion of my life." Educated at the University of Nebraska, she wrote as an art and drama critic for the school newspaper. After university, Cather moved to Pittsburgh, where she edited a women's magazine, the Home Monthly, *before moving on to work as a newspaper writer and a high school teacher. Her first short story collection was published by S. S. McClure, the editor of* McClure's *magazine, in 1905, and she later worked as an editor at* McClure's *while she began her career as a novelist.*

Most of Cather's writing is steeped in the landscapes she knew best, particularly the Nebraska plains, the setting of My Ántonia. *The Nebraska novels are famously based in the history of the region; as Cather once reflected, "From the first chapter, I decided not to 'write' at all—simply to give myself up to the pleasure of recapturing in memory people and places I had believed forgotten." Cather's breakthrough novel,* O Pioneers! *begins with a description of the Nebraska Divide in the 1880s, when the Cather family first arrived in the region. The book evokes the openness and the challenge of the prairie, acknowledging that "the great fact was the land itself, which seemed to overwhelm the little beginnings of human society that struggled in its sombre wastes." Her own most cherished novel,* My Ántonia, *is an unflinching yet luminous evocation of the Nebraska landscape and the farm and small-town experiences of her childhood. Often considered autobiographical, the story is told through the character of Jim Burden, whom Cather paints as a lifelong admirer of the title character, Ántonia Shimerda.*

Our selection begins when Jim and Ántonia are children on neighboring homesteads. Her father, a gentle and cultured Bohemian immigrant, has committed suicide—his spirit crushed by the desolate rawness of the frontier. In later excerpts the adolescent Jim and his family have moved to town, and Ántonia is a servant girl in a neighboring household. The novel closes decades later, with Jim a world-weary railroad lawyer and Ántonia a thriving farmwife and mother. Like Hamlin Garland, Cather loathed the narrow confines of small-town life, but from her memories of the land and its farmers she created a work of powerful agrarian affirmation.

From *My Ántonia* (1918)

Mr. Shimerda lay dead in the barn four days, and on the fifth they buried him. All day Friday Jelinek was off with Ambrosch digging the grave, chopping out the frozen earth with old axes. On Saturday we breakfasted before daylight and got into the wagon with the coffin. Jake and Jelinek went ahead on horseback to cut the body loose from the pool of blood in which it was frozen fast to the ground. . . .

When grandmother and I went into the Shimerdas' house, we found the women-folk alone; Ambrosch and Marek were at the barn. Mrs. Shimerda sat crouching by the stove, Ántonia was washing dishes. When she saw me she ran out of her dark corner and threw her arms around me. "Oh, Jimmy," she sobbed, "what you tink for my lovely papa!" It seemed to me that I could feel her heart breaking as she clung to me.

Mrs. Shimerda, sitting on the stump by the stove, kept looking over her shoulder toward the door while the neighbors were arriving. They came on horseback, all except the postmaster, who brought his family in a wagon over the only broken wagon-trail. The Widow Steavens rode up from her farm eight miles down the Black Hawk road. The cold drove the women into the cave-house, and it was soon crowded. A fine, sleety snow was beginning to fall, and every one was afraid of another storm and anxious to have the burial over with. . . .

Mrs. Shimerda came out and placed an open prayer-book against the body, making the sign of the cross on the bandaged head with her fingers. . . .

At a look from grandfather, Fuchs and Jelinek placed the lid on the box, and began to nail it down over Mr. Shimerda. I was afraid to look at Ántonia. She put her arms around Yulka and held the little girl close to her.

The coffin was put into the wagon. We drove slowly away, against the fine, icy snow which cut our faces like a sand-blast. When we reached the grave, it looked a very little spot in that snow-covered waste. The men took the coffin to the edge of the hole and lowered it with ropes. We stood about watching them, and the powdery snow lay without melting on the caps and shoulders of the men and the shawls of the women. Jelinek spoke in a persuasive tone to Mrs. Shimerda, and then turned to grandfather.

"She says, Mr. Burden, she is very glad if you can make some prayer for him here in English, for the neighbors to understand."

Grandmother looked anxiously at grandfather. He took off his hat, and the other men did likewise. I thought his prayer remarkable. I still remember it. He began, "Oh, great and just God, no man among us knows what a sleeper knows, nor is it for us to judge what lies between him and Thee." He prayed that if any man there had been remiss toward the stranger come to a far country, God would forgive him and soften his heart. He recalled the promises to the widow and the fatherless, and asked God to smooth the way before the widow and her children, and to "incline the hearts of men to deal justly with her." In closing, he said we were leaving Mr. Shimerda at "Thy judgment seat, which is also Thy mercy seat."

All the time he was praying, grandmother watched him through the black fingers of her glove, and when he said "Amen," I thought she looked satisfied with him. She turned to Otto and whispered, "Can't you start a hymn, Fuchs? It would seem less heathenish."

Fuchs glanced around to see if there was general approval of her suggestion, then began, "Jesus, Lover of my Soul," and all the men and women took it up after him. Whenever I have heard the hymn since, it has made me remember that white waste and the little group of people; and the bluish air, full of fine, eddying snow, like long veils flying:—

> While the nearer waters roll,
> While the tempest still is high.

Years afterward, when the open-grazing days were over, and the red grass had been ploughed under and under until it has almost disappeared from the prairie; when all the fields were under fence, and the roads no longer ran about like wild things, but followed the surveyed section-lines, Mr. Shimerda's grave was still there, with a sagging wire fence around it, and an unpainted wooden cross. As grandfather had predicted, Mrs. Shimerda never saw the roads going over his head. The road from the north curved a little to the east just there, and the road from the west swung out a little to the south; so that the grave, with its tall red grass that was never mowed, was like a little island; and at twilight, under a new moon or the clear evening star, the dusty roads used to look like soft gray rivers flowing past it. I never came upon the place without emotion, and in all that country it was the spot most dear to me. I loved the dim superstition, the propitiatory intent, that had put the grave there; and still more I loved the spirit that could not carry out the sentence—the error from the surveyed lines, the clemency of the soft earth roads along which the home-coming wagons rattled after sunset. Never a tired driver passed the wooden cross, I am sure, without wishing well to the sleeper. . . .

There was a curious social situation in Black Hawk. All the young men felt the attraction of the fine, well-set-up country girls who had come to town to earn a living, and, in nearly every case, to help the father struggle out of debt, or to make it possible for the younger children of the family to go to school.

Those girls had grown up in the first bitter-hard times, and had got little schooling themselves. But the younger brothers and sisters, for whom they made such sacrifices and who have had "advantages," never seem to me, when I meet them now, half as interesting or as well educated. The older girls, who helped to break up the wild sod, learned so much from life, from poverty, from their mothers and grandmothers; they had all, like Ántonia, been early awakened and made observant by coming at a tender age from an old country to a new.

I can remember a score of these country girls who were in service in Black Hawk during the few years I lived there, and I can remember something unusual and engaging about each of them. Physically they were almost a race apart, and out-of-door work had given them a vigor which, when they got over their first shyness on coming to town, developed into a positive carriage and freedom of movement, and made them conspicuous among Black Hawk women. . . .

The daughters of Black Hawk merchants had a confident, unenquiring belief that they were "refined," and that the country girls, who "worked out," were not. The American farmers in our country were quite as hard-pressed as their neighbors from other countries. All alike had come to Nebraska with little capital and no knowledge of the soil they must subdue. All had borrowed money on their land. But no matter in what straits the Pennsylvanian and Virginian found himself, he would not let his daughters go out into service. Unless his girls could teach a country school, they sat at home in poverty. The Bohemian and Scandinavian girls could not get positions as teachers, because they had had no opportunity to learn the language. Determined to help in the struggle to clear the homestead from debt, they had no alternative but to go into service. Some of them, after they came to town, remained as serious and discreet in behavior as they had been when they ploughed and herded on their father's farm. Others, like the three Bohemian Marys, tried to make up for the years of youth they had lost. But every one of them did what she had set out to do, and sent home those hard-earned dollars. The girls I knew were always helping to pay for ploughs and reapers, brood-sows, or steers to fatten.

One result of this family solidarity was that the foreign farmers in our county were the first to become prosperous. After the fathers were out of debt, the daughters married the sons of neighbors,—usually of like nationality,—and the girls who once worked in Black Hawk kitchens are to-day managing big farms and fine families of their own; their children are better off than the children of the town women they used to serve.

I thought the attitude of the town people toward these girls very stupid. If I told my schoolmates that Lena Lingard's grandfather was a clergyman, and much respected in Norway, they looked at me blankly. What did it matter? All foreigners were ignorant people who could 'nt speak English. There was not a man in Black Hawk who had the intelligence or cultivation, much less the personal distinction, of Ántonia's father. Yet people saw no difference between her and the three Marys; they were all Bohemians, all "hired girls."

I always knew I should live long enough to see my country girls come into their own, and I have. To-day the best that a harassed Black Hawk merchant can hope for is to sell provisions and farm machinery and automobiles to the rich farms where that first crop of stalwart Bohemian and Scandinavian girls are now the mistresses. . . .

I had only one holiday that summer. It was in July. I met Ántonia downtown on Saturday afternoon, and learned that she and Tiny and Lena were going to the river next day with Anna Hansen—the elder was all in bloom now, and Anna wanted to make elder-blow wine.

"Anna's to drive us down in the Marshalls' delivery wagon, and we'll take a nice lunch and have a picnic. Just us; nobody else. Couldn't you happen along, Jim? It would be like old times."

I considered a moment. "Maybe I can, if I won't be in the way."

On Sunday morning I rose early and got out of Black Hawk while the dew was still heavy on the long meadow grasses. It was the high season for summer flowers. The pink bee-bush stood tall along the sandy roadsides, and the cone-flowers and rose mallow grew everywhere. Across the wire fence, in the long grass, I saw a clump of flaming orange-colored milkweed, rare in that part of the State. I left the road and went through a stretch of pasture that was always cropped short in summer, where the gaillardia came up year after year and matted over the ground with the deep, velvety red that is in Bokhara carpets. The country was empty and solitary except for the larks that Sunday morning, and it seemed to lift itself up to me and come very close.

The river was running strong for midsummer; heavy rains to the west of us had kept it full. I crossed the bridge and went upstream along the wooded shore to a pleasant dressing-room I knew among the dogwood bushes, all overgrown with wild grapevines. I began to undress for a swim. The girls would not be along yet. For the first time it occurred to me that I would be homesick for that river after I left it. The sandbars, with their clean white beaches and their little groves of willows and cottonwood seedlings, were a sort of No Man's Land, little newly-created worlds that belonged to the Black Hawk boys. Charley Harling and I had hunted through these woods, fished from the fallen logs, until I knew every inch of the river shores and had a friendly feeling for every bar and shallow.

After my swim, while I was playing about indolently in the water, I heard the sound of hooves and wheels on the bridge. I struck downstream and shouted, as the open spring wagon came into view on the middle span. They stopped the horse, and the two girls in the bottom of the cart stood up, steadying themselves by the shoulders of the two in front, so that they could see me better. They were charming up there, huddled together in the cart and peering down at me like curious deer when they come out of the thicket to drink. I found bottom near the bridge and stood up, waving to them.

"How pretty you look!" I called.

"So do you!" they shouted together, and broke into peals of laughter. Anna Hansen took the reins and they drove on, while I zigzagged back to my inlet and clambered up behind an overhanging elm. I dried myself in the sun, and dressed slowly, reluctant to leave that green enclosure where the sunlight flickered so bright through the grape-vine leaves and the woodpecker hammered away in the crooked elm that trailed out over the water. As I went along the road back to the bridge I kept picking off little pieces of scaly chalk from the dried water gullies, and breaking them up in my hands.

When I came upon the Marshalls' delivery horse, tied in the shade, the girls had already taken their baskets and gone down the east road which wound through the sand and scrub. I could hear them calling to each other. The elder bushes did not grow back in the shady ravines between the bluffs, but in the hot, sandy bottoms along the stream, where their roots were always in moisture and their tops in the sun. The blossoms were unusually luxuriant and beautiful that summer.

I followed a cattle path through the thick underbrush until I came to a slope that fell away abruptly to the water's edge. A great chunk of the shore had been bitten out by some spring freshet, and the scar was masked by elder bushes, growing down to the water in flowery terraces. I did not touch them. I was overcome by content and drowsiness and by the warm silence about me. There was no sound but the high, sing-song buzz of wild bees and the sunny gurgle of the water underneath. I peeped over the edge of the bank to see the little stream that made the noise; it flowed along perfectly clear over the sand and gravel, cut off from the muddy main channel by a long sandbar. Down there, on the lower shelf of the bank, I saw Ántonia, seated alone under the pagoda-like elders. She looked up when she heard me, and smiled, but I saw that she had been crying. I slid down into the soft sand beside her and asked her what was the matter.

"It makes me homesick, Jimmy, this flower, this smell," she said softly. "We have this flower very much at home, in the old country. It always grew in our yard and my papa had a green bench and a table under the bushes. In summer, when they were in bloom, he used to sit there with his friend that played the trombone. When I was little I used to go down there to hear them talk—beautiful talk, like what I never hear in this country."

"What did they talk about?" I asked her.

She sighed and shook her head. "Oh, I don't know! About music, and about the woods, and about God, and when they were young." She turned to me suddenly and looked into my eyes. "You think, Jimmy, that maybe my father's spirit can go back to those old places?"

I told her about the feeling of her father's presence I had on that winter day when my grandparents had gone over to see his dead body and I was left alone in the house. I said I felt sure then that he was on his way back to his own country, and that even

Solomon D. Butcher, "Custer County, Nebraska," 1887. Reproduced by
permission of the Nebraska State Historical Society, Lincoln.

now, when I passed his grave, I always thought of him as being among the woods
and fields that were so dear to him.

Ántonia had the most trusting, responsive eyes in the world; love and credulous-
ness seemed to look out of them with open faces.

"Why didn't you ever tell me that before? It makes me feel more sure for him. . . ."

It was noon now, and so hot that the dogwoods and scrub-oaks began to turn
up the silvery under-side of their leaves, and all the foliage looked soft and wilted.
I carried the lunch-basket to the top of one of the chalk bluffs, where even on the
calmest days there was always a breeze. The flat-topped, twisted little oaks threw
light shadows on the grass. Below us we could see the windings of the river, and
Black Hawk, grouped among its trees, and, beyond, the rolling country, swelling
gently until it met the sky. We could recognize familiar farmhouses and windmills.
Each of the girls pointed out to me the direction in which her father's farm lay, and
told me how many acres were in wheat that year and how many in corn.

"My old folks," said Tiny Soderball, "have put in twenty acres of rye. They get it
ground at the mill, and it makes nice bread. It seems like my mother ain't been so
homesick, ever since father's raised rye flour for her."

"It must have been a trial for our mothers," said Lena, "coming out here and
having to do everything different. My mother had always lived in town. She says
she started behind in farm-work, and never has caught up."

"Yes, a new country's hard on the old ones, sometimes," said Anna thoughtfully. "My grandmother's getting feeble now, and her mind wanders. She's forgot about this country, and thinks she's at home in Norway. She keeps asking mother to take her down to the waterside and the fish market. She craves fish all the time. Whenever I go home I take her canned salmon and mackerel."

"Mercy, it's hot!" Lena yawned. She was supine under a little oak, resting after the fury of her elder-hunting, and had taken off the high-heeled slippers she had been silly enough to wear. "Come here, Jim. You never got the sand out of your hair." She began to draw her fingers slowly through my hair.

Ántonia pushed her away. "You'll never get it out like that," she said sharply. She gave my head a rough touzling and finished me off with something like a box on the ear. "Lena, you oughtn't to try to wear those slippers any more. They're too small for your feet. You'd better give them to me for Yulka."

"All right," said Lena good-naturedly, tucking her white stockings under her skirt. "You get all Yulka's things, don't you? I wish father didn't have such bad luck with his farm machinery; then I could buy more things for my sisters. I'm going to get Mary a new coat this fall, if the sulky plough's never paid for!" . . .

In the afternoon, when the heat was less oppressive, we had a lively game of "Pussy Wants a Corner," on the flat bluff-top, with the little trees for bases. Lena was Pussy so often that she finally said she wouldn't play any more. We threw ourselves down on the grass, out of breath.

"Jim," Ántonia said dreamily, "I want you to tell the girls about how the Spanish first came here, like you and Charley Harling used to talk about. I've tried to tell them, but I leave out so much."

They sat under a little oak, Tony resting against the trunk and the other girls leaning against her and each other, and listened to the little I was able to tell them about Coronado and his search for the Seven Golden Cities. At school we were taught that he never got so far north as Nebraska, but had given up his quest and turned back somewhere in Kansas. But Charley Harling and I had a strong belief that he had been along this very river. A farmer in the county north of ours, when he was breaking sod, had turned up a metal stirrup of fine workmanship, and a sword with a Spanish inscription on the blade. He lent these relics to Mr. Harling, who brought them home with him. Charley and I scoured them, and they were on display in the Harling office all summer. Father Kelly, the priest, had found the name of the Spanish maker on the sword and an abbreviation that stood for the city of Cordova.

"And that I saw with my own eyes," Ántonia put in triumphantly. "So Jim and Charley were right, and the teachers were wrong!"

The girls began to wonder among themselves. Why had the Spaniards come so far? What must this country have been like, then? Why had Coronado never gone back to Spain, to his riches and his castles and his king? I couldn't tell them. I only knew the school books said he "died in the wilderness, of a broken heart."

"More than him has done that," said Ántonia sadly, and the girls murmured assent.

We sat looking off across the country, watching the sun go down. The curly grass about us was on fire now. The bark of the oaks turned red as copper. There was a shimmer of gold on the brown river. Out in the stream the sandbars glittered like glass, and the light trembled in the willow thickets as if little flames were leaping among them. The breeze sank to stillness. In the ravine a ringdove mourned plaintively, and somewhere off in the bushes an owl hooted. The girls sat listless, leaning against each other. The long fingers of the sun touched their foreheads.

Presently we saw a curious thing: there were no clouds, the sun was going down in a limpid, gold-washed sky. Just as the lower edge of the red disk rested on the high fields against the horizon, a great black figure suddenly appeared on the face of the sun. We sprang to our feet, straining our eyes toward it. In a moment we realized what it was. On some upland field, a plough had been left standing in the field. The sun was sinking just behind it. Magnified across the distance by the horizontal light, it stood out against the sun, was exactly contained within the circle of the disk; the handles, the tongue, the share—black against the molten red. There it was, heroic in size, a picture writing on the sun.

Even while we whispered about it, our vision disappeared; the ball dropped and dropped until the red tip went beneath the earth. The fields below us were dark, the sky was growing pale, and that forgotten plough had sunk back to its own littleness somewhere on the prairie.

L. L. POLK

Leonidas Lafayette Polk, born in Anson County, North Carolina, in 1837, was a member of an extended family descended from Robert Bruce Polk, who had settled in Maryland in the seventeenth century. Relatives, distant or otherwise, included President James K. Polk and General Leonidas Polk, the "Fighting Bishop" of the Confederate army.

Orphaned at fourteen, L. L. Polk took up farming for his living. He married and had six daughters as well as a son who died in infancy. Polk served in the North Carolina state legislature in 1860, fought in the Civil War, and returned to the legislature in 1864. He continued farming as his political interests developed, and his political and agricultural pursuits coincided in many respects. Polk advocated the establishment of a state department of agriculture, then served as its first commissioner in 1877.

Polk was also a prominent editor and journalist. Beginning in 1880 he edited the Raleigh News, *and in 1886 he founded the* Progressive Farmer. *At first the* Progressive Farmer *focused on disseminating improved techniques to its readers. Before long, however, Polk used the journal to encourage his farm audience to organize to further their political interests. When the Farmers' Alliance came to North Carolina, spreading rapidly from its birthplace in Texas, Polk took up the cause and the* Progressive Farmer *became an official publication of the Alliance. In 1887 Polk became the organization's vice president, and by 1889 he was its president.*

In the early 1890s, Polk's political engagements changed in another critical way. A longtime Southern Democrat, he switched his allegiance to the People's Party, whose policies were closely aligned with the Farmers' Alliance. A man of great political skill, he appeared to be the leading contender for nomination as the Populist candidate for president in 1892, but died shortly before the party's convention in that year.

Despite his turn away from the Democratic Party and toward the more radical positions of the People's Party, Polk remained attuned to the conservative sensibilities of his native South. In his speeches and writing he sought to allay the fears of his core audience that the growing farmers' movement was too radical. When a convention of South Carolina farmers in 1887 set out a list of political demands, Polk commended them: "There is nothing inflammatory, or hot-blooded about them. Though pertinent and vigorous they are thoughtful and conservative. They voice the sentiments of the farmers of South Carolina strongly, but respectfully, to the men who are to compose the next legislature of South Carolina."

Polk gave the following address to the Inter-States Convention of Farmers in Atlanta, August 16–18, 1887. In it, with rhetorical finesse, he turns what starts like a standard Southern protest against domination by Northern industrial and banking interests into a call for alliance with farmers in other regions, and then to a critical inward look at the desolation taking shape in the South with the rise of an extractive system of sharecropping cotton monoculture.

From "Address before the Inter-States Convention of Farmers" (1887)

Mr. President, and gentlemen of the Convention.—I beg to express my very high appreciation of the distinguished honor done me through the kind partiality of your committee, in assigning me a place as an active participant in the deliberations of this Convention. And I would be untrue to myself and to my colleagues on this floor and false to the people whom we represent, did I fail to assure you of their earnest sympathy and co-operation in all that you may do to promote the great and important objects of this convention.

By the ordeal of common suffering, by the glories of common triumph, the cherished deeds and traditions of the past, the opportunities and resources of the present and the inviting promise of the future, North Carolina hails her Southern sister States here assembled, and pledges her most earnest endeavor and most loyal and fraternal support, in any and all measures for the advancement and development of the industrial and agricultural interests of the South and of the whole country.

Her people are profoundly impressed with the belief that we should no longer ignore the fact that in the great race of material progress—agriculture—and especially in the cotton States, is in a languishing condition; hence, in a large mass convention of our farmers, which met in the city of Raleigh, on the 26th day of January last, resolutions, suggesting and strongly commending a convention of the farmers of the Southern States, were unanimously adopted. We rejoice at the thrift and prosperity of all our great industrial enterprises, the growth and development of which, during the past two decades, has no parallel in all history.

At no period has this progressive country witnessed such active and substantial progress in its great railroad interests; never has it enjoyed so healthy and flourishing condition of manufacturing enterprise, in all its departments; never has it witnessed so rapid growth of towns and cities, or an easier accumulation of colossal fortunes by individuals and corporations. The whole country, from ocean to ocean, is charged and electrified with the exhilerating [sic] pulsations of active, animated industry, and yet agriculture, the great bed-rock on which rests all other interests, the source and basis of our material wealth, languishes.

Of all the busy millions of our population, engaged in the various occupations, 7,670,493, or 51 per cent. of the whole, are engaged in agriculture. They represent $10,000,000,000, in lands, $1,500,000,000, in live stock, $400,000,000, in implements and machinery, and $4,000,000,000, in the annual products of their labor. Of the millions and millions of dollars in public taxes which flow into our burdened and overflowing treasury, eighty cents of every dollar of it comes from the hard earnings of the agriculturalists of the country. The great propelling power which drives our ships of commerce to the ports of the world, is the muscle of the American farmer. It has supplied over seventy per cent. of our domestic exports, each year, since 1850, with the exception of the years 1864, and 1865. Without it, national progress would be paralyzed, our ships would rot at their docks; our railroads would be enveloped in grass and weeds and our towns and cities would become the dismal abode of the genius of ruin and desolation. Representing in part this gigantic power in our national organism, may not this convention appropriately ask to what extent does the government foster and encourage it? What voice has it in the council and cabinet of the nation?

And how is it with the States represented in this body? In a population of 4,186,235, employed in the various occupations, over ten years of age in these ten States, 3,007,493 are engaged in agriculture, or seventy-one per cent. of the whole, and representing thirty-eight per cent. of the entire agricultural population of the United States. The value of the aggregate agricultural products of these States, within the past twenty-one years, amounts to the enormous sum of $20,000,000,000! During that time we have produced 93,409,794 bales of cotton, realizing $7,357,000,000! By what process of computation are we to realize the vastness of this mighty sum?

It is enough money to pay for every foot of railway in this great country, and every dollar of their equipments, and have over $300,000,000 left. It is enough money to pay for every foot of real estate in these ten States, with the public debt of the United States and of Great Britain added. It is enough money to give $2,500 to each person over ten years of age in these States, engaged in agriculture. Where is this colossal pile of wealth? What has become of it? Is it to be found in beautified and improved homes, enriched and improved farms, improved public roads, improved implements and machinery, improved stock, in increasing and multiplying agencies for the comfort, happiness and independence of Southern farmers? Why is it that with all the wonderful and marvelous productive capacity of this God-favored South-land, States are represented here today, whose farmers are in as bad condition, financially, as they were in the dark days of 1866? Why is it, that they have contributed faithfully, the penury of their humble mite to the wealth of this great abundance, that they have worked hard and lived hard, have furnished the land, furnished the stock, furnished the labor, furnished the implements for the last twenty years to produce this vast sum of money; and today find themselves entangled in the meshes of merciless debt, and are bound to the earth with heaviness of spirit? Are our industries properly adjusted and equally fostered and protected? . . .

The causes which retard our progress and weigh down our energies, the serious exigency of the situation, to us as farmers, demands the most candid and honest investigation. The questions I have propounded should enlist the serious thought and most patriotic consideration, not only of the farmers of the South, but of the whole country. For, Mr. President, retrogression in American agriculture, means national decline, national decay and ultimate and inevitable ruin. . . .

Are we of the cotton States laboring under the aggregate evil of accumulated individual error, or the more important and serious trouble of inherent defects in our agricultural system? Representing a country of such vast and immense capabilities as are found within the limit of these ten States with its great variety of climate and fertility of soils, with its wonderful adaptability to the successful cultivation of all the leading agricultural products known to the civilized world, with its admirable and inviting conditions for diversified husbandry, with its accessibility to the markets of the world, with all the inherent conditions to make it indeed a mighty agricultural and industrial empire; why should any industrious owner of its fruitful soil work hard, live hard and die poor?

Where is the trouble? I mention first:

THE "ONE CROP," OR "ALL COTTON" SYSTEM.

The close of the war found us with our sons slain, our fair fields devastated, our homes desolate and household gods destroyed, without money, without food, without implements with which to work, our credit gone, labor utterly destroyed, our systems wiped out, the accumulated wealth of generations swept away as by a breath, and we were left friendless and unaided to depend on those high qualities of true manhood which are always evoked by terrible emergencies. Ours was the painful, but noble and heroic task of removing from our fair and beautiful Southland the wreck which marked the feast of the war-gods who had revelled here for four long years in the high carnival of blood, of carnage and of death. And if Southern men did win glory in arms, how grandly and resplendantly [*sic*] sublime that glory becomes when crowned with their nobler achievements as citizens in peace. In war their manhood, their unswerving devotion, their iron-like endurance, their superb heroism challenged the admiration of the whole world; in peace their forbearance, their patience and their indomitable perseverance, modeled by the heroic spirit and lofty patriotism of our peerless women, have wreathed their names with chaplets of imperishable honor.

We "beat our spears into pruning-hooks and our swords into plowshares," and turned, with a faith and devotion that was sublime, to the solution of problems so graphically described by your eloquent Grady [presumably Henry Grady, advocate for modernization in the South and economic integration with the rest of the country] this morning, and which would have appalled any hearts but those that had been educated in the terrible ordeal through which we had passed. For four years our ports had been locked against the world. The demands of commerce were hushed amid the thunderings of internal war, and our fleecy cotton fields were appropriated for supplies of bread for our struggling and beleagured [*sic*] armies. The world wanted and needed the cotton of the South, and the South in turn wanted and need[ed] the money of the world, and in that supreme hour of our extremity we yielded to the seductive and delusive promise of speedy relief, if not fabulous wealth, and turned a deaf ear to the wholesome lessons and admonitions of a prosperous and independent past. Every interest of the farm was subordinated to the production of cotton. We said to the Northern and Northwestern States, "make everything we need—our meat, corn, flour, hay, fertilizers, fruits, butter, potatoes, and ten thousand things we don't need—put your price on them, and we will raise one crop with which to pay for it all and you may price that crop." Magnanimous, but disastrously unwise proposition. Never was the connection between cause and effect more forcibly illustrated or traced than in the results which most naturally followed. With that old-fashioned, wise and only safe policy for the Southern farmer abandoned, viz.: the raising of our home supplies on our own farms—we drifted away from the sheet anchor of our safety and independence. Habits of reckless extravagance and improvidence grew on us, debt, the iron-gloved tyrant, was robbed of his terror, the wretched slavery of liens and mortgages was invoked and welcomed, ruinous profits on the necessaries as well as the luxuries of life were freely promised if not paid, stock was neglected, diversification of crops abandoned, rates were paid on money and supplies that would have wrecked the whole financial system of this government in ninety days, and finally the lordly gambler, under the sanction of legal authority, appears on this tempting and fruitful scene of wild and extravagant speculation and absolutely and arbitrarily fixes the price of the whole crop. Our independence as farmers surrendered, a system of genteel tenantry established—that we may have the privilege of producing the leading commercial crop of the world, the price of which we can no more control than can the shepherds in the highlands of Scotland.

Why should the Southern farmer pay one dollar, or more, per bushel for corn, when he can produce it at home at a cost not to exceed fifty cents per bushel?

Why should the Southern farmers pay ten to fourteen cents per pound for bacon, when he can produce it at home at five cents?

Why should the Southern farmer pay millions of dollars annually for commercial fertilizers, when for most of our soils and crops they may produce a better article for half the money?

Why should the Southern farmer pay twenty to twenty-five dollars per ton for hay, when he may produce it on his farm at two dollars per ton? Fighting grass all summer and buying it all winter!

Why is it that thousands and thousands of freight trains come southward loaded with the products of Northern farms and so many of them are returned empty?

My heart was moved this morning with pride as I listened to the gifted Grady in his thrilling and eloquent portrayal of the magnificent resources and capabilities of this Southern land. But the crowning argument, in my judgment, to sustain the claim that it is the grandest country on the globe, is the fact that it has undergone this stupendous drain on its energies and resources for these long years and we are alive! It could have been accomplished nowhere on the globe except on Southern soil and by Southern men.

The next defect, I notice, is our "Tenant system," by which I mean that practice among the farmers in the South of renting or leasing lands to tenants to be managed and cultivated according to their own methods. In my judgment, no system could be devised for more unremunerative and unprofitable crops, or for the more certain exhaustion and destruction of our lands. With our cheap and naturally fertile lands, if we had intelligent, skilled and reliable labor, subjected to wholesome rules of labor-discipline and labor-system, as in England or other old and advanced European countries, then indeed might we hope for the best results. But I hazard nothing in saying that with this practice engrafted on our agricultural system, neither the tenant nor landlord may expect remunerative results, and the farmer who persists in it will leave to his children a legacy of old fields and gullies. For if there be any one phase of the labor system of the South which is settled beyond all question among intelligent Southern farmers, it is that we may hope for no improvement of lands or crops or methods if left to the control of free-negro tenants. The wage-labor system, established on the equitable principle of giving a just price for labor, and demanding in return a just equivalent from labor, would give to the employee and those dependent on him the full benefit of a fair value for his labor and would keep the care and cultivation of the lands where it should always be,—under the direct control and supervision of the owner.

THIRDLY—*The "Broad Acre" System.* Is it not harder to cultivate a two hundred acre farm poorly, than it is to cultivate a one hundred acre farm well? Can that system of farming be right which impoverishes the soil and reduces the yield? Can that system of farming be right and profitable which does not improve the soil and increase the yield?

The average farmer in the South is attempting to cultivate too much land. The average farmer in the South owns more land than he can manage profitably. The time, labor and money expended on twenty acres should be concentrated on ten. But it was one of those venerable and reverend customs of the past, which, although un-profitable and wasteful, and unsuited to those conditions imposed by the destruction of our slave-labor system, we still cling to with singular tenacity. The slave-owner of the South was rich in proportion as he was able to increase the number of his slaves and to multiply his broad acres. And what though, by his system of farming, fair and fertile fields were exhausted, laid waste and abandoned? He had but to slay and destroy his magnificent forests to replace them with fresher and richer fields, to be in their turn depleted, exhausted and ruined. Thus was this goodly heritage, which the God of Nature had so bountifully provided, wasted and positively robbed, and today all over the beautiful hills and plains of this fair land, abandoned, exhausted fields and gullied hillsides greet the eye as the mournful footprints of slavery. Shall this process of depletion, exhaustion, waste and ruin be continued? Shall the genera-tions which are to follow us continue to be robbed?

Mr. President, slavery is gone. In the name of progress, in the name of our chil-dren, let those systems and practices born of its existence, go with it.

FOURTHLY—*Want of unity of action or co-operation among our farmers.* Every inter-est of any magnitude in the civilized world is protected and advanced by the fostering care and power of co-operative effort except that of agriculture. It is exposed as a helpless prey to all, and all prey upon it. Co-operation, organization, consolidation are the watch-words of the hour in all departments of progress, and yet the farmers of the land are slow to discover its mighty and restless power. "No man liveth to himself" is a law that is as unyielding in its demands and operations on the farmer as the one which declares that he "shall eat bread in the sweat of his face."

Do the farmers of neighborhoods, of townships, of counties, of States, meet together, counsel together, act together for their common advancement and protec-tion and for their mutual good? Do they complain of imperfect and unprofitable methods and systems? Do they complain of unjust laws, and discriminating legisla-tion? Where is the remedy? Belonging to 51 per cent. of the entire population of the United States engaged in the various occupations and constituting 71 per cent. of that of cotton States—where is the remedy?

FIFTHLY—*A want of practical training and agricultural education for the masses.*

Our system of education, not only in the South, but throughout the country, has hitherto tended to the development of only one department of human effort, viz: Conquest on the forum or renown in professional life. In our system of education no conquests in the great field of agriculture were planned—no avenues for distinction and renown were opened up to the aspiring young mind in the great field of industry.

The great and imperative need of our people and of our time, is the practical education of the masses. Wherever attainable, give the youth of the South brightly burnished mental machinery, but we owe it to ourselves, to our posterity and to the world to see to it that they are provided with the weapons of practical knowledge with which to fight life's battles. Let us teach them how to work and teach them to be proud of it. Let us teach them how to make a dollar and how to live on 90 cents of it. Establish institutions for the practical, industrial education of the masses of our people. It will be a glorious day for the South, when our young men shall not be ashamed to hang their diplomas in their workshops, their factories, their laboratories, their school rooms, their counting rooms and their farm houses. It will be a glorious day for the South when her young ladies, educated in all the higher and refined arts of life, shall boast and without blushing of equal proficiency in the management of the household and the flower-garden. Teach them the important principles involved in the great science of agriculture, the nobility of honorable industry and thus lift our boys to a higher plane of thought, of aspiration and of manhood. The most pitiable of all the examples of worthless manhood, is that young man who is too proud to be poor and who is too lazy to work. Ah, the towering spirits that have been stripped of their bright plumage by this unmanly and cowardly dread of honorable labor! Let us have institutions for the special practical training of the brain and the hand, and thus bring forth an army of young men, inspired with a just appreciation of the dignity and nobility of honorable labor, and whose skill and ability shall command success. We need and must have skilled artisans, skilled mechanics, skilled scientists, skilled agriculturists, skilled brain and skilled labor in all departments of industry, if we would develop and utilize all the varied and wonderful capabilities of this highly favored section. Many of our children will live to be numbered among one hundred and fifty millions of American people. That young man, who shall be able to write before his maturity, the date, January 1, A. D., 2,000, will look out upon an era fraught with the grandest and most stupendous achievement and glorious triumph of intellect and genius, that has marked any period in all the world's history. Shall we not equip and train him for the responsibilities which he must inevitably encounter? . . .

Young men of the South, let us turn our faces toward the rising sun and with hopeful energy, with willing hearts and ready hands, rejuvenate the grand old homestead of our ancestors, and by fidelity to principle and unswerving devotion to duty, transmit it, with all its endearing associations and cherished traditions, to those who are to come after us, with the lamp of liberty, burning in undimmed splendor on its sacred altars.

IGNATIUS DONNELLY, FOR THE PEOPLE'S PARTY

The political ferment that simmered in the latter decades of the nineteenth century peaked in the early 1890s. This ferment had deep roots in agrarian unrest in the Great Plains and South but was also fed by urban and industrial discontent. In both cities and countryside, there was a strong feeling that the American system was failing ordinary people in the face of economic exploitation by powerful corporations, aided and abetted by political leaders who were unwilling to address the needs of the "producing" classes. Leaders in the Farmers' Alliance movement and other agrarian organizations reached out to the Knights of Labor in the late 1880s, which led to the founding of the coalition that formed the People's Party. This union held out the promise that the concerns of rural and urban workers could be met by a third party, free of the need to compromise with either the Democrats or the Republicans. In the summer of 1892 the party convened at Omaha, Nebraska, and on July 4 adopted a platform that laid out its political agenda. The Omaha Platform is remembered today for planks that once seemed radical but would later become familiar parts of the American political landscape. These include the popular election of U.S. senators and the graduated income tax.

Ignatius Donnelly, author and longtime political activist, drafted the platform's preamble. Originally from Philadelphia, Donnelly moved to Minnesota, where he attempted to found a cooperative agricultural community, Nininger City. The community failed in 1857 during an economic downturn, and Donnelly entered political life, serving as lieutenant governor of Minnesota, as a state senator, and as a representative to the U.S. Congress. He was known for his rhetorical flair, honed by writing highly popular literary works. Donnelly, who founded and edited the Anti-Monopolist newspaper, was drawn to many reform movements, including women's suffrage, and helped organize the Minnesota Farmers' Alliance.

From the Omaha Platform of the People's Party (1892)

Assembled upon the 116th anniversary of the Declaration of Independence, the People's Party of America, in their first national convention, invoking upon their action the blessing of Almighty God, put forth in the name and on behalf of the people of this country, the following preamble and declaration of principles:

PREAMBLE

The conditions which surround us best justify our co-operation; we meet in the midst of a nation brought to the verge of moral, political, and material ruin. Corruption dominates the ballot-box, the Legislatures, the Congress, and touches even the ermine of the bench. The people are demoralized; most of the States have been compelled to isolate the voters at the polling places to prevent universal intimidation and bribery. The newspapers are largely subsidized or muzzled, public opinion silenced, business prostrated, homes covered with mortgages, labor impoverished, and the land concentrating in the hands of capitalists. The urban workmen are denied the right to organize for self-protection, imported pauperized labor beats down their wages, a hireling standing army, unrecognized by our laws, is established to shoot them down, and they are rapidly degenerating into European conditions. The fruits of the toil of millions are boldly stolen to build up colossal fortunes for a few, unprecedented in the history of mankind; and the possessors of those, in turn, despise the republic and endanger liberty. From the same prolific womb of governmental injustice we breed the two great classes—tramps and millionaires.

The national power to create money is appropriated to enrich bondholders; a vast public debt payable in legal tender currency has been funded into gold-bearing bonds, thereby adding millions to the burdens of the people.

Silver, which has been accepted as coin since the dawn of history, has been demonetized to add to the purchasing power of gold by decreasing the value of all forms of property as well as human labor, and the supply of currency is purposely abridged to fatten usurers, bankrupt enterprise, and enslave industry. A vast conspiracy against mankind has been organized on two continents, and it is rapidly taking possession of the world. If not met and overthrown at once it forebodes terrible social convulsions, the destruction of civilization, or the establishment of an absolute despotism.

We have witnessed for more than a quarter of a century the struggles of the two great political parties for power and plunder, while grievous wrongs have been inflicted upon the suffering people. We charge that the controlling influences dominating both these parties have permitted the existing dreadful conditions to develop without serious effort to prevent or restrain them. Neither do they now promise us any substantial reform. They have agreed together to ignore, in the coming campaign, every issue but one. They propose to drown the outcries of a plundered people with the uproar of a sham battle over the tariff, so that capitalists, corporations, national banks, rings, trusts, watered stock, the demonetization of silver and the oppressions of the usurers may all be lost sight of. They propose to sacrifice our homes, lives, and children on the altar of mammon; to destroy the multitude in order to secure corruption funds from the millionaires.

Assembled on the anniversary of the birthday of the nation, and filled with the spirit of the grand general and chief who established our independence, we seek to restore the government of the Republic to the hands of "the plain people," with which class it originated. We assert our purposes to be identical with the purposes of the National Constitution; to form a more perfect union and establish justice, insure domestic tranquillity, provide for the common defence, promote the general welfare, and secure the blessings of liberty for ourselves and our posterity.

We declare that this Republic can only endure as a free government while built upon the love of the whole people for each other and for the nation; that it cannot be pinned together by bayonets; that the civil war is over, and that every passion and resentment which grew out of it must die with it, and that we must be in fact, as we are in name, one united brotherhood of free men.

Our country finds itself confronted by conditions for which there is no precedent in the history of the world; our annual agricultural productions amount to billions of dollars in value, which must, within a few weeks or months, be exchanged for billions of dollars worth of commodities consumed in their production; the existing currency supply is wholly inadequate to make this exchange; the results are falling prices, the formation of combines and rings, the impoverishment of the producing class. We pledge ourselves that if given power we will labor to correct these evils by wise and reasonable legislation, in accordance with the terms of our platform.

We believe that the power of government—in other words, of the people—should be expanded (as in the case of the postal service) as rapidly and as far as the good sense of an intelligent people and the teachings of experience shall justify, to the end that oppression, injustice, and poverty shall eventually cease in the land.

While our sympathies as a party of reform are naturally upon the side of every proposition which will tend to make men intelligent, virtuous, and temperate, we nevertheless regard these questions, important as they are, as secondary to the great issues now pressing for solution, and upon which not only our individual prosperity but the very existence of free institutions depend; and we ask all men to first help us to determine whether we are to have a republic to administer before we differ as to the conditions upon which it is to be administered, believing that the forces of reform this day organized will never cease to move forward until every wrong is remedied and equal rights and equal privileges securely established for all the men and women of this country.

PLATFORM

We declare, therefore—

First.—That the union of the labor forces of the United States this day consummated shall be permanent and perpetual; may its spirit enter into all hearts for the salvation of the Republic and the uplifting of mankind.

Second.—Wealth belongs to him who creates it, and every dollar taken from industry without an equivalent is robbery. "If any will not work, neither shall he eat." The interests of rural and civic labor are the same; their enemies are identical.

Third.—We believe that the time has come when the railroad corporations will either own the people or the people must own the railroads, and should the government enter upon the work of owning and managing all railroads, we should favor an amendment to the Constitution by which all persons engaged in the government service shall be placed under a civil-service regulation of the most rigid character, so as to prevent the increase of the power of the national administration by the use of such additional government employes.

FINANCE.—We demand a national currency, safe, sound, and flexible, issued by the general government only, a full legal tender for all debts, public and private, and that without the use of banking corporations, a just, equitable, and efficient means of distribution direct to the people, at a tax not to exceed 2 per cent. per annum, to be provided as set forth in the sub-treasury plan of the Farmers' Alliance, or a better system; also by payments in discharge of its obligations for public improvements.

1. We demand free and unlimited coinage of silver and gold at the present legal ratio of 16 to 1.
2. We demand that the amount of circulating medium be speedily increased to not less than $50 per capita.
3. We demand a graduated income tax.
4. We believe that the money of the country should be kept as much as possible in the hands of the people, and hence we demand that all State and national revenues shall be limited to the necessary expenses of the government, economically and honestly administered.
5. We demand that postal savings banks be established by the government for the safe deposit of the earnings of the people and to facilitate exchange.

TRANSPORTATION—Transportation being a means of exchange and a public necessity, the government should own and operate the railroads in the interest of the people. The telegraph, telephone, like the post-office system, being a necessity for the transmission of news, should be owned and operated by the government in the interest of the people.

LAND.—The land, including all the natural sources of wealth, is the heritage of the people, and should not be monopolized for speculative purposes, and alien ownership of land should be prohibited. All land now held by railroads and other corporations in excess of their actual needs, and all lands now owned by aliens should be reclaimed by the government and held for actual settlers only.

Expression of Sentiments

Your Committee on Platform and Resolutions beg leave unanimously to report the following:

Whereas, Other questions have been presented for our consideration, we hereby submit the following, not as a part of the Platform of the People's Party, but as resolutions expressive of the sentiment of this Convention.

1. RESOLVED, That we demand a free ballot and a fair count in all elections and pledge ourselves to secure it to every legal voter without Federal Intervention, through the adoption by the States of the unperverted Australian or secret ballot system.
2. RESOLVED, That the revenue derived from a graduated income tax should be applied to the reduction of the burden of taxation now levied upon the domestic industries of this country.
3. RESOLVED, That we pledge our support to fair and liberal pensions to ex-Union soldiers and sailors.
4. RESOLVED, That we condemn the fallacy of protecting American labor under the present system, which opens our ports to the pauper and criminal classes of the world and crowds out our wage-earners; and we denounce the present ineffective laws against contract labor, and demand the further restriction of undesirable emigration.
5. RESOLVED, That we cordially sympathize with the efforts of organized workingmen to shorten the hours of labor, and demand a rigid enforcement of the existing eight-hour law on Government work, and ask that a penalty clause be added to the said law.
6. RESOLVED, That we regard the maintenance of a large standing army of mercenaries, known as the Pinkerton system, as a menace to our liberties, and we demand its abolition. . . .
7. RESOLVED, That we commend to the favorable consideration of the people and the reform press the legislative system known as the initiative and referendum.
8. RESOLVED, That we favor a constitutional provision limiting the office of President and Vice-President to one term, and providing for the election of Senators of the United States by a direct vote of the people.
9. RESOLVED, That we oppose any subsidy or national aid to any private corporation for any purpose.
10. RESOLVED, That this convention sympathizes with the Knights of Labor and their righteous contest with the tyrannical combine of clothing manufacturers of Rochester, and declare it to be a duty of all who hate tyranny and oppression to refuse to purchase the goods made by the said manufacturers, or to patronize any merchants who sell such goods.

Luna Kellie

Luna Sanford Kellie was born in 1857 in Pipestone, a settlement in southwestern Min-
nesota. Her family was itinerant, struggling to make a living on farms in various locations
before moving to St. Louis, Missouri. There Luna met and soon married J. T. Kellie.

Luna and J. T. later moved to south-central Nebraska to homestead. The young couple's
initial enthusiasm was undermined in time by the disasters that seem to have afflicted
so many homesteaders: severe weather, grasshopper infestations, infant mortality, and
financial difficulties. Drought was less of a problem during the Kellies' tenure as home-
steaders than in other periods on the Great Plains, but even abundant crops failed to
provide them a sound living. As J. T. Kellie put it, "We have raised enough grain this
year to feed us and all our descendants for a hundred years yet have to sell every bushel
of it to pay expenses and our expenses did not include a salary for ourselves or anything
not necessary to produce the crop."

Luna Kellie noted years later that she was attracted to homesteading in part by the
railroad advertising campaigns that portrayed the West as an Eden for farmers. In time
she came to denounce the role of the railroads in the life of the Plains farmer, "for the
minute you crossed the Missouri River, your fate both soul and body was in their hands.
What you should eat and drink, what you should wear, everything was in their hands
and they robbed us of all we produced except enough to keep body and soul together and
many times not that."

In response to their troubles homesteading, the Kellies became ardent supporters of the
Farmers' Alliance. By 1890 Luna had emerged as an effective speaker and songwriter in
Alliance campaigns. She prepared the following speech for the group's 1894 Nebraska state
meeting, when the fortunes of the Farmers' Alliance were beginning to decline.

From "Stand Up for Nebraska" (1894)

There are those who think the work of the Nebraska Farmers' Alliance is ended;
that while the bankers of the state keep up their organization with the avowed purpose of "better influencing legislation" in their behalf, while the merchants, manufacturers, lawyers, doctors, men of every trade or profession, find it to their interest
to keep up organizations to aid each other and look after their political welfare, the
agriculturalists of the state and nation have no interest in common sufficient for
the existence of an organization, but should leave their financial and political business for office seeking politicians to look after. It grieves us to think how little has
been accomplished by the Alliance compared with all that is necessary to be done
before the farmers of the state obtain anything like justice. At times we grow weary

Millions of Acres, 1872. Courtesy Library of Congress, Washington, DC.

and discouraged when we realize that the work of the Alliance is hardly begun, and that after the weary years of toil of the best men and women of the state we have hardly taken a step on the road to industrial freedom. We know that although we may not arrive there *our children will* enter into the promised land, and we can make their trials fewer and lighter, even if we live not to see the full light of freedom for mankind. We work in the knowledge that our labor of education is not in vain, some one, sometime, will arise and call the Alliance blessed. Meanwhile to us who have learned "to labor and to wait" there come sometimes sweet glimpses of the land beyond, and it seems so near, the road so short, that we can not have long to wait to enter and possess the land.

> There's a land where the toiler is free,
>> Where no robber of labor can come,
> Where wealth gives not power to oppress,
>> Nor another man's labor to own.

> In that sweet by and by
>> Which has been for long ages foretold,
> In that sweet by and by
>> Moral worth will rank higher than gold.

> We can dwell in that land of the free,
>> If we will, in the near by and by;
> We can soon wrest the scepter from gold.
>> We can make labor free if we try.

> Vote no interest whatever to gold,
>> Vote for naught which will favor a class,
> Make an injury offered to one
>> The most vital concern of the mass.

It does not seem as if that would be hard to do, nor that the road to the promised land of freedom need be long; yet there is a shorter one given by a noted guide hundreds of years ago. But they say, he was visionary, and his way impracticable. It was simply "Do unto others as ye would they should do unto you." The Christian way closely followed at the ballot box would soon right every legalized injustice, and yet the majority of the voters pretend to be his followers. Had they been so in deed and in truth how different would be the condition of our country. We have annually seen the greater part of the wealth produced in the state legislated out of the hands of the rightful owners and into the pockets of those who are allowed to eat, although they *will not* work.

The condition of the farmers of the state has changed greatly in the last three years.

Then the abolishment of high rates of interest on money and reduction of freight rates was all the average Alliance member desired. Thousands of farmers who would have preserved their homes if they could have obtained that relief at that time have now had the mortgage cleared off their farms by the sheriff and are today without a home, and they now demand that *occupancy and use shall be the sole title of land.*

So with the transportation question. While a slight reduction would have satisfied three years ago, the people now know that they have the constitutional right to take the railroads, under right of eminent domain and run them at cost in the interest of *all the people*; never again will any party arouse any enthusiasm among them who advocate less.

Of course the renter does not care greatly for anything which does not free him from the servitude of giving one third or one half his labor for the *chance* to work on the earth. The farmers comprising this organization are the wealthier class of the farmers of the state, and doubtless most of them own land and a home; but if we do unto others as we would they should do unto us we must look out for the interest of our neighbors, who are mostly renters. This is now a state of renters, and the politicians will find they have a new factor to deal with, and that the rapidly increasing number of renters is proportioned very like that of the stay at home vote. And it is reasonable that any man should stay at home unless he sees some hope of benefiting himself by going to the polls. A renter does not care greatly for transportation charges. He who owns the land owns the man who works it, and as soon as freight rates go down the prices rise [and] the renter is raised in proportion. So also he regards the money question. If the value of his products is increased by increasing money volume the rent is raised in proportion so as barely to allow him to exist to produce more. He has no hope of education for his children, or of giving them a better chance in life than he has until he is permitted to go upon the unoccupied land of the state and make for himself a home while adding yearly to the state's productive capacity and wealth. It will soon be necessary for any organization political or social that wishes the renters' allegiance, to advocate occupancy and use as the sole title to land. And if they desire the allegiance of those who, owing to an insufficient money volume, have become debtors, they must advocate a sufficient medium of exchange so that no usury interest will be exacted for its use. The Alliance must not ask if an idea is popular, but rather is it right? If right advocate it, agitate it, write it, speak it, vote it. We can make it popular. If we wish the farmers to join and keep up this society we must convince them each and every one that it will benefit him individually. We should take a decided step forward in co-operative work. We can compel the building of a co-operative road to the Gulf. We can get an agent to contract the crops of the state at foreign markets for better prices. We can by ordering machinery, flour, coal, etc., in large quantities get greatly reduced prices, and we ought to place ourselves on a level with the Grange and F.M.B.A. [the Farmers' Mutual Benefits Association] in these respects, then each member can soon receive a benefit and a new impetus be given.

Some think the People's party has taken the place of the Alliance. It has to some extent, but cannot entirely.

Leaving out business co-operation which a political organization will not touch, the Alliance has an educational work to perform which no political party can do. Politicians are notoriously cowardly, and not over truthful, especially the law-interpreting class which make speeches for them, and the people will not put faith in them or be taught by them.

A farmer can teach his brother farmers much better the principles of political economy and what he needs to better his condition than the most silvery-tongued office-seeking lawyer that ever lived in any party. There is a large class (yearly becoming larger) who put no faith in political organizations of any class, as regards benefiting the toilers. They think as soon as the party attains power politicians will crowd to the front who care only for the "spoils of office," and the wishes of the voters will be ignored. The Alliance must make it its future work to educate this class to demand the Referendum and direct legislation. It is an excellent time to show the folly of placing one-sixth of the legislative power in the hands of a corrupt governor and president.

If this is to become a government by the people, they must have the right to initiate new laws and not have important questions tabled by a committee appointed by some scoundrel in the shape of a speaker. No power higher than the vote or veto of the people can exist in a free country. The Nebraska farmers and toilers whose productive labor has made the state all it is, whose labor will make it all it ever will become, should stand up for Nebraska by showing what wealth has been produced from her fertile soil and the vast amount paid by her each year to foreigners for the privilege of using the highways of our own state, and as interest money borrowed to replace that legislated from the pockets of our farmers.

Had the farmers of Nebraska obtained justice ten years ago not a dollar of foreign capital would now be drawing interest in the state. That is the sole reason why the loan agents oppose every effort to increase the price of Nebraska's products.

> Stand up for Nebraska! from the hand of her God
> She came forth, bright and pure as her own golden rod.
> Sweet peas and wild roses perfumed all the air.
> Her maker pronounced her both fertile and fair.
> Not a boodler or pauper disgraced the state then;
> Stand up for Nebraska and cleanse her again. . . .

Stand up for Nebraska! Let no foot of her soil
Be held by the idlers to tax rent from toil.
Bid the hard-working tenants of other states come,
And build on each wild quarter section a home.
And soon the world over the watchword will be,
Stand up for Nebraska, the home of the free.

Leo Breslau, *Plowing*, 1934. Reproduced by permission of the Smithsonian American Art Museum, Washington, DC/Art Resource, New York.

5. Agrarians in an Industrial Nation, 1900–1945

By 1900 the United States was poised to realize an urban, industrialized future. Yet coupled with the growing nation's optimism was a deep and pervasive anxiety about the American way of life. The nineteenth century had closed with the resounding defeat of the Populist program, and the age of agricultural movements appeared at its end. Many Americans still wondered how a democracy founded upon independent producers should adapt to an increasingly centralized economy of wage laborers. How would the new industrial paradigm affect the American countryside and its people? If the number of city-dwellers continued to increase while the number of farmers did not, could the nation even feed itself?

In 1890 the superintendent of the U.S. Census had announced that the nation no longer contained a frontier of settlement, with the West having surpassed an average population density of at least two people per square mile. Therefore, according to demographers, the period of internal colonization was over, and the nation had entered a new era of development. In 1893 historian Frederick Jackson Turner seized upon this watershed as the key to a grand reinterpretation of American history, arguing that the nation's identity had been shaped primarily by the pioneers' encounter with the raw natural environment. At bottom, this was another version of the old agrarian claim that the ideal "middle landscape" lay between the rude frontier of the West and the corrupt civilization of the East. Turner, however, relocated the democratic heart of the story from a place to a process—from the innate moral virtue of the yeoman farmer in a world of improved cultivation to the dynamic encounter between pioneers and the frontier. By 1890, alarmingly, the frontier that had supposedly given the country its exceptional vigor had closed.

Turner's "frontier thesis" heralded a preoccupation of American culture throughout the twentieth century: what new frontiers, internal or external, could be found to replace the one that had already been conquered? The Jeffersonian idea that access to land would provide a place for upright, independent freehold farmers had been transmuted within a few generations into an appetite for boundless economic and cultural territory to sustain the growth of an industrial nation. As the national economy (and mainstream agriculture along with it) plowed on toward a beckoning horizon of ever-increasing production and consumption, a new agrarianism began to emerge in response. It was an amalgamation of traditional rural values, such as self-reliance, frugality, neighborliness, and the centrality of the family farm, with newer Romantic ideals of closeness with nature and ecological ideals of maintaining balance with nature. It sought a return to what it took to be a core agrarian virtue of becoming rooted in particular places and communities. American agrarianism in the twentieth century was no longer "Go West, young man!" It was about staying on the land, or going *back* to the land in order to create a "permanent agriculture."

The century began with a hard question: Could America's farmers continue to increase their production to meet the country's needs? Turner's frontier thesis mirrored the pragmatic concerns of many policymakers. With the end of the period of "free land" came serious doubts about the ability of American agriculture to grow, built as it had been upon an unlimited supply of land and natural resources. Over the preceding decades the low price of basic commodities had fueled the nation's industrial growth and the resultant boom in urban population. Conversely, the closing of the frontier posed a direct challenge for economic and political leaders: With no more easy expansion of prime farmland available on the continent, how could the nation ensure a supply of cheap food? Many wondered whether production could even hold steady, let alone increase. In the early twentieth century crop prices began to rise: good news for some farmers, maybe, but not so good for an industrial nation. Federal officials confronted this predicament by pushing a new science-based agriculture that sought to increase crop yields by relying on chemical inputs and new technologies.

As state and federal governments turned their research energies toward improving yields, the increased attention to the countryside led American policymakers to a renewed sense that the country faced a "rural problem." Farmers appeared to be lagging behind the modern era, both economically and culturally: they lacked electricity, running water, good roads, adequate schools. Meanwhile, the nation was going to need educated, progressive farmers to manage new scientific farms. The drive to improve efficiency, expertise, and methods of production generated pressure to weed out the old mossbacks on their "marginal" farms. In the long run these policies would succeed only too well. Improvements in technology, particularly the gasoline-powered tractor, meant that fewer farmers were needed to produce an ever-expanding supply of food and fiber, which spurred farm consolidation and rural outmigration. Generations of twentieth-century policymakers such as economist Edwin Nourse treated the steady erosion of small farmers as nothing but a cause for celebration and did their utmost to push the trend along, thus emptying large portions of the American heartland. Consequently, many a rural county reached its population peak in the decades around 1900.

Farmers wondered about the future of their communities during these transformative years. Following the defeat of Populism and the broken dream of an agrarian nation, those living in the countryside searched for an updated cultural identity, which meant shedding the "hayseed" stigma even as they remained devoted to the special quality of rural life. Farmers' pride and anxiety both endured, creating a culture that became simultaneously more modern and in many ways more defensive and conservative. As industrial capacity continued to expand, American agriculture moved steadily toward the sale of cash crops, and at the same time rural people increasingly became reliant upon mass-produced goods. The influx of mail order catalogues allowed farm families to begin to share in the consumer culture that defined middle-class urban life. Isolated farmers were among the first Americans to take up the automobile, and early rural telephone party lines were strung along interconnected barbed-wire fences. Industrialization shaped both the production and the consumption patterns of farm families as they became fully integrated into the nation's economy.

Farmers increasingly found themselves swept along by a modernizing and industrializing nation. As Kenyon Butterfield observed in his 1908 *Chapters in Rural Progress*, "The old farmer was a pioneer, and he had all the courage, enterprise, and resourcefulness of the pioneer. He owned and controlled everything in sight." By contrast, "The new farmer lives in a day when the nation is not purely an agricultural nation, but is also a manufacturing and a trading nation. But he realizes that out of this seeming decline of agriculture grow his best opportunities." Did they? The completion of the long transition away from a culture of agrarian independence to a partnership in industrial production marked a new age in American rural development, rife with both opportunities and the constant threat of ruin. The choices farm people faced were incremental but over time could amount to a complete transformation: Should they hold to time-honored farming practices or invest in new techniques and machines? Should they strive to remain diversified and self-sufficient, still growing most of their own food alongside an increasing surplus for market, or should they concentrate entirely on high-value commodity crops that promised big gains but also risked crippling losses?

Urban and rural thinkers alike confronted this transitional moment, from the highest levels of government to the smallest community organizations. In 1908 President Theodore Roosevelt appointed a National Commission on Country Life, composed of proponents of rural improvement including Liberty Hyde Bailey, Walter Hines Page, Henry Wallace, Kenyon Butterfield, and Gifford Pinchot. The commission's report captured both these Progressive reformers' optimism and their sense that large cultural and economic advances remained to be made in rural areas. "Broadly speaking, agriculture in the United States is prospering and the conditions in many of the great farming regions are improving. . . . There has never been a time when the American farmer was as well off as he is today, when we consider not only his earning power, but the comforts and advantages he may secure. Yet the real efficiency in farm life, and in country life as a whole, is not to be measured by historical standards, but in terms of its possibilities. Considered from this point of view, there are very marked deficiencies." The Country Life Commission proposed social reforms and economic reorganization as a means of rectifying the cultural disadvantages of rural communities. In the view of these experts, real rural prosperity could rest only on increased productivity, not on rising crop prices that were a by-product of backwardness. A national cheap food policy was being born—but would that really bring rural prosperity?

With most of the best land already claimed and put to use, increased productivity eventually had to mean more bushels per acre and more acres per farmer. This necessitated a new system of production based upon chemical fertilizers, hybrid seeds, pesticides, expanded irrigation networks, and mechanization. As these changes took hold after World War I, rural communities faced dramatic economic and cultural shifts—most fundamentally in the number of farms needed to supply the nation and in the proportion of people who lived in rural areas. While the total number of farms had expanded from about 2 million in 1860 to over 6 million in 1910, a rate that kept pace with urban growth, from that peak the numbers began to decline—slowly at first, but then rapidly after 1940. By 1920 urban population in America had surpassed the rural. Since then the number of people in the countryside has remained essentially flat, while the urban and suburban population has skyrocketed; by midcentury two-thirds of the population was urbanized.

These demographic changes were tied to a technological revolution. What Lewis Mumford termed the nineteenth-century "paleotechnic era" of coal, iron, and steam gave way during these decades to the "neotechnic era" of oil, electricity, and chemistry—new energy sources, materials, and production processes that had been developed through the application of science. The face of agriculture changed dramatically as more and more farms converted from muscle power to machines. Producers steadily embraced the gasoline-powered tractors, combines, trucks, and automobiles that promised to liberate them from dependence on horsepower. Meanwhile, the size of fields, the size of farms, what was being grown, and how it was being grown all changed as a result of the new technologies. The acres and hours once required to feed work stock were freed up to be devoted to cash crops, but in exchange farmers became dependent on purchased inputs, particularly gasoline, fertilizers, and commercial seeds. As these products became increasingly central to increased production, the calculus of successful farming changed.

The drive toward monoculture did not occur by accident. Following the advice of the Country Life Commission, state and federal governments began to invest in agricultural research, education, and infrastructure. Many of the early efforts were aimed at the young. State agricultural leaders worked to form "corn clubs," in which boys competed to produce spectacularly high yields on small plots of land. Thanks to the support of the USDA and state agricultural colleges, these clubs evolved into the 4-H movement. With the 1914 Smith-Lever Act, state university cooperative extension services began to reach out directly to farmers. In 1916 the federal government started to invest in rural transportation using matching funds through the Federal Aid Road Act, while land banks offered government-supported low-interest loans to farmers under the Federal Farm Loan Act. Crop prices were high during the first decades of the century, and many farmers responded by investing time and money to expand their operations. Much of the scientific advice for rural producers that promoted new methods, implements, and crops was sound enough. But it tended to come wrapped in a cultural and economic package that encouraged farmers to abandon old methods altogether in favor of expensive new ways of doing business, and over the course of the century many a family bet the farm on the new ways, and lost.

Not everyone agreed that the new regime was the best course for American farming. While many economists and government planners called for the most efficient, large-scale agriculture possible, other thinkers envisioned an updated rural culture founded on the values of community, conservation, and what we now call sustainability. These neoagrarians believed that the future viability of farming would be best served by perfecting time-honored traditions of mixed husbandry and integrated rotations. They chose to keep their dependence on outside inputs as limited as possible, often still relying on animals and small, diversified farms, rather than fully embracing the technological ideal that was sweeping the nation. Many of the writers who critiqued the new industrial system in the first decades of the twentieth century took issue with the growing specialization of the American economy and strove to detach themselves from complete dependence on markets and consumer goods. Instead, they celebrated the multifaceted experience of rural people, masters of all trades. As Louis Bromfield argued in *Malabar Farm,* "The *good* farmer or livestock man is no longer a 'hick,' as indeed he never was. He must always be an intelligent man of parts, knowing perhaps more about more things than any other citizen. He must know and understand something of markets, the weather, distribution, machinery, economics, history, ecology, disease, bacteriology, and many other things, but most of all he must understand the earth and the laws of God and nature which govern its maintenance and productivity." In a form of protest against the homogenizing pattern of American life, Bromfield and his peers held up the old agrarian virtues of competence and independence, hoping to modernize and improve farming without reducing it to an unsentimental business of efficient resource extraction.

Widely read and historically minded, the neoagrarians saw clear connections between their vision and those who had come before them. Ralph Borsodi and Louis Bromfield constructed their agrarianism upon historical practices of self-sufficiency, including ideas drawn from the earliest descriptions of agriculture. David Grayson (the pen name of journalist Ray Stannard Baker) picked up the theme of rural contentment from Andrew Jackson Downing, focusing on designing an aesthetic and satisfying country life, just as Helen and Scott Nearing would do decades later. Similarly, Luigi Ligutti and John Rawe reiterated some of the same themes as the Mormons and other religious communities as they integrated ecclesiastical visions of agricultural virtue and independence into designs for Catholic agrarian settlements. Often raised off the farm, these writers embraced the landscapes of their adopted regions and integrated a critique of urbanism into their endorsement of rural life. These rural immigrants used the tools at their disposal—intellectual, cultural, and communal—to become native to their chosen places. In fact, their enthusiasm for pastoral experience often set them apart as comparative newcomers to the farm. This was in sharp contrast to mainstream agricultural thinkers like economist Edwin Nourse (farm born but primarily the product of an urban upbringing), who promoted the application of industrial principles on the farm and mocked romanticized portrayals of rural life that might undermine agricultural efficiency.

Other neoagrarians, like horticulturalist Liberty Hyde Bailey and botanist Henry Agard Wallace—the sons and grandsons of innovative farmers—were steeped in the culture of agricultural improvement and used their experience on their family farms as the grounds for experiments in conservation, crop breeding, and economics. A (not uncritical) sensitivity to rural culture suffused the work of writers such as Louis Bromfield, who chose to return to rural Ohio where his grandparents had farmed, seeking to update common agricultural practices for ecologically sound profits; or Ralph Borsodi, who created a new model of sustainable, self-sufficient production in rural New York.

In spite of their optimism about rural independence, these neoagrarians acknowledged the challenges facing small-scale cultivators during the twentieth century. By adopting traditional methods, they often accepted more laborious, difficult work than their mechanizing peers. As Borsodi observed in a 1974 interview, the rural reality was far more taxing than many urbanites imagined. "Living in the country has been called 'the simple life.' This is not true. It's much more complex than city life. City life is the one that's simple. You get a job and earn money and you go to a store and buy what you want and can afford. The decentralist life in the country, on the other hand, is something else again. When you design your own things and make plans about what you're going to produce and really live in a self-sufficient manner, you've got to learn . . . you've got to master all sorts of crafts and activities that people in the city know nothing about." Borsodi wholeheartedly embraced the complexity of farm life and widely advertised its virtues, in contradiction of the dominant urban and consumerist ethos of the era.

The agrarian writers of the first decades of the twentieth century feared that reducing farming to just another industry would lead to economic and ecological disaster. Their attempt to ensure a sustainable, socially conscious agriculture was a significant, and yet distinctly minority, counterbalance to the industrializing trend in farms nationwide. As agriculture modernized, technological innovations offered the possibility of more efficient production, while new chemicals and methods of cultivation raised the threat of increased environmental disruption. Most of the neoagrarian thinkers were devoted to the idea that scientific reform was needed in American agriculture, but they were also alert to the danger that science narrowly applied to production might make the problem of land degradation worse. As soil scientist Hugh Hammond Bennett warned in 1928, "If we continue to twiddle thumbs and warn our brothers not to be stampeded by the wailing of those apostles and prophets, who . . . recurrently wave red flags of fear and point to direful dangers lurking around the corner, then we of to-day are going to be the recipients of much tongue-lashing by our children a few generations hence." Bennett's fears were realized even sooner than he had imagined, as images of the Dust Bowl on the Plains and monstrous gullies in the cottoned-out South came to symbolize the degradation of American farmland.

The years surrounding World War I are remembered by American farmers as the time of "parity," the peak of farmer purchasing power. But good prices for crops stimulated production, and soon the economic consequences of market saturation again convulsed rural areas. Not only were new technologies increasingly applied, but the lure of quick profit also drew farmers into vulnerable environments such as the High Plains, thus flooding the market in another round of boom and bust. The Roaring Twenties marked a return to sluggish hard times on the farm, marked by low crop prices, foreclosures, and forced sales. With the 1930s came ruin, as expanding supply met crumpling demand. Economic crisis forced many farmers to turn inward in search of independence or even sheer survival on the land. Total farm income dropped from a high of over $15 billion in 1920 to just under $2 billion in 1932, a devastating decline of over 92 percent. The price of a bushel of wheat fell from $1.03 in 1929 to $.36 in 1931. No longer able to rely on a regular income from crop sales, many farmers looked for alternatives to staple-crop agriculture, and once again sought greater stability through diversified production. In the wake of this disaster in agriculture, the emergence of the nationwide Great Depression in the 1930s called the entire modern industrial economy into question.

The economic crisis brought unprecedented attention to writers like Borsodi, Bromfield, and Rawe and Ligutti, who enthusiastically shared their vision of self-sufficiency with readers around the world. Many of the land economists, planners, and conservationists within Franklin Delano Roosevelt's administration were working toward a vision of stable rural communities that looked similar in some respects. The federal government (along with many states) initiated vigorous programs to acquire what they classified as submarginal farmlands in the Appalachians, the dry Plains, the Deep South, and the cutover of the North Woods in a bid to reduce overproduction, protect the land from erosion and waste, and attack the worst of rural poverty. But while the government was working to relocate some small farmers, it was also promoting conservation methods on better farmland, establishing rural public works programs, and pushing rural electrification to improve the quality of farm life. New Dealers also experimented with the idea of resettling poor farmers on subsistence homesteads where they might divide their family labor between factory work and farming. At least for the duration of the Depression, government policy seemed to support the emergence of a new vision of "permanent agriculture" designed to reinvigorate the American countryside.

But the temporary vindication of rural self-reliance brought on by the 1930s would not long prevail. The material demands of world war soon returned the nation to a commodity-driven productionist mentality. During the 1940s research scientists and agricultural suppliers developed even more new machinery, crop varieties, chemical fertilizers, and pesticides to boost agricultural yields, paving the way for increasingly capital-intensive and consolidated farming during the second half of the century. The transformation of American agriculture that ensued built upon deep changes that had been under way since the middle of the nineteenth century. It would take yet another era of agricultural crisis for the dissenting voices of agrarians to once again receive wide attention.

LIBERTY HYDE BAILEY JR.

Among the leaders in early twentieth-century American agricultural thought was Liberty Hyde Bailey, one of the nation's foremost horticulturalists, and professor and dean at the Cornell University College of Agriculture. Bailey was an influential advisor to policymakers and presidents, and he chaired Theodore Roosevelt's 1908 Commission on Country Life, the first advisory panel to formulate recommendations about the needs of American agriculture.

Bailey was born in South Haven, Michigan, in 1858. The son of a progressive orchardist, he grew up attuned to the caprices of markets and weather, the potential rewards of experimenting with new varieties and techniques, and the need for conservation-mindedness among farmers and country people. Bailey attended Michigan State College, then studied with botanist Asa Gray at Harvard before returning to Michigan State as a professor of horticulture and landscape gardening. After receiving his master's degree in 1885, he accepted a professorship of horticulture at Cornell University in 1888, which he held until his retirement in 1913.

Bailey remained active in research, writing, and teaching until his death in 1954. He was a leader in the fields of experimental botany and horticulture, an outspoken advocate of nature study as a technique for educating children, and a spokesperson for a better rural life. As chairman, Bailey framed the Country Life Commission's sweeping statement on the possibilities and challenges of rural life. In his 1915 The Holy Earth, *parts of which are excerpted below, Bailey observed, "A good part of agriculture is to learn how to adapt one's work to nature, to fit the crop-scheme to the climate and to the soil and the facilities. To live in right relation to the natural conditions is one of the first lessons that a wise farmer or any other wise man learns." Optimism about the future of American agriculture was a recurrent theme in Bailey's work, and he mixed a concern with the conservation of natural resources with a firm commitment to husbanding agricultural land that was shared by few others at the time. In this Bailey was particularly prescient, and many of the points he made in* The Holy Earth *will resonate with modern readers.*

From *The Holy Earth* (1915)

The earth is holy

Verily, then, the earth is divine, because man did not make it. We are here, part in the creation. We cannot escape. We are under obligation to take part and to do our best, living with each other and with all the creatures. We may not know the full plan, but that does not alter the relation. When once we set ourselves to the pleasure of our dominion, reverently and hopefully, and assume all its responsibilities, we shall have a new hold on life.

We shall put our dominion into the realm of morals. It is now in the realm of trade. This will be very personal morals, but it will also be national and racial morals. More iniquity follows the improper and greedy division of the resources and privileges of the earth than any other form of sinfulness.

If God created the earth, so the earth is hallowed; and if it is hallowed, so we must deal with it devotedly and with care that we do not despoil it, and mindful of our relations to all beings that live on it. We are to consider it religiously: Put off thy shoes from off thy feet, for the place whereon thou standest is holy ground.

The sacredness to us of the earth is intrinsic and inherent. It lies in our necessary relationship and in the duty imposed upon us to have dominion, and to exercise ourselves even against our own interests. We may not waste that which is not ours. To live in sincere relations with the company of created things and with conscious regard for the support of all men now and yet to come, must be of the essence of righteousness. . . .

I do not mean all this, for our modern world, in any vague or abstract way. If the earth is holy, then the things that grow out of the earth are also holy. They do not belong to man to do with them as he will. Dominion does not carry personal ownership. There are many generations of folk yet to come after us, who will have equal right with us to the products of the globe. It would seem that a divine obligation rests on every soul. Are we to make righteous use of the vast accumulation of knowledge of the planet? If so, we must have a new formulation. The partition of the earth among the millions who live on it is necessarily a question of morals; and a society that is founded on an unmoral partition and use cannot itself be righteous and whole. . . .

The habit of destruction

The first observation that must be apparent to all men is that our dominion has been mostly destructive.

We have been greatly engaged in digging up the stored resources, and in destroying vast products of the earth for some kernel that we can apply to our necessities or add to our enjoyments. We excavate the best of the coal and cast away the remainder; blast the minerals and metals from underneath the crust, and leave the earth raw and sore; we box the pines for turpentine and abandon the growths of limitless years to fire and devastation; sweep the forests with the besom of destruction; pull the fish from the rivers and ponds without making any adequate provision for renewal; exterminate whole races of animals; choke the streams with refuse and dross; rob the land of its available stores, denuding the surface, exposing great areas to erosion.

We do not exercise the care and thrift of good housekeepers. We do not clean up our work or leave the earth in order. The remnants and accumulation of mining-camps are left to ruin and decay; the deserted phosphate excavations are ragged, barren, and unfilled; vast areas of forested lands are left in brush and waste, unthoughtful of the future, unmindful of the years that must be consumed to reduce the refuse to mould and to cover the surface respectably, uncharitable to those who must clear away the wastes and put the place in order; and so thoughtless are we with these natural resources that even the establishments that manufacture them—the mills, the factories of many kinds—are likely to be offensive objects in the landscape, unclean, unkempt, displaying the unconcern of the owners to the obligation that the use of the materials imposes and to the sensibilities of the community for the way in which they handle them. The burden of proof seems always to have been rested on those who partake little in the benefits, although we know that these non-partakers have been real owners of the resources; and yet so undeveloped has been the public conscience in these matters that the blame—if blame there be—cannot be laid on one group more than on the other. Strange it is, however, that we should not have insisted at least that those who appropriate the accumulations of the earth should complete their work, cleaning up the remainders, leaving the areas wholesome, inoffensive, and safe. How many and many are the years required to grow a forest and to fill the pockets of the rocks, and how satisfying are the landscapes, and yet how desperately soon may men reduce it all to ruin and to emptiness, and how slatternly may they violate the scenery!

All this habit of destructiveness is uneconomic in the best sense, unsocial, unmoral.

Society now begins to demand a constructive process. With care and with regard for other men, we must produce the food and the other supplies in regularity and sufficiency; and we must clean up after our work, that the earth may not be depleted, scarred, or repulsive. . . .

In times past we were moved by religious fanaticism, even to the point of waging wars. To-day we are moved by impulses of trade, and we find ourselves plunged into a war of commercial frenzy; and it has behind it vaster resources and more command of natural forces, so is it the most ferocious and wasteful that the race has experienced, exceeding in its havoc the cataclysms of earthquake and volcano. Certainly we have not learned how to withstand the prosperity and the privileges that we have gained by the discoveries of science; and certainly the morals of commerce has not given us freedom or mastery. Rivalry that leads to arms is a natural fruit of rivalry in trade.

Man has dominion, but he has no commission to devastate: And the Lord God took the man, and put him into the garden of Eden to dress it and to keep it.

Verily, so bountiful hath been the earth and so securely have we drawn from it our substance, that we have taken it all for granted as if it were only a gift, and with little care or conscious thought of the consequences of our use of it.

The new hold

We may distinguish three stages in our relation to the planet,—the collecting stage, the mining stage, and the producing stage. These overlap and perhaps are nowhere distinct, and yet it serves a purpose to contrast them.

At first man sweeps the earth to see what he may gather,—game, wood, fruits, fish, fur, feathers, shells on the shore. A certain social and moral life arises out of this relation, seen well in the woodsmen and the fishers—in whom it best persists to the present day—strong, dogmatic, superstitious folk. Then man begins to go beneath the surface to see what he can find,—iron and precious stones, the gold of Ophir, coal, and many curious treasures. This develops the exploiting faculties, and leads men into the uttermost parts. In both these stages the elements of waste and disregard have been heavy.

Finally, we begin to enter the productive stage, whereby we secure supplies by controlling the conditions under which they grow, wasting little, harming not. Farming has been very much a mining process, the utilization of fertility easily at hand and the moving-on to lands unspoiled of quick potash and nitrogen. Now it begins to be really productive and constructive, with a range of responsible and permanent morals. We rear the domestic animals with precision. We raise crops, when we will, almost to a nicety. We plant fish in lakes and streams to some extent but chiefly to provide more game rather than more human food, for in this range we are mostly in the collecting or hunter stage. If the older stages were strongly expressed in the character of the people, so will this new stage be expressed; and so is it that we are escaping the primitive and should be coming into a new character. We shall find our rootage in the soil.

This new character, this clearer sense of relationship with the earth, should express itself in all the people and not exclusively in farming people and their like.

It should be a popular character—or a national character if we would limit the discussion to one people—and not a class character. Now, here lies a difficulty and here is a reason for writing this book: the population of the earth is increasing, the relative population of farmers is decreasing, people are herding in cities, we have a city mind, and relatively fewer people are brought into touch with the earth in any real way. So is it incumbent on us to take special pains—now that we see the new time—that all the people, or as many of them as possible, shall have contact with the earth and that the earth righteousness shall be abundantly taught.

I hasten to say that I am not thinking of any back-to-the-farm movement to bring about the results we seek. Necessarily, the proportion of farmers will decrease. Not so many are needed, relatively, to produce the requisite supplies from the earth. Agriculture makes a great contribution to human progress by releasing men for the manufactures and the trades. In proportion as the ratio of farmers decreases it is important that we provide them the best of opportunities and encouragement: they must be better and better men. And if we are to secure our moral connection with the planet to a large extent through them, we can see that they bear a relation to society in general that we have overlooked.

Even the farming itself is changing radically in character. It ceases to be an occupation to gain sustenance and becomes a business. We apply to it the general attitudes of commerce. We must be alert to see that it does not lose its capacity for spiritual contact. . . .

We are not to look for our permanent civilization to rest on any species of robber-economy. No flurry of coal-mining, or gold-fever, or rubber-collecting in the tropics, or excitement of prospecting for new finds or even locating new lands, no ravishing of the earth or monopolistic control of its bounties, will build a stable society. So is much of our economic and social fabric transitory. It is not by accident that a very distinct form of society is developing in the great farming regions of the Mississippi Valley and in other comparable places; the exploiting and promoting occupancy of those lands is passing and a stable progressive development appears. We have been obsessed of the passion to cover everything at once, to skin the earth, to pass on, even when there was no necessity for so doing. It is a vast pity that this should ever have been the policy of government in giving away great tracts of land by lottery, as if our fingers would burn if we held the lands inviolate until needed by the natural process of settlement. The people should be kept on their lands long enough to learn how to use them. But very well: we have run with the wind, we have staked the lands; now we shall be real farmers and real conquerors. Not all lands are equally good for farming, and some lands will never be good for farming; but whether in Iowa, or New England, or old Asia, farming land may develop character in the people.

My reader must not infer that we have arrived at a permanent agriculture, although we begin now to see the importance of a permanent land occupancy. Probably we have not yet evolved a satisfying husbandry that will maintain itself century by century, without loss and without the ransacking of the ends of the earth for fertilizer materials to make good our deficiencies. All the more is it important that the problem be elevated into the realm of statesmanship and of morals. Neither must he infer that the resources of the earth are to be locked up beyond contact and use (for the contact and use will be morally regulated). But no system of brilliant exploitation, and no accidental scratching of the surface of the earth, and no easy appropriation of stored materials can suffice us in the good days to come. City, country, this class and that class, all fall and merge before the common necessity.

It is often said that the farmer is our financial mainstay; so in the good process of time will he be a moral mainstay, for ultimately finance and social morals must coincide.

The gifts are to be used for service and for satisfaction, and not for wealth. Very great wealth introduces too many intermediaries, too great indirectness, too much that is extrinsic, too frequent hindrances and superficialities. It builds a wall about the man, and too often does he receive his impressions of the needs of the world from satellites and sycophants. It is significant that great wealth, if it contributes much to social service, usually accomplishes the result by endowing others to work. The gift of the products of the earth was "for meat": nothing was said about riches.

Yet the very appropriation or use of natural resources may be the means of directing the mind of the people back to native situations. We have the opportunity to make the forthcoming development of water-power, for example, such an agency for wholesome training. Whenever we can appropriate without despoliation and loss, or without a damaging monopoly, we tie the people to the backgrounds. . . .

I hope my reader now sees where I am leading him. He sees that I am not thinking merely of instructing the young in the names and habits of birds and flowers and other pleasant knowledge, although this works strongly toward the desired end; nor of any movement merely to have gardens, or to own farms, although this is desirable provided one is qualified to own a farm; nor of rhapsodies on the beauties of nature. Nor am I thinking of any new plan or any novel kind of institution or any new agency; rather shall we do better to escape some of the excessive institutionalism and organization. We are so accustomed to think in terms of organized politics and education and religion and philanthropies that when we detach ourselves we are said to lack definiteness. It is the personal satisfaction in the earth to which we are born, and the quickened responsibility, the whole relation, broadly developed, of the man and of all men,—it is this attitude that we are to discuss.

The years pass and they grow into centuries. We see more clearly. We are to take a new hold.

THE FARMER'S RELATION

The surface of the earth is particularly within the care of the farmer. He keeps it for his own sustenance and gain, but his gain is also the gain of all the rest of us.

At the best, he accumulates little to himself. The successful farmer is the one who produces more than he needs for his support; and the overplus he does not keep; and, moreover, his own needs are easily satisfied. It is of the utmost consequence that the man next the earth shall lead a fair and simple life; for in riotous living he might halt many good supplies that now go to his fellows.

It is a public duty so to train the farmer that he shall appreciate his guardianship. He is engaged in a quasi-public business. He really does not even own his land. He does not take his land with him, but only the personal development that he gains from it. He cannot annihilate his land, as another might destroy all his belongings. He is the agent or the representative of society to guard and to subdue the surface of the earth; and he is the agent of the divinity that made it. He must exercise his dominion with due regard to all these obligations. He is a trustee. The productiveness of the earth must increase from generation to generation: this also is his obligation. He must handle all his materials, remembering man and remembering God. A man cannot be a good farmer unless he is a religious man. . . .

Into [the farmer's] secular and more or less technical education we are now to introduce the element of moral obligation, that the man may understand his peculiar contribution and responsibility to society; but this result cannot be attained until the farmer and every one of us recognize the holiness of the earth.

The farmer and every one of us: every citizen should be put right toward the planet, should be quickened to his relationship to his natural background. The whole body of public sentiment should be sympathetic with the man who works and administers the land for us; and this requires understanding. We have heard much about the "marginal man," but the first concern of society should be for the bottom man.

If this philosophy should really be translated into action, the farmer would nowhere be a peasant, forming merely a caste, and that a low one, among his fellows. He would be an independent co-operating citizen partaking fully of the fruits of his labor, enjoying the social rewards of his essential position, being sustained and protected by a body of responsive public opinion. The farmer cannot keep the earth for us without an enlightened and very active support from every other person, and without adequate safeguards from exploitation and from unessential commercial pressure.

This social support requires a ready response on the part of the farmer; and he must also be developed into his position by a kind of training that will make him quickly and naturally responsive to it. The social fascination of the town will always be greater than that of the open country. The movements are more rapid, more picturesque, have more color and more vivacity. It is not to be expected that we can overcome this fascination and safeguard the country boy and girl merely by introducing more showy or active enterprises into the open country. We must develop a new background for the country youth, establish new standards, and arouse a new point of view. The farmer will not need all the things that the city man thinks the farmer needs. We must stimulate his moral response, his appreciation of the worthiness of the things in which he lives, and increase his knowledge of all the objects and affairs amongst which he moves. The backbone of the rural question is at the bottom a moral problem.

We do not yet know whether the race can permanently endure urban life, or whether it must be constantly renewed from the vitalities in the rear. We know that the farms and the back spaces have been the mother of the race. We know that the exigencies and frugalities of life in these backgrounds beget men and women to be serious and steady and to know the value of every hour and of every coin that they earn; and whenever they are properly trained, these folk recognize the holiness of the earth. . . .

THE SUBDIVIDING OF THE LAND

The question then arises whether lands and other natural resources shall now be divided and redistributed in order that the share-and-share of the earth's patrimony shall be morally just. Undoubtedly the logic of the situation makes for many personal points of very close contact with the mother earth, and contact is usually most definite and best when it results from what we understand as ownership. This, in practice, suggests many small parcels of land—for those who would have their contact by means of land, which is the directest means—under personal fee. But due provision must always be made, as I have already indicated, for the man who makes unusual contribution to the welfare of his fellows, that he may be allowed to extend his service and attain his own full development; and moreover, an established order may not be overturned suddenly and completely without much damage, not only to personal interests but to society. Every person should have the right and the privilege to a personal use of some part of the earth; and naturally the extent of his privilege must be determined by his use of it.

It is urged that lands can be most economically administered in very large units and under corporate management; but the economic results are not the most important results to be secured, although at present they are the most stressed. The ultimate good in the use of land is the development of the people; it may be better that more persons have contact with it than that it shall be executively more efficiently administered. The morals of land management is more important than the economics of land management; and of course my reader is aware that by morals I mean the results that arise from a right use of the earth rather than the formal attitudes toward standardized or conventional codes of conduct.

If the moral and economic ends can be secured simultaneously, as eventually they will be secured, the perfect results will come to pass; but any line of development founded on accountant economics alone will fail.

Here I must pause for an explanation in self-defense, for my reader may think I advise the "little farm well tilled" that has so much captured the public mind. So far from giving such advice, I am not thinking exclusively of farming when I speak of partitioning the land. One may have land merely to live on. Another may have a wood to wander in. One may have a spot on which to make a garden. Another may have a shore, and another a retreat in the mountains or in some far space. Much of the earth may never be farmed or mined or used for timber, and yet these supposed waste places may be very real assets to the race: we shall learn this in time. I am glad to see these outlying places set aside as public reserves; and yet we must not so organize and tie up the far spaces as to prevent persons of little means from securing small parcels. These persons should have land that they can handle and manipulate, in which they may dig, on which they may plant trees and build cabins, and which they may feel is theirs to keep and to master, and which they are not obliged to "improve." In the parks and reserves the land may be available only to look at, or as a retreat in which one may secure permission to camp. The regulations are necessary for these places, but these places are not sufficient.

If it were possible for every person to own a tree and to care for it, the good results would be beyond estimation.

Now, farming is a means of support; and in this case, the economic possibilities of a particular piece of land are of primary consequence. Of course, the most complete permanent contact with the earth is by means of farming, when one makes a living from the land; this should produce better results than hunting or sport; but one must learn how to make this connection. It is possible to hoe potatoes and to hear the birds sing at the same time, although our teaching has not much developed this completeness in the minds of the people. . . .

THE GROUP REACTION

. . . The real farmer, the one whom we so much delight to honor, has a strong moral regard for his land, for his animals, and his crops. These are established men, with highly developed obligations, feeling their responsibility to the farm on which they live. No nation can long persist that does not have this kind of citizenry in the background.

I have spoken of one phase of the group reaction, as suggested in the attitude of the farmer. It may be interesting to recall, again, the fact that the purpose of farming is changing. The farmer is now adopting the outlook and moral conduct of commerce. His business is no longer to produce the supplies for his family and to share the small overplus with society. He grows or makes a certain line of produce that he sells for cash, and then he purchases his other supplies in the general market. The days of homespun are gone. The farmer is as much a buyer as a seller. Commercial methods and standards are invading the remotest communities. This will have far-reaching results. Perhaps a fundamental shift in the moral basis of the agricultural occupations is slowly under way.

The measuring of farming in terms of yields and incomes introduces a dangerous standard. It is commonly assumed that State moneys for agriculture-education may be used for "practical"—that is, for dollar-and-cents—results, and the emphasis is widely placed very exclusively on more alfalfa, more corn, more hogs, more fruit, on the two-blades-of-grass morals; and yet the highest good that can accrue to a State for the expenditure of its money is the raising up of a population less responsive to cash than to other stimuli. The good physical support is indeed essential, but it is only the beginning of a process. I am conscious of a peculiar hardness in some of the agriculture-enterprise, with little real uplook; I hope we may soon pass this cruder phase.

Undoubtedly we are at the beginning of an epoch in rural affairs. We are at a formative period. We begin to consider the rural problem increasingly in terms of social groups. The attitudes that these groups assume, the way in which they react to their problems, will be determined in the broader aspects for some time to come by the character of the young leadership that is now taking the field.

DAVID GRAYSON

David Grayson was the pen name of journalist Ray Stannard Baker, one of the foremost investigative journalists of the early twentieth century. Baker was born in 1870 in Lansing, Michigan, the son of a middle-class businessman. When he was five, his family moved to rural St. Croix Falls in northwestern Wisconsin. Baker took his degree at Michigan Agricultural College and, after briefly studying law at the University of Michigan, moved to Chicago to write for the Chicago Record. *Like his contemporaries Theodore Dreiser and Frank Norris, he often wrote about the underbelly of industrial society in Chicago. In 1898 Baker moved to New York City to write for* McClure's *magazine, a position that placed him in the company of the finest journalists in the nation. Within a few years, however, he suffered from nervous exhaustion. Granted a leave of absence with salary, he left the city for a small farm in Amherst, Massachusetts, which became his primary home.*

Baker found solace and new inspiration in the country, and for the rest of his career he balanced his work in journalism with a passion for the countryside. In his autobiography, he reflected: "It was fortunate, perhaps beyond anything else I ever did, that I had set up a home in the country. It was only a small house with a bit of land, but it was my own. I had paid for it; I held it in fee simple; it seemed to me in those days an extraordinarily beautiful place." In 1907, writing as David Grayson, he published the first of his rural volumes, Adventures in Contentment, *a critically acclaimed work that became one of the key texts of the turn-of-the-century "back-to-the-land" movement.*

Baker retained a foot in the worlds of both country and city, writing on agrarian themes while continuing his work as a hard-hitting journalist. A close friend of Theodore

Roosevelt and Woodrow Wilson during their presidencies, Baker had the ear of powerful people, and both as a writer and as an advisor to policymakers he sought to correct the manifold injustices he saw in society. The following selection from Adventures in Contentment *is both a celebration of the virtues of rural practical knowledge and a call for the recognition of hard work and intelligence in all walks of life. In this excerpt, Grayson asserts the farmer's social equality, but he makes this case from a very different angle than do the fiery populists of the Farmers' Alliance.*

From *Adventures in Contentment* (1907)

AN ARGUMENT WITH A MILLIONAIRE

I have been hearing of John Starkweather ever since I came here. He is a most important personage in this community. He is rich. Horace especially loves to talk about him. Give Horace half a chance, whether the subject be pigs or churches, and he will break in somewhere with the remark: "As I was saying to Mr. Starkweather—" or, "Mr. Starkweather says to me—" How we love to shine by reflected glory! Even Harriet [the narrator's sister] has not gone unscathed; she, too, has been affected by the bacillus of admiration. She has wanted to know several times if I saw John Starkweather drive by: "the finest span of horses in this country," she says, and "*did* you see his daughter?" Much other information concerning the Starkweather household, culinary and otherwise, is current among our hills. We know accurately the number of Mr. Starkweather's bedrooms, we can tell how much coal he uses in winter and how many tons of ice in summer, and upon such important premises we argue his riches.

Several times I have passed John Starkweather's home. It lies between my farm and the town, though not on the direct road, and it is really beautiful with the groomed and guided beauty possible to wealth. A stately old house with a huge end chimney of red brick stands with dignity well back from the road; round about lie pleasant lawns that once were cornfields: and there are drives and walks and exotic shrubs. At first, loving my own hills so well, I was puzzled to understand why I should also enjoy Starkweather's groomed surroundings. But it came to me that after all, much as we may love wildness, we are not wild, nor our works. What more artificial than a house, or a barn, or a fence? And the greater and more formal the house, the more formal indeed must be the nearer natural environments. Perhaps the hand of man might well have been less evident in developing the surroundings of the Starkweather home—for art, dealing with nature, is so often too accomplished!

But I enjoy the Starkweather place and as I look in from the road, I sometimes think to myself with satisfaction: "Here is this rich man who has paid his thousands to make the beauty which I pass and take for nothing—and having taken, leave as much behind." And I wonder sometimes whether he, inside his fences, gets more joy of it than I, who walk the roads outside. Anyway, I am grateful to him for using his riches so much to my advantage.

On fine mornings John Starkweather sometimes comes out in his slippers, bare-headed, his white vest gleaming in the sunshine, and walks slowly around his garden. Charles Baxter says that on these occasions he is asking his gardener the names of the vegetables. However that may be, he has seemed to our community the very incarnation of contentment and prosperity—his position the acme of desirability.

What was my astonishment, then, the other morning to see John Starkweather coming down the pasture lane through my farm. I knew him afar off, though I had never met him. May I express the inexpressible when I say he had a rich look; he walked rich, there was richness in the confident crook of his elbow, and in the positive twitch of the stick he carried: a man accustomed to having doors opened before he knocked. I stood there a moment and looked up the hill at him, and I felt that profound curiosity which every one of us feels every day of his life to know something of the inner impulses which stir his nearest neighbour. I should have liked to know John Starkweather; but I thought to myself as I have thought so many times how surely one comes finally to imitate his surroundings. A farmer grows to be a part of his farm; the sawdust on his coat is not the most distinctive insignia of the carpenter; the poet writes his truest lines upon his own countenance. People passing in my road take me to be a part of this natural scene. I suppose I seem to them as a partridge squatting among dry grass and leaves, so like the grass and leaves as to be invisible. We all come to be marked upon by nature and dismissed—how carelessly!—as genera or species. And is it not the primal struggle of man to escape classification, to form new differentiations? . . .

I assumed that he was out for a walk, perhaps to enliven a worn appetite (do you know, confidentially, I've had some pleasure in times past in reflecting upon the jaded appetites of millionaires!), and that he would pass out by my lane to the country road; but instead of that, what should he do but climb the yard fence and walk over toward the barn where I was at work.

Perhaps I was not consumed with excitement: here was fresh adventure!

"A farmer," I said to myself with exultation, "has only to wait long enough and all the world comes his way."

I had just begun to grease my farm wagon and was experiencing some difficulty in lifting and steadying the heavy rear axle while I took off the wheel. I kept busily at work, pretending (such is the perversity of the human mind) that I did not see Mr. Starkweather. He stood for a moment watching me; then he said:

"Good morning, sir."

I looked up and said:

"Oh, good morning!"

"Nice little farm you have here."

"It's enough for me," I replied. I did not especially like the "little." One is human.

Then I had an absurd inspiration: he stood there so trim and jaunty and prosperous. So rich! I had a good look at him. He was dressed in a woollen jacket coat, knee-trousers and leggins; on his head he wore a jaunty, cocky little Scotch cap; a man, I should judge, about fifty years old, well-fed and hearty in appearance, with grayish hair and a good-humoured eye. I acted on my inspiration:

"You've arrived," I said, "at the psychological moment."

"How's that?"

"Take hold here and help me lift this axle and steady it. I'm having a hard time of it."

The look of astonishment in his countenance was beautiful to see.

For a moment failure stared me in the face. His expression said with emphasis: "Perhaps you don't know who I am." But I looked at him with the greatest good feeling and my expression said, or I meant it to say: "To be sure I don't: and what difference does it make, anyway!"

"You take hold there," I said, without waiting for him to catch his breath, "and I'll get hold here. Together we can easily get the wheel off."

Without a word he set his cane against the barn and bent his back, up came the axle and I propped it with a board.

"Now," I said, "you hang on there and steady it while I get the wheel off"—though, indeed, it didn't really need much steadying.

As I straightened up, whom should I see but Harriet standing transfixed in the pathway half way down to the barn, transfixed with horror. She had recognised John Starkweather and had heard at least part of what I said to him, and the vision of that important man bending his back to help lift the axle of my old wagon was too terrible! She caught my eye and pointed and mouthed. When I smiled and nodded, John Starkweather straightened up and looked around.

"Don't, on your life," I warned, "let go of that axle."

He held on and Harriet turned and retreated ingloriously. John Starkweather's face was a study!

"Did you ever grease a wagon?" I asked him genially.

"Never," he said.

"There's more of an art in it than you think," I said, and as I worked I talked to him of the lore of axle-grease and showed him exactly how to put it on—neither too much nor too little, and so that it would distribute itself evenly when the wheel was replaced.

"There's a right way of doing everything," I observed.

"That's so," said John Starkweather: "if I could only get workmen that believed it."

By that time I could see that he was beginning to be interested. I put back the wheel, gave it a light turn and screwed on the nut. He helped me with the other end of the axle with all good humour.

"Perhaps," I said, as engagingly as I knew how, "you'd like to try the art yourself? You take the grease this time and I'll steady the wagon."

"All right!" he said, laughing, "I'm in for anything."

He took the grease box and the paddle—less gingerly than I thought he would.

"Is that right?" he demanded, and so he put on the grease. And oh, it was good to see Harriet in the doorway!

"Steady there," I said, "not so much at the end: now put the box down on the reach."

And so together we greased the wagon, talking all the time in the friendliest way. I actually believe that he was having a pretty good time. At least it had the virtue of unexpectedness. He wasn't bored!

When he had finished we both straightened our backs and looked at each other. There was a twinkle in his eye: then we both laughed. "He's all right," I said to myself. I held up my hands, then he held up his: it was hardly necessary to prove that wagon-greasing was not a delicate operation.

"It's a good wholesome sign," I said, "but it'll come off. Do you happen to remember a story of Tolstoi's called 'Ivan the Fool'?"

("What is a farmer doing quoting Tolstoi!" remarked his countenance—though he said not a word.)

"In the kingdom of Ivan, you remember," I said, "it was the rule that whoever had hard places on his hands came to table, but whoever had not must eat what the others left."

Thus I led him up to the back steps and poured him a basin of hot water—which I brought myself from the kitchen, Harriet having marvellously and completely disappeared. We both washed our hands, talking with great good humour.

When we had finished I said:

"Sit down, friend, if you've time, and let's talk."

So he sat down on one of the logs of my woodpile: a solid sort of man, rather warm after his recent activities. He looked me over with some interest and, I thought, friendliness.

"Why does a man like you," he asked finally, "waste himself on a little farm back here in the country?"

For a single instant I came nearer to being angry than I have been for a long time. *Waste* myself! So we are judged without knowledge. I had a sudden impulse to demolish him (if I could) with the nearest sarcasms I could lay hand to. He was so sure of himself! "Oh well," I thought, with vainglorious superiority, "he doesn't know," So I said:

"What would you have me be—a millionaire?"

He smiled, but with a sort of sincerity.

"You might be," he said: "who can tell!"

I laughed outright: the humour of it struck me as delicious. Here I had been, ever since I first heard of John Starkweather, rather gloating over him as a poor suffering millionaire (of course millionaires *are* unhappy), and there he sat, ruddy of face and hearty of body, pitying *me* for a poor unfortunate farmer back here in the country! Curious, this human nature of ours, isn't it? But how infinitely beguiling!

So I sat down beside Mr. Starkweather on the log and crossed my legs. I felt as though I had set foot in a new country.

"Would you really advise me," I asked, "to start in to be a millionaire?"

He chuckled:

"Well, that's one way of putting it. Hitch your wagon to a star; but begin by making a few dollars more a year than you spend. When I began—" he stopped short with an amused smile, remembering that I did not know who he was. . . .

"I suppose I might," I said, "but do you think I'd be any better off or happier with fifty thousand a year than I am now? You see, I like all these surroundings better than any other place I ever knew. That old green hill over there with the oak on it is an intimate friend of mine. I have a good cornfield in which every year I work miracles. I've a cow and a horse, and a few pigs. I have a comfortable home. My appetite is perfect, and I have plenty of food to gratify it. I sleep every night like a boy, for I haven't a trouble in this world to disturb me. I enjoy the mornings here in the country: and the evenings are pleasant. Some of my neighbours have come to be my good friends. I like them and I am pretty sure they like me. Inside the house there I have the best books ever written and I have time in the evenings to read them—I mean *really* read them. Now the question is, would I be any better off, or any happier, if I had fifty thousand a year?"

John Starkweather laughed.

"Well, sir," he said, "I see I've made the acquaintance of a philosopher."

"Let us say," I continued, "that you are willing to invest twenty years of your life in a million dollars." ("Merely an illustration," said John Starkweather.) "You have it where you can put it in the bank and take it out again, or you can give it form in houses, yachts, and other things. Now twenty years of my life—to me—is worth more than a million dollars. I simply can't afford to sell it for that. I prefer to invest it, as somebody or other has said, unearned in life. I've always had a liking for intangible properties."

"See here," said John Starkweather, "you are taking a narrow view of life. You are making your own pleasure the only standard. Shouldn't a man make the most of the talents given him? Hasn't he a duty to society?"

"Now you are shifting your ground," I said, "from the question of personal satisfaction to that of duty. That concerns me, too. Let me ask you: Isn't it important to society that this piece of earth be plowed and cultivated?"

"Yes, but—"

"Isn't it honest and useful work?"

"Of course."

"Isn't it important that it shall not only be done, but well done?"

"Certainly."

"It takes all there is in a good man," I said, "to be a good farmer."

"But the point is," he argued, "might not the same faculties applied to other things yield better and bigger results?"

"That is a problem, of course," I said. "I tried money-making once—in a city—and I was unsuccessful and unhappy; here I am both successful and happy. I suppose I was one of the young men who did the work while some millionaire drew the dividends." (I was cutting close, and I didn't venture to look at him). "No doubt he had his houses and yachts and went to Europe when he liked. I know I lived upstairs—back—where there wasn't a tree to be seen, or a spear of green grass, or a hill, or a brook: only smoke and chimneys and littered roofs. Lord be thanked for my escape! Sometimes I think that Success has formed a silent conspiracy against Youth. Success holds up a single glittering apple and bids Youth strip and run for it; and Youth runs and Success still holds the apple."

John Starkweather said nothing.

"Yes," I said, "there are duties. We realise, we farmers, that we must produce more than we ourselves can eat or wear or burn. We realise that we are the foundation: we connect human life with the earth. We dig and plant and produce, and having eaten at the first table ourselves, we pass what is left to the bankers and millionaires. Did you ever think, stranger, that most of the wars of the world have been fought for the control of this farmer's second table? Have you thought that the surplus of wheat and corn and cotton is what the railroads are struggling to carry? Upon our surplus run all the factories and mills; a little of it gathered in cash makes a millionaire. But we farmers, we sit back comfortably after dinner, and joke with our wives and play with our babies, and let all the rest of you fight for the crumbs that fall from our abundant tables. If once we really cared and got up and shook ourselves, and said to the maid: 'Here, child, don't waste the crusts: gather 'em up and to-morrow we'll have a cottage pudding,' where in the world would all the millionaires be?"

Oh, I tell you, I waxed eloquent. I couldn't let John Starkweather, or any other man, get away with the conviction that a millionaire is better than a farmer. "Moreover," I said, "think of the position of the millionaire. He spends his time playing not with life, but with the symbols of life, whether cash or houses. Any day the symbols may change; a little war may happen along, there may be a defective flue or a western breeze, or even a panic because the farmers aren't scattering as many crumbs as usual (they call it crop failure, but I've noticed that the farmers still continue to have plenty to eat) and then what happens to your millionaire? Not knowing how to produce anything himself, he would starve to death if there were not always, somewhere, a farmer to take him up to the table."

"You're making a strong case," laughed John Starkweather.

"Strong!" I said. "It is simply wonderful what a leverage upon society a few acres of land, a cow, a pig or two, and a span of horses gives a man. I'm ridiculously independent. I'd be the hardest sort of a man to dislodge or crush. I tell you, my friend, a farmer is like an oak, his roots strike deep in the soil, he draws a sufficiency of food from the earth itself, he breathes the free air around him, his thirst is quenched by heaven itself—and there's no tax on sunshine."

I paused for very lack of breath. John Starkweather was laughing.

"When you commiserate me, therefore" ("I'm sure I shall never do it again," said John Starkweather)—"when you commiserate me, therefore, and advise me to rise, you must give me really good reasons for changing my occupation and becoming a millionaire. You must prove to me that I can be more independent, more honest, more useful as a millionaire, and that I shall have better and truer friends!"

John Starkweather looked around at me (I knew I had been absurdly eager and I was rather ashamed of myself) and put his hand on my knee (he has a wonderfully fine eye!).

"I don't believe," he said, "you'd have any truer friends."

"Anyway," I said repentantly, "I'll admit that millionaires have their place—at present I wouldn't do entirely away with them, though I do think they'd enjoy farming better. And if I were to select a millionaire for all the best things I know, I should certainly choose you, Mr. Starkweather."

He jumped up.

"You know who I am?" he asked.

I nodded.

"And you knew all the time?"

I nodded.

"Well, you're a good one!"

We both laughed and fell to talking with the greatest friendliness. I led him down my garden to show him my prize pie-plant [rhubarb], of which I am enormously proud, and I pulled for him some of the finest stalks I could find.

"Take it home," I said, "it makes the best pies of any pie-plant in this country."

He took it under his arm.

"I want you to come over and see me the first chance you get," he said. "I'm going to prove to you by physical demonstration that it's better sport to be a millionaire than a farmer—not that I am a millionaire: I'm only accepting the reputation you give me."

So I walked with him down to the lane.

"Let me know when you grease up again," he said, "and I'll come over."

So we shook hands: and he set off sturdily down the road with the pie-plant leaves waving cheerfully over his shoulder.

EDWIN G. NOURSE

Edwin Griswold Nourse was born on a farm in Lockport, New York, near Buffalo, but the family soon moved to Chicago, where his father taught music in the public schools. Nourse began college at Lewis Institute (now the Illinois Institute of Technology), before completing his BA at Cornell University. He received his PhD in economics from the

University of Chicago in 1915, with a doctoral thesis that studied marketing in Chicago, published in 1918 as The Chicago Produce Market.

A theoretician of cooperative agriculture, Nourse wrote widely on the practice of cooperative marketing. He frequently bemoaned the antiquated practices of twentieth-century farmers, critiquing the state of agriculture early in the century as a throwback to the seventeenth century. Nourse endorsed the growing efficiencies in agriculture that emerged in the 1910s and 1920s, praising farmers who applied modern business principles to farming.

Over the course of his long career, Nourse held academic appointments at the Wharton School of the University of Pennsylvania, the University of South Dakota, and the University of Arkansas. He also served as a director of the Institute of Economics (later the Brookings Institution) and as the first chairman of the Council of Economic Advisors, from 1946 to 1949, under President Truman.

A critic of romantic agrarianism and the propensity of farmers to hew to "old sentimental and habitual patterns," Nourse instead celebrated the capacity of farmers to integrate scientific agriculture "into their everyday farming methods with a speed and effectiveness which are truly remarkable." Nourse's approach resonated with other Midwestern proponents of agricultural economic reform such as the American Farm Bureau, and his writings helped mold the postwar approach to high-production farming. This excerpt demonstrates how different Nourse's vision of an improved agriculture was from that of the neoagrarians included in this section.

From "The Place of Agriculture in Modern Industrial Society" (1919)

That any faction still argues that agriculture does by innate right and should in our social arrangements occupy a position superior to other industries simply shows that the more adequate economic theories of today have not yet superseded in some quarters the quaint patterns of thought which belong to the time of Louis XV. Such a philosophy it is that animates the extravagent [sic] utterances which, like thunder, precede the storm of each new farmers' "movement." On it can be based any amount of denunciation of bankers, of produce exchanges, of transportation companies, and of industrial concerns in general. Such an argument is implicit in most of the attacks upon the middleman and has prejudiced many a rural-credits discussion. It adds fuel to the Non-Partisan League's damning of "Big Business" and in part made the issue upon which farmers were attracted to the Populist party. Just in proportion as our rural interests approach other economic classes with this obsession of divine right or of natural preferment, by so much are they unfitted to deal fairly and effectively with the social and economic adjustments in which these several industries are mutually concerned. . . .

Certainly the United States has outgrown the idea of being merely a nation of

McCanless Brothers, "Harvesting in Washington State," 1903.
Courtesy Library of Congress, Washington, DC.

cowboys and cotton planters, wheat growers and hog raisers, truck farmers and orchardists. We are already a nation of iron smelters, machine builders, spinners and shoemakers, bankers, and scholars and artists. No one in his right mind would deny the fundamental and residual importance of agriculture in our scheme of economic life. However, instead of stopping our national building with the completion of that good foundation, we are concerned to rear upon its broad support the noblest conceivable structure of modern civilization. In this we are concerned not to have the largest possible proportion of our labor employed directly upon the land but rather to relieve as many as can be from the soil to fabricate raw materials into finished forms and to enrich our civilization with the greatest possible diversity and perfection of culture employments. The ruthless preacher of "Back to the land" would "kill the thing he loves," because the decline of non-agricultural callings means the impoverishment of agriculture by striking from under it many of the most essential props of its efficiency.

The maintenance and expansion of agriculture can be effectively provided for only by developing at a suitable rate the agencies of transportation, trade, and industry. Not for a moment should we forget the extent to which the modern farmer's performances are conditioned by the scientist from whom he has learned his present technique, the manufacturer who provides his equipment, or the financial institutions through which his business operations are transacted. Thus, for example, the farmer cannot hope to make his labors on the land achieve the maximum productivity possible to the modern industrial state of his art unless there are great factories turning out labor-saving implements and fertilizers and mill feeds. American farming would lapse back into an ineffectual past if it were deprived of building materials and potash by the closing of cement mills; or if it found the supply of slag phosphate or the raw materials of its implements, fencing, etc., curtailed by the decline of our steel industry; if it did not find that, under an effective division of

Montana wheat fields, Campbell Farms. Courtesy Phoebe Knapp Warren.

labor, rock phosphate and ground limestone and hog cholera serums and electric light plants and automobiles and papers, magazines, and books were being produced for it by specialized but non-farming groups. It prospers by and depends upon a system of rail and water transportation which brings it jute and sisal hemp from tropical lands, nitrate from Chile, and countless greater or lesser wares from the four corners of the earth. Most of all, however, the country depends, and must continue to depend, upon the town for the creation and distribution of those products of a high civilization without which our whole life would slip back to a lower and more primitive level, and which are produced only in the centers of population, where certain intensive forms of social activity are possible, and where the highly specialized ability and equipment which are needful can be supplied. In other words, the maintenance of our general civilization is essential to the farmer not less than to other classes of our society.

This is a fact apparently forgotten by certain dangerous friends of agriculture. They seem blind to the fact that if our life is to rise above the level of mere belly-filling and back-covering, a considerable fraction of our population must be freed from the soil to pursue the not less essential callings of the manufacturer, trades-man, engineer, scientist, teacher, writer, statesman, and artist; and that they must be fed and clothed and, in general, given economic rewards commensurate with the importance of their service. How many men and women can be spared and how adequately they can be equipped to pursue the labor necessary to endow our common life with better machines or better government or greater spiritual goods depends first and last upon how large a surplus of food and clothing is produced in our agricultural industry. Hence their reciprocal interest in the farmer and in the maintenance of agricultural productivity. If we are to multiply the conveniences of life, add to its beauties, and study its natural and spiritual mysteries, this rising standard of life must be supported by an enlarging flow of subsistence goods. Urban dwellers will follow their own true interests in safeguarding the prosperity and hence the efficiency of our rural population, and, equally, our country folk will advance their own well-being by a careful consideration of the true interests of towns and industries.

The day when men seek to draw invidious and wholly imaginary distinctions of greater and less, more honorable and less honorable, between the different classes of our productive population should have passed away long since. Such futile notions hark back to the dark ages of economic thought or rather to the metaphysical wranglings and theologic quibbles which preceded the coming of natural and social science. The real problem is that of economic expediency or industrial equilibrium, and the obvious truth that our civilization is "one and inseparable" should be kept vividly and steadily in our consciousness.

HENRY A. WALLACE

Henry Agard Wallace was born into a position of influence within Midwestern agriculture. During his career he became one of the most influential secretaries of agriculture in the twentieth century, vice president in 1940, and Progressive candidate for president in 1948.

Wallace was born in 1888 near the town of Orient in Adair County, Iowa. He was particularly close to his grandfather, "Uncle Henry" Wallace, the first editor of the family's influential farm paper Wallace's Farmer. *As an adolescent, Henry benefited from his father Henry C. Wallace's teaching position at Iowa State College, where he was mentored by one of his father's most promising students, a botanist named George Washington Carver. Carver taught the young Wallace his research methodologies, and the two remained close friends as Carver went on to develop his research in the South. Wallace's ambition to improve corn yields led to his successful first career, as a plant geneticist, and the hybrids that he developed in his gardens in the early 1920s led to the creation of one of the largest seed companies in the United States, now known as Pioneer Hi-Bred.*

Wallace replaced his father as editor of the family's paper in 1921 when the latter was named secretary of agriculture by Warren Harding. In 1933 Wallace again followed in his father's footsteps, ascending to the position of secretary of agriculture under Franklin Roosevelt. A principal architect of New Deal agricultural policy, Wallace had an incalculable influence on the development of modern agriculture. While serving as chief of the USDA, Wallace was responsible for implementing the Agricultural Adjustment Act, which recalibrated American farming through price supports and acreage reductions. It was Wallace who called for the notorious "Kill" and "Plow Up" policies, which caused over 6 million sows and piglets and 10 million acres of cotton to be destroyed in an attempt —ultimately successful—to control surpluses and raise agricultural prices.

Wallace was also a renowned economist and conservationist. The leading agrarian in the Roosevelt administration, he sought throughout his career to secure a stable place for agriculture in the American economy and to ensure the viability of rural areas in the modern nation. The following passage demonstrates his vision during the tough middle years of the 1930s—a period when the fate of the nation was most in jeopardy and his influence was at its peak. Here Wallace links the removal of farmers from submarginal lands to the provision of allotments of self-sufficient homesteads to the same people, arguing that both were central to the renewal of the American economy.

From "Putting Our Lands in Order" (1934)

The Chinese are the greatest individualists on earth. They cut their forests, silted up their streams, and destroyed millions of acres of their land by erosion gullies. Thus, they became increasingly subject to floods and drought. Their soil, exposed without cover to high winds, blew around in raging dust storms. The Chinaman's individualistic treatment of the land has exposed the Chinese again and again to famine.

They have been there a long time. Destructive as they have been, we in the United States, during the past mere 150 years, have handled our land in a way that indicates even more destructive possibilities. Over large areas we are even worse than the Chinese, because we made no real effort to restore to the soil the fertility which has been removed.

We have permitted the livestock men of the West to overgraze the public domain and so expose it to wind and water erosion. Much of the grass land of the great plains has been plowed, exposed, and allowed to blow away. Timber land under private ownership has been destructively logged off, without proper provision for leaving seed trees. All of this has been careless, thoughtless, wanton and to the disadvantage of nearly every one, immediately and in the future.

Early in the century, the conscience of our people began to awaken, under the leadership of Gifford Pinchot and Theodore Roosevelt. One hundred and sixty million acres of western public land were withdrawn from entry and set aside as great national forests. Places of great scenic beauty were set aside as National Parks. More recently we have appropriated money year by year to buy new land to incorporate in the national forests.

There are now 230 million acres of land in private ownership which are in serious danger of being exploited in ways harmful to the public welfare, and which should be purchased as rapidly as possible. During the 22 years of the Weeks' Forest Purchase Act, previous to this Administration, a total of 4,700,000 acres of forest land had been acquired. During the first year and a half of this Administration, this 22-year total has been slightly exceeded, and if this rate of acquisition is continued, the forest resources of the United States will be under adequate supervision and the headquarters of streams will be protected from erosion and unduly rapid run-off within twenty years.

Under the leadership of Franklin Roosevelt, the whole land problem has received an emphasis such as it never had before. As a young man, he was a close student of the Pinchot conservation policies, and he set out thousands of trees on his own farm. As Governor of New York, he had inaugurated a policy of buying poor land and reforesting it. As President, he soon saw the foolishness of spending millions in public works money to irrigate new land while at the same time, the AAA [the Agricultural Adjustment Administration] was taking land out of use. His suggested solution was to unify the program by allocating money to buy sub-marginal land to off-set the new irrigated land. To carry out this policy, 25 million dollars were allocated by Public Works in 1934 to buy poor land. . . .

Nevertheless government land purchase of poor and eroded land should be pushed with all possible speed. Human beings are ruining land, and bad land is ruining human beings, especially children. There are certain poor land regions so remote that it is impossible to maintain decent schools, roads and churches. It would be economy not to permit such areas to be settled except perhaps by childless adults who do not expect ever to lead civilized lives. Some areas are so unproductive that they will actually produce more food per acre if returned to natural game cover and restocked with wild life.

The human waste on poor land is even more appalling than the soil waste. The Federal Emergency Relief has approached the land problem from the standpoint of farmers now on relief. Several hundred thousands of these have been trying to accomplish the impossible, but they didn't know it until their endurance was sapped, and their case made plainly hopeless by the depression. With these families, the smaller part of the problem is buying the miserable land where they have been trying to make a living. The difficult thing is to find a new and better place for them to go—a place that will not mean that the government, in placing them there, is simply subsidising trouble for someone else. While no hard and fast rule can be drawn, it would seem that in the eastern half of the United States, the ideal location for many of these poor land farmers who are now on relief would be on self-subsistence homesteads where part of the family can work in industry. This may also be the destiny of some of the unemployed in our cities. It costs only one-third as much to take care of a farm family on relief as a city family, and if the Federal Relief has to support several million families for several years, it will try to make the money go as far as possible by getting several hundred thousand of them out on the land, establishing them in part-time farming and part-time industry. . . .

Land planning is no longer an academic question. A wise use of our land is intimately related to future of industry and the unemployed. . . .

The repeated droughts of the last five years demand special attention to another critical maladjustment. During the last 70 years several generations of suckers have been enticed by real estate men to the western great plains in years of good rainfall, only to be burned out later. This sort of robbery should not be perpetrated. No state can build a permanent prosperity on the falsehoods of realtors and pro-

Arthur Rothstein, "Fighting the Drought and Dust
with Irrigation, Cimarron County, Oklahoma," 1936.
Courtesy Library of Congress, Washington, DC.

moters. Wisconsin has recognized this by adopting a zoning law which divides the
state in such a way as definitely limits the field of these predators. Their game can
be rather definitely limited to areas where a man has at least a gambling chance to
make a living at farming.

Persons interested in ducks, pheasants, deer, fish and other forms of game, and
who lament the passing of these native species, want the government to acquire
the poorer types of land, especially low-land pastures and meadows near streams
and lakes, for game refuges. By planting the right kind of shrubs and grasses and
protecting nestlings from pasturing and mowing, we can work wonders in restoring
wild life over considerable areas unfit to farm.

In the eastern half of the United States, we need national recreational parks fifteen or twenty miles from the larger cities. These parks might well be located where
most of the land is so rough that farmers there are making a miserable living now.

Because of white encroachments, many of the Indian reservations are now terribly
short of land. An increase in Indian population has contributed to this jamming-in
of too many Indians on too little land. Indians are already subsistence farmers. To

transfer to certain of their land-hungry tribes the sub-marginal land of the neighborhood, would seem to be both fair and wise.

Other poor land might well be turned over to the Erosion Service to see what can be done to restore it. If nothing else can be done with it, it can be put into long-time use under the Forest Service, the National Parks Service, or restored to the Public Domain.

Certain large areas of the great plains now plowed, should be put down to grass again. Some of these regrassed areas might be grazed by cattle, under controlled conditions. Other areas of regrassed land might be best restored to wild life, including antelope, deer and buffalo.

But these are fragments. Fortunately the whole land question will probably be debated at length in 1935. In June of 1934, President Roosevelt appointed a special natural resources committee—the Secretary of the Interior, Secretary of War, Secretary of Agriculture, Secretary of Labor, Frederick Delano, Wesley C. Mitchell, Charles Merriam, and Harry Hopkins.

Frederick Delano, serving as chairman of a special advisory committee of this board, has prepared a report which will be submitted to Congress when it meets in January of 1935.

When this report is submitted we shall have had the benefit of some of the preliminary experience gained by spending a part of the 25 million dollars for purchase of sub-marginal land. We shall know more definitely the nature of the obstacles and how fast we can go. . . .

In the past, our land policies have often contradicted and cancelled each other. The government had no agency to unify such policies. Both the land and its people have suffered.

This year we are really beginning to build on the foundation nobly laid by the forest acquisition policy of Theodore Roosevelt and Gifford Pinchot. The report submitted by the natural resources committee to Congress in January of 1935 will, in all probability, if the Congress and the American people are willing, furnish the blueprint for putting our lands in order. In many parts of our social structure the blueprint method of approach is not advisable, but land is so fundamental and precious a heritage, that we should outline a policy to continue over many administrations, and stick to it for the sake of our children and their great grandchildren. The alternative is to maim and misuse our basic heritage, as have the Chinese.

RALPH BORSODI

One of the leading advocates of a return to self-sufficient farming in the early twentieth century, Ralph Borsodi was born in New York City in 1886 or 1887 (he asserted that he did not know which). As he grew up he frequently visited his father's printing shop, where the elder Borsodi published several classics of the early twentieth-century back-to-the-land movement.

From 1908 Ralph Borsodi experimented with his own ideas about independent living, but he did not put his thoughts fully into practice until 1920 when he committed to a self-sufficient life in the countryside. As he reported in one interview late in his life, his theory was simple: "that it was possible to live more comfortably in the country than in the city." Books chronicling his experiments on the farm, This Ugly Civilization (1928) and Flight from the City (1933), became the inspiration for thousands of people to try a new and independent life in the country. In response to the growing interest in self-sufficiency, in 1934 Borsodi created the School of Living at his Rockland County, New York, farm. This center institutionalized self-sufficient farming techniques and supported the establishment of enthusiastic but capital-poor farmers on their own land.

Borsodi was also an outspoken critic of land grant schools and the U.S. Department of Agriculture. In an article published in The Land in 1947–48, entitled "The Case against Farming as a Big Business," he laid out most clearly his thoughts on modern agriculture, issuing a stinging denunciation of the agricultural establishment:

> I accuse the United States Department of Agriculture, the Departments of Agriculture of our various states, our Agricultural and Mechanical Colleges, and the Association of Land Grant Colleges, of treason to the land!
>
> Blind leaders of the blind, I accuse them of deliberately commercializing and industrializing agriculture; of subordinating the real interest of agriculture to that of the fertilizer industry, the seed industry, the milk-distributing industry, the meat-packing industry, the canning industry, the agricultural implement industry, the automotive and petroleum industry, and all the other industries and interests which prosper upon a commercialized agriculture.
>
> I accuse them of teaching the rape of the earth and the destruction of our priceless heritage of land.
>
> I accuse them of impoverishing our rural communities, wiping out our rural schools, closing our rural churches, destroying our rural culture, and depopulating the countryside upon which all these are dependent.
>
> This is strong language. But it is not too difficult to prove that it may not be strong enough.

Borsodi lived to see state and federal agricultural agencies gather ever more influence and adhere ever more closely to industrial agriculture, and yet when he died in 1977 at his home in Exeter, New Hampshire, he was still living the independent practices that he had advocated for so long.

From *Agriculture in Modern Life* (1939)

The dictionary defines agriculture as the *science and art* of cultivating the soil, including the gathering in of crops and the rearing of livestock. In considering the problems of agriculture today, it is important to bear this definition in mind. Particularly important to note is the fact that in agriculture we are presumably concerned with a *science and art*, because one of the first things which this definition requires of us is abandonment of the idea, assiduously inculcated for nearly a century, that agriculture is a *business and industry* and that every problem connected with it should be approached in the same manner in which we would approach all other businesses and industries. This latter conception of agriculture is not merely modern, it is distinctly American. Not only the leaders and teachers of agriculture in America, but most of the farmers of America today, consider agriculture a business similar in all its essentials to the business of mining, of manufacturing, of trade, and of finance.

Yet it may prove to be the case that in these two conflicting conceptions of agriculture will be found the clue to the unsatisfactory condition of agriculture today. We moderns may be treating agriculture as a business, instead of a way of life. When it is too late, we may find that it is no more possible to treat agriculture as a business (without utter disregard of its intrinsic nature), than to treat art or religion in that manner.

It is necessary, therefore, in connection with this matter of definition, to record the probability that while nearly everybody will be willing to accept the definition here given to the term "modern life," only a few students of the agricultural problem will be willing to accept the definition given to the term "agriculture." Some students of the subject will insist, as I do, that agriculture is necessarily and by its nature a vocation. They will maintain that the true agricultural problem today is, "How can this particular way of life absorb what modern science and invention have to contribute to enrich it without surrendering itself to modern commercialism and industrialism?"

Others will insist that agriculture is a business, and that it is pure romanticism not to recognize that the real problem today is how to make the farmer as prosperous as other businessmen. They will maintain that self-sufficient family farming has been made into an anachronism by the modern world and that the sooner all farming is commercialized, the sooner the agricultural problem will disappear.

Still other authorities will maintain that both kinds of agriculture have existed side by side in the past and must continue to exist side by side in the future, even though the proportion of commercial to subsistence farming may continue to be increased by the developments of modern life.

The first group will maintain that agriculture is intrinsically a way of life with an incidental business aspect; the second group will claim that it must be treated as a business pure and simple; the third, that it is becoming—and may have already become—a business but with peculiarities arising from the fact that for many farmers it is also a way of life.

Those who belong to the first of these three schools of thought will insist that concern about agriculture today should not be narrowed to the counting-house concern to which most present studies of modern agriculture have accustomed us. The Farm Board of the Hoover administration; the AAA of the Roosevelt administration, the American Farm Bureau Federation; the Dairymen's Leagues and other cooperative marketing associations of farmers; the designers and manufacturers of modern agricultural implements such as the rubber-tired tractor, the cotton-picker, and the combine, all devote themselves to this counting-house approach to the problem.

I, however, feel that it is a mistake to concern ourselves primarily with the problem of how to make money out of farming. I do not feel that we ought to concern ourselves so much with the technical problems which increase the quantity and lower the cost of producing agricultural products and so make the business of farming as profitable as that of other businesses and industries. Even less do I believe that we should absorb ourselves with the problem of how to lower prices for the ultimate buyers of agricultural products so as to improve the conditions of life for our urban population and for foreigners. The commercial "profit and loss" approach to agriculture seems to me an approach in the interest of modern industry and finance-capitalism, while the technical and engineering approach seems to me an approach in the interest of urbanism and the development ultimately of a socialized state.

What I think really needs consideration is the problem presented by modern life to those who practice the art and science of agriculture. To me, the great need is for consideration of the problems of *agriculturalists* rather than of the agricultural *industry.*

In spite of the great development of mechanized farming, the distinctively commercialized agriculturalists are still only a minority of all the population of the nation which practices the art and science of agriculture. Commercial agriculture, in spite of its dominance in terms of production for the market, is only one phase of the life of enormous numbers of American agriculturalists, of the millions who are still engaged in general farming and who own family-sized farms. It plays practically no part at all in the life of the part-time working population which lives in the country—which draws part of its sustenance and support from agriculture, but which is not even considered a part of the farming population by proponents of modern commercial and industrial farming. Yet all these part-time farmers and all the sub-marginal agriculturalists (often farming sub-marginal land with sub-marginal capital) whom the advocates of a commercialized or socialized agriculture would "liquidate" in the interests of what they call progress, are human beings who still support their families at least in part from the farming of land. They are practicing the art and science of agriculture just as truly as are those farmers who support themselves altogether from the farming business, who have capital enough to operate a modern, specialized, one-crop farm, and who secure a cash income from the sale of crops large enough to buy most of the goods which their families consume and most of the supplies their farms utilize.

The real question to which it is high time we gave consideration is how *both* the millions of commercial and the millions of non-commercial agriculturalists should either adjust themselves to modern life—to a life scientific, industrial, commercial, and urban—*or how modern life should be adjusted to what is inherent and inescapable in the art and science of cultivating the land*. It is possible that if we ask this question, we shall find out that there is not only something wrong with modern agriculture but that there is *also* something wrong with modern life. We may even find out that what is wrong with agriculture today is caused by the effort which we have made for over a century to modernize it by commercializing, by industrializing, and by urbanizing it.

This suspicion—which for myself has long since become a profound conviction—is based upon certain aspects of modern agriculture which are almost entirely neglected in present-day prescriptions for ailing agriculture. These matters are neglected in the prescriptions mainly because they are accepted as fixed and unchangeable. Yet it seems to me that their appearance dooms the prescriptions to failure and the prescribers to futility. Among these basic aspects of the problem of agriculture and modern life which it is usual to ignore, I would place our system of land tenure with its accompaniment of speculation in land values; our mechanization of farming and misuse of specialized farming; our acceptance of the idea that soil should be treated as mineral or chemical capital to be converted into wealth, and our devotion to farming for the market with its accompanying high cost of distribution.

No phase of modern agriculture is more typically modern than that very fundamental one represented by our present system of land tenure—by our present methods of capitalizing land values and of treating the fertility of the land itself as commercial capital.

In a democratic society such as our own, in which social well-being and economic prosperity are to be achieved without any denial of the principle of individual freedom—without any denial of the right to life, liberty, and the pursuit of happiness—the population must consist mainly of families living in the country and owning their own homesteads. The system of land tenure in such a society must therefore be one which furnishes every family the opportunity to acquire land and to establish a homestead.

Our present system of land tenure, which is based upon the legal theory of freehold ownership, offers such an opportunity in appearance, but denies it in practice. A century of trial of the present system in this country is the proof of this statement.

A hundred years ago probably 80 per cent of the farmers of the United States were landowners. Since the establishment of the republic, and in spite of the fact that up to the beginning of this century, agricultural land was free under the Homestead Law to any American family, the proportion of farmers owning their own land has steadily declined. At the present time more than half the farmers of the United States do not own their own land.

The principal institution responsible for this, in my opinion, has been the system of alodial land tenure [that is, unfettered ownership] with its accompaniment of speculation in land values. Under the present system of freehold ownership, the American farmer is steadily losing the ownership of the land which he is farming. The reason for this is the fact that freehold ownership is a system of land ownership which has persistently tempted the individual American farmer to rely upon land speculation, rather than upon crops, livestock, woodland, and pasturage for his prosperity.

Yet the whole history of American farm and home ownership has been a record of . . . alternate speculative booms and years of depression, followed by foreclosures during the period of liquidation. No socially and economically sound agriculture can be maintained under the conditions which have produced this record. Even if one were momentarily to be established by a revolutionary redistribution of land, but without any change in the present system of land tenure, it would only be a question of time until present conditions of absentee-ownership, farm tenancy, mortgage exploitation, and urban concentration of the population would again develop.

The question which must, in my opinion, be answered is what sort of system of land tenure would furnish the bulk of the population of the nation an opportunity to acquire homesteads, and prevent farmers and owners of land and homes from losing them.

My own answer to this question begins with an acceptance of certain basic ideas upon the land question which were first expressed in their fullness by Henry George in his "Progress and Poverty." Land, George pointed out, ought not to be considered capital. Land differs from all other kinds of property because it is a gift of nature. It is unlike property such as buildings, machinery, agricultural and manufactured commodities, all of which are produced by the labor of men. Its value, as distinguished from the improvements made in or upon it, is due either to its utility as a natural resource—its fertility, its timber, its coal, oil, or other minerals—or to the demand for it created by its location. Neither of these values, George made clear, was individually created by the owners of land. When not gifts of nature, land values are produced by the activities of the entire community.

It is the tragedy of American society that we have used our land, not to establish a nation of independent farmers and home owners, but to indulge in gigantic speculations in real estate.

But it is not only by speculating in land that we have treated agriculture predominantly as a means of making money. The history of American agriculture is a history not only of wave after wave of farmers (each of whom lived on land which he had acquired at a low price until it rose in value and he was tempted to repeat the process further west), but also one of converting the fertility of our virgin soil into money, sometimes even more quickly than was possible by land-speculation itself. At the beginning we started by converting the riches of our soil, slowly created by natural forces over a period of thousands of years, into money by shipping it in the form of tobacco, cotton, and grain to Europe. Every ship conveying produce across the Atlantic, carried some of the fertility of our soil with it.

Inventors like Deere and McCormick perfected, and the manufacturers of agricultural implements distributed to every region of the nation, all sorts of ingenious labor-saving devices for hastening this process.

As soon as signs of exhaustion began to appear, agricultural scientists came to the rescue of commercialized agriculture. Chemical fertilizers were developed to continue the process of exhausting the remaining humus from the land.

Today we are shipping in some regions almost the last vestiges of some of our original resources to the cities. In many sections the denuded land, deprived of a proper proportion of trees, and without its organic binder of humus, is being eroded by wind and water and permanently destroyed.

So long as we insist upon being *modern* in this way, no permanently good way of life for the agriculturalist is possible. It is only by renouncing both land-speculation and land-mining, and by ceasing to be modern as the term modern is customarily defined, that we can lay the foundations for a permanently secure and prosperous life for the agriculturalist.

Agricultural land is a trust inherited by those who possess it today, to be used while they live, and to be bequeathed in at least the same, if not better, condition to those who follow them. The idea that the present-day possessors of land hold it absolutely, and that it is theirs to monopolize its value and exhaust its fertility, is not only ethically and esthetically false, but economically unsound. Each generation of farmers consists of tenants for life only. It is impossible for any generation to speculate in land or to exhaust the soil entrusted to it without depriving its own children and all future inheritors of the earth of their birthrights. Yet having accepted the idea that both land and farming are primarily methods of making money, that is precisely what we are doing, and what we are today being urged to do in an ever greater measure by most of the agricultural leaders who are telling us what is essential in order to modernize agriculture.

John C. Rawe and Luigi G. Ligutti

From the 1930s through the 1950s, a vocal minority of Catholics engaged in their own push for environmental and social renewal through a back-to-nature movement, born in the rural parishes of the Midwest. Catholic clerics called for a "green revolution," to be marked by sustainable practices and self-sufficiency. As John Rawe described it: "It is a green revolution because it takes place out in the green fields where the land, owned by the patient, productive, profitable, democratic, free, personal laborer is blessed and gladdened with the divine benedictions of life-giving moisture and smiling sunshine." This organic vision of agricultural change contrasts sharply with the technological and input-intensive innovations that would characterize the more famous Green Revolution in the developing world during the 1960s and 1970s.

John Rawe and Luigi Ligutti presented the most focused, comprehensive presentation of their green revolution in the 1940 Rural Roads to Security. *Rawe and Ligutti, like the other members of the National Catholic Rural Life Conference (NCRLC), viewed Catholic agrarianism as a new way of life embracing independence through work. Similar in many respects to the other neoagrarians of the 1930s, the NCRLC stressed the importance of life on the land for a viable citizenry and promoted farm living as a healthy and ethical way to raise a family. Their work was distinguished from other neoagrarianisms by a more specifically sacramental reverence for the natural world and the work of the farmer as well as by their debt to Catholic social doctrine.*

Whereas many of the other agrarians of the 1930s and 1940s sought freedom from the constraints of society on the farm, Rawe and Ligutti envisioned that rural life could be oriented around vibrant Catholic communities. The New Deal community of subsistence homesteads in Granger, Iowa, was cosponsored by the NCRLC and embodied the organization's ideas of how to support viable rural communities. There was significant overlap with non-Catholic thinkers, however, and both the Southern Agrarians and Ralph Borsodi had an important influence on the thinking and writing of both Rawe and Ligutti. In fact, each taught at Borsodi's School of Living during the 1940s.

After Rawe died in 1947 at the age of forty-seven, Ligutti continued to lead the NCRLC until the late 1950s, even though the movement foundered after the transformative years of World War II.

From *Rural Roads to Security: America's Third Struggle for Freedom* (1940)

When in 1775 British oppression threatened the human rights and happiness of Americans, the Minute Men united for the defense of liberty. That war gave us political independence. Ultimate independence, however, with liberty for all, Black and White, cost America another war. It then became a truly independent nation. Henceforth a future of freedom was to become the birthright of every American.

By a tragic paradox, at the very time when wars were being waged for political independence, there was injected into the veins of American industry a deadly virus which would tend more and more to paralyze the mind and spirit of American manhood. In a word, American life fell heir to the liberalistic system which Europe had fostered. When the renaissance individualism cast off moral restraint through the influence of the Reformation, the road was paved for a materialistic philosophy. Liberalism saw only good in the ambitions of men, demanded fullest liberty for the satisfaction of personal aggrandizement without hindrance of law, or organization, or any effort to safeguard one man against the greed of another. As the Rev. Joseph C. Husslein, S.J., points out in *The Christian Social Manifesto*, this liberalistic dream was taken seriously for more than a century and a half and it has not yet been dissipated.

The discoveries of science in the eighteenth and nineteenth centuries coincided with this liberalistic stream of philosophy, ever deepening and expanding the glittering sea of modern Capitalism. Heavy machinery, power, crowded factories, congested cities, large-scale production, greater and still greater profits and investments complemented each other, and accelerated the evolution of liberalistic industrialism. And so, with every new stage of the swirling cycle, liberty retreated farther from the wage earner, as economic necessity left him ever more helplessly at the mercy of the capital which he served.

The factory system, which today signifies concentrated mass production, took root in our country during the War of 1812. Its early growth was comparatively slow, for as late as 1850, the bulk of American goods was still produced in the household, the shop, or the small factory.

While the open frontier still continued in competition with the factory, the demand for labor was never filled. Gradually women and children were themselves drawn into the factory. Although hours were long for all alike, and although children were growing up illiterate and without the normal experiences of childhood, yet deluded Americans flocked to the factory as though it were the gate to prosperity.

The entrance of women into the industrial field tended to reduce the wages of men, since men were no longer the sole support of a family the idea of a family wage for the head of the family was slipping to that of a mere individual wage in competition with women and children. Still labor was not at once shackled by this condition. There was still a possibility of escape, and when escape is possible, liberty is not dead.

Harold Faulkner gives the alternative when he writes:

> As long as public land could be had at nominal cost, "wage slavery," in the sense that there was no escape, did not exist. If times were hard and wages low, the worker could always go West.

After 1850, transportation underwent marked improvements. Steam railroads increased 300 per cent between 1850 and 1860. With steam transportation established, the factory system began that forward leap which continued, with but brief lulls during the great panics, through the remainder of the century.

This twofold development, growth of factories and improvement in transportation, was directly instrumental in changing from bad to worse the conditions of labor. Wages tended to become standardized at a minimum, since goods from one city were brought into competition with the same type of goods from another city. Price plus quality capture the market. By established custom the necessary curtailment was taken from wages. Transportation and growth of factories also made profitable the sub-division of labor, thereby creating vast numbers of detail jobs, simple enough to be classed with unskilled labor and each paid the correspondingly lower wage.

CITY CONCENTRATIONS AND THEIR SOCIAL PROBLEMS

The specialized capitalist, alert to the possibilities of saving by division of production, concentrated industry in fewer and larger plants. Labor, long below the ability of housing itself in health and decency, huddled more densely in the industrial tenements. This urbanization of population paralleled the concentration of industry and was, in greater part, due directly to it.

Labor declined rapidly, losing not only ownership of tools, productive property, and control of conditions of labor, but also home ownership as well. Company tenements, company stores, company commodities were being provided, but in a very inadequate manner, and under circumstances that left only a shadow of liberty or recognition of rights on the side of the working people.

Another factor that greatly stimulated urbanization of population was the rapid disappearance, since 1880, of desirable western land obtainable on easy terms. During the first half of the nineteenth century public land of rare quality was limitless and given on terms that were meant to be an invitation and reward for settlement. Little or no capital was required to secure and work a claim. The disappearance of such public land closed a safety valve of escape from the city and dammed the floods of immigrants in the already close confines of industrial cities.

Urbanization, so rapid and so concentrated, created a host of social and economic problems. Of these the most tragic to human freedom was the increasing depth of helpless surrender to which an ever greater and greater portion of the nation's citizens was reduced, succumbing to the unscrupulous and liberalistically sanctioned avarice of the "robber barons." Labor had become depersonalized as regards the relations of employer and employee. Corporate ownership and control lodged in the hands of a relatively few. These few, interested primarily in greater profits, better business, and more production, neither saw nor cared to see the laborers, nor still less the slums in which they existed. Public opinion protested, and government took action again and again, but the philosophy of wealth continued unconquered and almost unquestioned except in subconscious thought, and the conditions of labor, even though improved, lagged behind that of the favorites of fortune as far as ever.

Keeping pace with economic changes, the destructive influence of these new conditions on the home and family life now made themselves sadly felt. With the appearance of the factory system home occupations decreased quite generally. Competition with factory products forced members of the family to seek employment in the large-scale industries. Spacious homes were abandoned for dingy, unsanitary shacks near the noisy, smoking factory. The health and well-being of women were menaced. Low wages for men, brought lower by the competition of women in industry, left marriage to be deferred or renounced. After marriage both husband and wife were often compelled to continue in their former outside employment so that the family might be able to subsist. Ideals of family life were thus shattered by the absence of one or both parents. Children were left without care and childless homes became common. Self-interest was created by the separate purses of husband and wife. Divorce increased. Submarginal living was almost the rule for larger families. Commercialized amusements, the movie, dance halls, soft-drink parlors, saloons, and the automobile assisted in drawing people out of the home. . . .

Depression and the Present Struggle

Materialistic philosophers taught that Liberalism, free and unrestrained, would bring the nation to the peak of prosperity. Capital falsely claimed that efficiency and security for any industry was attained by way of complete monopoly. Labor wondered wistfully, and submitted and hoped with infinite patience, but the peaks of prosperity seldom ever broke the level horizon of day-by-day toils.

Today, with consumption deadlocked for years and the unfulfilled desires of the masses soured to envy, distrust, and destructiveness, capital too has paused to calculate whether its excessively high profits were really so excessively profitable.

Have the unhampered ambitions of the fortunate brought about the highest common good? Americans form the wealthiest nation in the world. Our productive potentiality is, so to say, limitless. But consumption has collapsed, and gone on relief for artificial stimulation. Overproduction and starvation are puzzling neighbors. We think our money gods have tricked us. We are unhappy. We seem to have sold our birthright of freedom for a mess of pottage. Our tenants and sharecroppers are numbered in millions. Few slaves lived lives as wretched and insecure. A sharecropper "in the red" is bound to remain on his land, not as a serf with recognized rights, but as an exploited commodity. He is an American robbed of freedom. . . .

We do not want to be "taken care of." We want our birthright back. Let those who despair use the methods of despair. We have our dream, our ideal. We dreamed of liberty, equality, precious freedom, our inalienable, God-given rights. We can give no more than a passing thought to anything less.

But, surely, all this is just another sour joke, they will tell us, that some day may come in handy to humor the crowd. An ideal! Americans with dreams? Observe those who pass, these dissatisfied, listless job holders with a haunting fear in their eyes lest the job that is theirs today will be no longer theirs when the sun rises tomorrow; migratory tenants always hoping for the better and getting the worse; dehumanized sharecroppers artificially always "in the red"; the army of unemployed frantically walking the streets till even the beautiful sunshine is cruel; spiritless men and women on relief, some wishing life could pass more quickly, others spinelessly content to remain as they are; old age filling in government blanks; youth staring into a blank future. Behold flourishing American cities grown tiresome, harsh, and cold; fertile American valleys commercialized, gullied, barren, and parched. . . .

Mammoth-scale industry, commercialized farms, human lives ground by marvelous machines, efficiency substituted for liberty, money codes displacing justice, with God and His spirit forgotten—these, too, are lifeless bones, well shaped and bleached to a beautiful, gleaming white, yet mere bones. Bones that need not be destroyed, but rather let sinews and spirit be given them. We want to remain modern and also be human and live. We shall do this when we come once more to know that we are the Lord's; that He opens our sepulchers and brings us from our graves, and puts His spirit into us, makes us live and brings us to rest upon our own land. If philosophy in the sixteenth and seventeenth centuries made mistakes, it can be set right in the twentieth, for the nineteenth already pointed out the error and the remedy. Rugged individualism must yield to love of fellow men. The striving to become rich, measuring success by fortunes, and seeing values in all things only by the dollars marked on the price tag, should give place to the joy of life. Human living must again be able to lift its thought above the passing things of earth to see the dignity and immortal destiny of man.

New Values—New Life

"We the people" are the power and authority that can make the change. Our first activity is to establish a new view of life. We want truly to live. All our strivings should aim to win, not a fortune, but a greater fullness of life, for happiness, comfort, and security by honest, God-pleasing work. Next we need a plan to make this possible. From our side, it is only made possible by breaking down this iron-clad wage slavery. Pope Leo XIII, the workingman's truest friend, wished to see every laborer owner of some productive property.

Ownership of productive property in the industrial world is available to labor through coupon clipping. Though such ownership, when practical steps are taken to establish it, is good and commendable, it is not an ownership which gives effective control. A way to both ownership and control, for the many, in some productive property is the way *by land*.

Ownership of Homes with Small Acreages

The movement for ownership of homes on the land is eminently human and satisfying whether it is the relief garden planned toward ownership for the needy, the subsistence homestead or part-time farm for the industrial family, or the full-time diversified family-unit farm for the rural people.

Modern comfort and modern, small, human-scale machines in many forms of production without any loss of efficiency, point the way to restored freedom and security through the ownership of a few acres. A blend of the rural and urban modes of life, in both part-time and full-time farming, a mode of life which modern technocracy makes possible, is the one which can accomplish the aim expressed by Alexis Carrel: "Restore to man his intelligence, his moral sense, his virility, and lead him to the summit of his development."

The major economic and social need in work for spiritual and cultural advancement and the preservation of liberty in American life is: *family-unit operation and fee-simple, family-basis ownership of land* based on religious principles and spiritually motivated. Landownership is a determining factor in human well-being. This is not a matter of selfish concern for rural life and rural people. It is a matter of great national importance. Perhaps the most tragic aspect of changing America is the general decline from landownership to mere factory work and tenancy. The number of landowners, tenants, and wage earners will in large measure determine whether the life of the nation is to be democratic or proletarian, and will ultimately decide the destiny of our civilization. Landownership goes hand in hand with a predisposition for education and the building of good communities and a democratic citizenship. The problem of this change from landownership to tenancy, the status of mere factory workers, demands the study and help of all those who are interested in the spiritual and democratic progress of America.

Extremes to Be Avoided

It is a wise policy, when discussing a complex problem, to avoid extremes. In the present instance, those who demand, for the good of society, a clean cleavage either in favor of agrarianism to the exclusion of industrialism, or industrialism to the exclusion of agrarianism, are extremists. Some zealots tell us to flee to the fields in order to avoid contamination of soul and body in the city, which is a blot on the face of the earth; others, equally foolish, extol commercialized, scientific progress and dream of the day when the countryside will be completely covered over with the sprawling works of big industry and big farming.

Both views are untenable. A nation cannot be prosperous, unless there is a proper balance between town and country, between the rural and urban way of life.

Why can't we develop a constructive economics and sociology? Why can't we restore some natural economic functions to the family, the natural economic unit? Why should all economic and social functions be swallowed up by corporations and states in a mad rush toward concentration and collectivism of one form or another? Why can't we plan and work for the proper preservation of many natural units in a food raising economy, a home-owning and home-building environment? We would soon discover that such work would result in the building of a new democratic nation at much less expense than it takes to give pensions and doles and maintain an artificial system that leaves human capabilities exposed to corruption and decay, and sets the mind in a groove of false thinking.

Some Family-Centered Production

The best way to restore the home is to provide for some family-centered production, family-centered activity where the child can soon become an economic asset instead of remaining an economic liability. That is why the food-producing homestead has economic, social, cultural, and ethical significance. That is why every housing program should be a homestead program. Nothing prevents the successful combination of industrial wage earning and part-time farming today save a certain spirit of narrow urban industrialism, an erroneous self-sufficiency, and a want of democratic vision. Many industrial workers would welcome the new type of living which homesteading embodies—a life which is neither strictly rural nor strictly urban, a life which is an intermediate type between the two, combining the benefits of both.

LOUIS BROMFIELD

Louis Bromfield was born near Mansfield, Ohio, in 1896, the son of a businessman and politician who nevertheless fondly remembered his family's agricultural roots. After leaving home to attend Cornell University's College of Agriculture in 1914, Louis was called back to Ohio to help work the family's ancestral farm. In 1916 he returned to college, at Columbia University, but left again to serve in the French Ambulance Service in the First World War. After the end of the war he wrote several popular novels, won a Pulitzer Prize, and enjoyed the revenues from book and film royalties.

Still attracted to agriculture, in 1938 Bromfield settled his family upon a thousand-acre farm in his native northern Ohio, "a desolate farm, ruined by some ignorant and evil predecessors." He argued that the mistreatment of farms was too common, writing that "a great many" farmers "actually hate the soil which they work, the very soil which, if

treated properly, could make them prosperous and proud and dignified and happy men."
Bromfield saw his own land as representative of the hundreds of millions of misused acres
in the United States and experimented upon Malabar Farm using his ideas about ecologi-
cal agriculture to promote the virtues of rural life.

Bromfield sought to educate the American people about agriculture and land use, and
after his move to Malabar he claimed that he wrote primarily to "lure readers who never
had any interest in agriculture or whose interest had been dulled or killed by pamphlets."
Bromfield enjoyed a wide audience, gained friends such as Humphrey Bogart (who mar-
ried Lauren Bacall at Malabar in 1945) and E. B. White along the way, and contributed
a great deal to the midcentury understanding of the potential of a permanent agriculture.

From *Pleasant Valley* (1945)

In northwestern Ohio the land is flat and one can see for miles across the country
without a perceptible roll in the land. It is the newest land in Ohio and some of the
most fertile in the world. Once the whole area had been lake bottom, a part of Lake
Erie, and then for a few centuries it had been a vast marsh inhabited by bear and
wolves and wild duck and geese.

A little before the Civil War the whole area was drained by a system of vast ditches
and the land sold off at a dollar an acre to settlers. It is cornland and the best baby
beef in America comes from it. Thousands of steers are "fed out" there every year
and the fertility has been kept high by the millions of tons of manure that goes
back into the soil every year. The yields of crops, particularly corn, are prodigious.
Almost any farmer in the region will tell you that his soil is ten or twelve feet deep
and that his land is so flat that there is no erosion. Yet the time is not too far off
when all that region may once again become marshland, for the Black Swamp area is
slowly returning to its old condition as the soil wears down to the level of the drainage
ditches and the big lake to the north. You cannot see it go; evidence of the change
might never occur to a farmer until he begins to plow up his drainage tiles and even
then he will try to persuade you and himself that this is not because the soil is wearing off but because the frost or the "working of the soil" has brought the drainage
tiles to the surface. But the terrible evidence is Sandusky Bay.

The whole area lies in the watersheds of the Maumee and Sandusky rivers. The
Sandusky flows into a marsh-bordered bay about thirty or forty miles long and
from four to five miles across. Within the memory of a man forty years old, some
of the best fishing in the world existed in that bay—bass and pickerel and all sorts
of native game fish. Today there are in it only the sluggard mud-loving carp and a
few perch because the water of the bay is seldom clear any longer. The soil of the
drained great swamp has been moving into Sandusky Bay since the Black Swamp

Kate Lord, drawing on scratchboard. From Louis Bromfield, *Pleasant Valley* (1945). Reproduced by permission of Wooster Book Company, Wooster, OH.

was drained and put to the plow. The bottom now is mud and the areas along the edge grow shallower each year as more and more silt is deposited. Meanwhile the game fish have left the bay. Once it was an important spawning area for the lake fish which provide Ohio with one of its important industries. The fish no longer spawn in the mud-filled bay and the effects are being felt in the fishing industry.

One day the farms of the rich Black Swamp country will have moved into Sandusky Bay, filling it up and leveling off the land of the rich watershed behind it. One day both bay and farm land will be level and the Black Swamp will return.

It is true that in this area there are good farmers who practice proper crop rotation and cover their bare cornfields in winter with root blankets of wheat or rye and they are doing a good job of anchoring their soil. Very little of it is going down into Sandusky Bay. But there are others who leave their fields bare to wind and frost and rain. They will tell you that you are crazy, that they have no erosion on their flat land. When a farmer in the South, the East or the Middle West, tells you that he has no erosion by wind or water on his bare fields and that it is unnecessary to take any precautions, you may put him down either as ignorant or a fool. The truth, which any man can see, is that a bare field is abhorrent to Nature and she sets about at once to blanket it with vegetation or to destroy it. Winter-bare corn, cotton and tobacco fields cost this country annually millions of dollars in the loss of soil.

On Malabar, in genuine hill country, where ugly gullies and denuded hilltops tell their own story, the evidence was all about us, yet there were farmers of the last generation, unlearning and unwilling to learn, who thought and said the measures we were taking that first year were crazy. It is true that they were not very good farmers and that in the case of most of them, they will probably be the last occupants of the land they were farming. Bankruptcy and the forest will move in again.

Behind that philosophy lies a large segment of the history of the United States, once a vast wilderness of incredible richness inhabited by a few hundred thousand half-savage redskins. White men from Europe came to it and set out to pilfer rather than to develop it. The riches seemed inexhaustible and a tradition of farming grew up among the frontier men which made of the American farmer one of the worst farmers in the world. The tradition and habit was simply that of mining the land.

The formula was simple. First you simply cut off or burned over the forest or prairie and then you went to work wresting the fertility from the soil in terms of crops as rapidly as possible. Sometimes the fertility or the topsoil lasted two or three generations, sometimes longer. Then when the soil was worn out you went west to Ohio or Indiana and repeated the formula. When land was exhausted there, Iowa and Kansas and Dakota lay ahead, and finally Oregon and Washington and California. The good land that could be had for little or no investment and could be "mined" seemed without limits. "The West" became a byword for opportunity—opportunity principally for more free, rich, virgin land. Very often men went west to take up land less good than the land they had recklessly destroyed. Sometimes they exchanged a mild climate for a harsher one, well-watered areas for country afflicted by drought or waterless land which had to be irrigated. They could have done better and been happier and more prosperous and comfortable if they had cherished the good land they destroyed and remained on it.

Like a plague of locusts they moved across the continent, leaving behind here and there men who found the soils so deep and the mining so inexhaustible that there was no necessity for migration. And here and there they left behind a good farmer, wiser than the rest, who cherished his soil and farmed well. But the good farmer was and is as a rule a "foreigner"—an American whose tradition and training in reality went far back into a Europe where there had been little or no cheap land for five centuries. By then their piece of land was regarded properly as their capital; and an intelligent or a wise man does not throw his capital out of the window. As a rule the more recently arrived the immigrants, the more they respected the piece of land they were able to acquire in the New World. Very often they acquired farms ruined by farmers of old American stock and restored them. Too many of the descendants of the older stock followed the wide-open reckless traditions of the American farmer in which land was not a capital and treasure but merely a speculation or a "mine." The principal exceptions among the old stock were the Pennsylvania "Dutch," the Amish, the Mennonites who lived closely among themselves, holding fast to their particular variety of religion, to their customs, even to their language. They stayed on the land they settled upon and made it richer and more valuable each year by farming well.

Not all of the fault lies with the farmer himself. As Americans, we are all immigrants to a new world and since most of our early stock came from central or northern Europe, the traditional agricultural methods of that region were brought here with them. The earlier stock not only came to a country of apparently inexhaustible resources but the agricultural methods it brought were hopelessly unsuited to the different climate and soils of America. The inexhaustible richness made our farmers of old American stock reckless and greedy exploiters; largely speaking, the older the American stock the poorer the farmer. Those immigrants who arrived more recently have managed to preserve, like the earlier religious sects, their reverence and respect for the soil.

The use of European agricultural methods in the American climate has been disastrous. In the temperate areas of northern and central Europe there rarely occur the cloudbursts, the thunderstorms, the violent winds, the seasonal droughts and floods, the temperatures ranging from 20 below zero in winter to 100 degrees Fahrenheit in summer which are commonplace in the American climate. In the European area the climate is more temperate and the rains fall gently, sometimes as in Normandy and in the Channel Islands, in a form of more or less steady drizzle. In all the eighteen years of my experience with agriculture in France, I saw only once a thunderstorm and cloudburst as violent as the kind of storm that happens a score of times every summer in our Ohio country.

As the wilderness was subdued, during the progress of the white man westward, forests were cut down, high prairie grass plowed under and millions of acres of the best mixed legume grazing land in the world was plowed up, overgrazed and burned over. Swamps were drained and streams straightened, sometimes senselessly. Within a period of from fifty to one hundred years the whole of the vast Mississippi Basin was changed almost beyond recognition by the hand of man. Much of the land once covered with forest, grass and marshland was left bare to cloudbursts, thawing snow, tornadoes and other violent manifestations of a climate much more like that of China than of Europe. The result was devastating floods, dust storms, droughts and lowered productions in some areas approaching desert status.

Dr. Hugh Bennett of the Soil Conservation Service estimates that if the soil lost annually by erosion in the United States was placed in ordinary railway gondola cars, it would fill a train reaching four times around the earth at the equator. At the Georgia State Agricultural College tests made upon one acre of ground farmed by the conventional method used in cotton cultivation in that state showed an average loss of 127 tons of topsoil a year over a period of five years.

About twenty-five years ago there began a spontaneous movement toward finding a system of agriculture more suited to the American climates and soils than that imported from Europe. It came none too soon. In this search countless people took part, most of them working individually, experimenting, watching the earth, working always toward a common end. There were market gardeners, garden club members, government bureau men, agricultural college professors, city farmers, dirt farmers, schoolteachers, all working individually toward the common end of finding an *American agriculture*. Only in the past few years has it become apparent that an actual revolution in agriculture had been in progress for a long time. I think that the publication of *Plowman's Folly* by Edward Faulkner did much to make the character of the revolution evident. It is certainly true that never before in the history of the nation has there been so intense and so widespread an interest in agriculture.

Slowly but certainly a system of agriculture suitable to the United States has been evolved. It has grown out of the remote past of Asia and the Near East, out of discoveries made in half the nations of the world, out of experiments and the brains of countless devoted and intelligent workers. The system includes terracing and cover crops, trash farming, proper drainage and forestry practices and pasture treatment, the use of legumes as green fertilizer, diversified farming and the rotation of crops. The revolution is still in progress, growing and expanding. It is not only important to the people of this nation, but to the people of other nations with similar problems of soil and climate. South Africa, Palestine, China, Mexico, even so new a country as Venezuela, have called for and received the aid and experience of Drs. Bennett and Lowdermilk and their staffs from our Soil Conservation Service. Perhaps of all the aid given so lavishly to foreign nations by our government, none will prove so valuable as that given, without fanfare, without special reward, by the men of our Soil Conservation Service. . . .

In our own country there is no more virgin, well-watered land available. Land that can be irrigated or drained is not always good agricultural land, and often enough in our West, the very water used to irrigate the land is so alkaline that it ruins the soil within a few years. Fully a quarter of our good agricultural land has been already reduced to the lowest status, fit only for reforestation and sometimes not even fit for such a purpose. Fully another quarter is in an intermediate stage of destruction. There has been much talk of settling returning veterans on the land, but what land? The good land is not for sale or if for sale it is expensive land. There are no great prairies with deep, black soil waiting for settlement to absorb returning soldiers as it absorbed them after the Civil War.

Meanwhile our population continues to increase while the productive capacity of our land decreases rapidly. It is probable that not 5 per cent of our agricultural land produces anywhere near the 100 per cent of its potentiality. It is also probable that about 70 per cent of our agricultural land produces not 30 per cent of its potentiality. The farmer's problem is not entirely one beyond his control. Much of the rural poverty and insecurity arises from the fact that there are too many bad and careless farmers and too many lazy ones who are content to live as their pioneer grandfathers and great-grandfathers lived, working during the crop season on single cash crops and sitting by the stove or in the village store all through the winter. Their forefathers had deep virgin soil which could support that sort of existence and when the land wore out they could go elsewhere. Today in order to succeed, no matter what the prices or the parity support, a farmer has to be both intelligent and informed and to work and work intelligently with good modern farm machinery. This is no longer a new country with limitless resources. Each day conditions come nearer and nearer to approximate those of Europe with respect to agriculture, population and economic security. This need not be so; it is so largely because we have been wasteful and reckless in the treatment of our natural resources.

Thomas Hart Benton, *The Wreck of the Ole '97 Train*, 1943. Egg tempera on gessoed masonite.
Hunter Museum, Chattanooga, TN. Art © T. H. Benton and R. P. Benton Testamentary Trusts/
UMB Bank Trustee/Licensed by VAGA, New York, NY.

6. Southern Agrarianism,

1925–1940

If, as economists say, there is no free lunch, then how do we fully account for economic growth? The benefits of modernization may be obvious, but the questions "What is the cost?" and "Who pays?" often prove more elusive. Cutting against the grain of American optimism about economic development, during the 1920s and 1930s the Southern Agrarians, a group of writers centered at Vanderbilt University, set out to calculate the cultural price paid for industrialization. In their manifesto *I'll Take My Stand,* the Agrarians attacked narrow economic thinking and centralized economic power, which they believed would ultimately destroy rural communities, distinctive regional cultures, self-reliance, enjoyment of good work, and affinity for the mystery of the natural creation. They saw industrialization as an immediate threat to the Southern society they loved, but also as a cancer that deformed the inner workings of Western civilization in general. As a sympathetic T. S. Eliot put it in commenting on the work of the Southern Agrarians, they addressed the troubling question of "how far [it is] possible for mankind to accept industrialization without spiritual harm."

Those concerns have been shared and elaborated upon by many neoagrarian thinkers since, most notably Wendell Berry—and by other critics of the industrial market economy who might not consider themselves agrarians. But in championing the rural South as a bulwark against corrosive modernization, in praising its very "backwardness," the Southern Agrarians all but invited critics to raise an uncomfortable question in return: How much hookworm and Jim Crow should the South endure as the price of keeping its traditions intact? For there is no free lunch in failing to modernize, either.

Industrialization, with its mixture of promise and peril, arrived with uneven speed in different parts of the country. Nowhere was it received, though, with the kind of ambivalence that greeted it in the South. As the nation's manufacturing power leapt forward during the latter half of the nineteenth century, the South lagged behind, retaining its largely agricultural economy. Modernization had its Southern advocates, however. By the turn of the century newspaper publisher Henry Grady and others were calling for a New South, in which a growing industrial sector would balance and strengthen the region's economy. Hopes for this New South began to bear fruit after World War I. United with the rest of the country in fighting the war, increasingly integrated into broader national and international economic systems, and with its own industries—especially textiles—growing rapidly, a New South did take shape. As Allen Tate put it, "With the war of 1914–1918, the South reentered the world—but gave a backward glance as it stepped over the border."

That backward glance was the inspiration, Tate believed, for a Southern literary renaissance during the interwar years. Tate himself was a leading figure in that renaissance, and he helped organize the collection of essays—or symposium, as it was called—published in 1930 as *I'll Take My Stand: The South and the Agrarian Tradition*. The essays in the collection were wide-ranging and the contributors varied in academic and professional background. The volume explored prospects for the regional economy, the place of the arts in an industrial society, education in a time of change, the South's religious legacy, and the tradition of Southern manners. The twelve essayists found they were united by an inclination to defend the value of the South's traditional society, rooted as it largely was in agriculture. They also shared a deep distrust of the growing power of American industrialism and distaste for the society that was shaped by that economy.

Although the contributors came from various disciplines, there was a strongly literary cast to the group. Most of the core authors were creative writers who are remembered as poets (in the case of John Crowe Ransom, Donald Davidson, Allen Tate, and Robert Penn Warren) or authors of fiction (notably Andrew Lytle and Warren). Many also enjoyed distinguished careers as literary critics, with Ransom, Tate, and Warren being founding spirits of the New Criticism that dominated literary theory during the middle of the twentieth century. These were prominent American intellectuals.

The core members had shared a long history before they worked together on *I'll Take My Stand*. Ransom, after studying at Vanderbilt as an undergraduate, had returned years later to become an influential teacher there. Davidson, Tate, and Warren were his students. These four were among the Fugitives, a literary group that gathered to discuss poetry, critique each other's poems, and eventually to publish a literary quarterly, *The Fugitive*, that won admirers far beyond the South.

When publication of *The Fugitive* ceased in 1925, Ransom and Davidson continued to correspond with Tate, who had moved to New York. From their different vantage points, each grew increasingly concerned about assaults on the South, which were coming from the likes of journalist H. L. Mencken. These attacks crested in the summer of 1925 during the Scopes "Monkey Trial," in which a high school science teacher, John Scopes, was prosecuted for teaching the theory of evolution in defiance of

Tennessee state law. The trial, argued in part by celebrity lawyers Clarence Darrow (for the defense) and William Jennings Bryan (for the prosecution), came to signify in the national press the conflict between Southern ignorance and modern, scientific thinking. Wrote Tate in 1927, "We must do something about Southern history and the culture of the South." Plans for what would become *I'll Take My Stand* grew out of this correspondence, and these core activists began to recruit other contributors.

In writing about the evolution of the ideas behind *I'll Take My Stand*, Donald Davidson remembered that as the project was taking shape, he and the others began to turn common assumptions about the South on their collective head. Perhaps, as Davidson put it, "in its very backwardness the South had clung to some secret which embodied, it seemed, the precise elements out of which its own reconstruction—and possibly even the reconstruction of America—might be achieved. With American civilization, ugly and visibly bent on ruin, before our eyes, why should we not explore this secret?" Eventually this line of reasoning brought them to the question of the South's distinctiveness as an agrarian region. "By this route we came at last to economics and so found ourselves at odds with the prevailing schools of economic thought. These held that economics determines life and set up an abstract economic existence as the governor of man's effort. We believed that life determines economics, or ought to do so, and that economics is no more than an instrument, around the use of which should gather many more motives than economic ones. The evil of industrial economics was that it squeezed all human motives into one narrow channel and then looked for humanitarian means to repair the injury."

Ransom associated industrialization with a restless acquisitiveness incompatible with healthy culture. He criticized what he called the pioneering spirit in the economy. It was in this spirit that newcomers to any land set out to conquer and subdue it for human use. In the normal course of affairs, the pioneering stage would be left behind for a more stable, settled phase, in which manners, cuisine, conversation, and the finer arts could evolve. But in America, Ransom suggested, the pioneering spirit seemed never to have loosened its hold on the imagination, with dire results: "The American progressive principle was like a ball rolling down the hill with an increasing momentum; and by 1890 or 1900 it was clear to any intelligent Southerner that it was a principle of boundless aggression against nature that could hardly offer much hospitality to a society devoted to the arts of peace."

The early ambitions of the Southern Agrarians, as they came to be called, went well beyond the publication of *I'll Take My Stand*. Tate, in 1929, proposed the formation of a kind of academy that could "set forth, under our leading idea, a complete social, philosophical, literary, economic, and religious system." The proposed academy would communicate its beliefs through a newspaper, a weekly magazine, and a quarterly journal. None of this came to pass, but the Agrarians did continue to press their case. In the wake of publication of *I'll Take My Stand*, Ransom and Davidson debated the merits of industrialization and agrarianism against proponents of the New South. These public debates drew large crowds—as many as thirty-five hundred in Richmond and a thousand or more in New Orleans and Atlanta. Core contributors to *I'll Take My Stand* also found a sympathetic editor, Seward Collins at the *American Review*, who published dozens of articles in the early and mid-1930s in which Ransom, Tate, Davidson, and others extended the arguments presented in the book. And several of the Agrarians, joined by others, contributed to the 1936 collection *Who Owns America?* which served as a kind of follow-up to *I'll Take My Stand*.

By the late 1930s, most of the core contributors to *I'll Take My Stand* had drifted away from social and political commentary to focus again on their literary work. Over time, their reflections on the development of the symposium and, more generally, on their agrarian activism shed even more light on what agrarianism meant to them and what they hoped to achieve. Their comments also highlight differences in the aims and understandings of key participants.

Allen Tate, who still called himself an Agrarian in a 1961 interview, emphasized the consistency between his agrarianism and his literary thought. In both, his primary concern was with the damage modernity was doing to the individual and to culture, considered in all their (potential) fullness. By contrast, Donald Davidson had hoped the Agrarian movement would have direct political results. In 1942, Tate wrote to Davidson: "You evidently believe that agrarianism was a failure; I think it was and *is* a very great success; but then I never expected it to have any political influence. It is a reaffirmation of the humane tradition, and to reaffirm that is an end in itself. Never fear: we shall be remembered when our snipers are forgotten. . . . We live in a bad age in which we cannot give our best; but no age is good."

Like Davidson, Andrew Lytle also believed their movement missed an opportunity to achieve practical results. The Agrarian with the most direct experience in agriculture, he believed that family farming was crucial for the health of a society, as the title of his essay for *Who Owns America?* "The Small Farm Secures the State," implied. Fifty years after the publication of *I'll Take My Stand,* Lytle still held to an agrarian vision that emphasized the importance of farms in a decentralized economy. "At the time we wrote there were enough families living on the land and enough privately owned businesses in small towns and cities to counterbalance the great industrial might, which was a fact and had to be reckoned with. If our proposal had been listened to, this necessary industry might have been contained, might not have grown into the only idea of the kind of life everybody must be forced to accept. A family, and I mean its kin and connections, too, thrives best on some fixed location which holds the memories of past generations by the ownership of farms or even family businesses."

John Crowe Ransom also viewed the Southern Agrarian movement in practical as well as aesthetic terms, but his thinking evolved in another direction. Around the time he worked on *I'll Take My Stand,* he also attempted a formal study of economics, hoping to advance the Agrarian cause on that front. He believed for a time that the movement could set the stage for policy changes that would improve the place of the small farmer in the nation's economy.

Only for a time, though. Unlike his closest Agrarian collaborators, Ransom would clearly distance himself from the movement, renouncing his earlier position in a 1945 essay, "Art and the Humane Economy." In it he commented on postwar proposals that aimed to reestablish the German economy on an agricultural basis. "Once I should have thought there could have been no greater happiness for a people, but now I have no difficulty seeing it for what it is meant to be: a heavy punishment. Technically it might be said to be an inhuman punishment, in the case where the people in the natural course of things have left the garden far behind."

Whatever the hopes of its authors may have been, the publication of *I'll Take My Stand* did not change the trajectory of Southern history in any measurable way. The movement did have its effects, however. One was semantic and, like so much of the Southern Agrarian project, controversial. Historian Thomas P. Govan wrote in 1964 about the "uses and abuses" of the terms *agrarian* and *agrarianism*. He was troubled by a growing imprecision and elasticity in their usage, especially during the twentieth century. These terms had once been narrowly legal, deriving from the Latin *lex agraria,* the Roman laws concerned with the uses of public land, such as the redistribution policies pursued by the Gracchi. Likewise, eighteenth- and nineteenth-century "agrarians"—most notably Thomas Paine in *Agrarian Justice*— espoused radical land reform, advocating systematically breaking up large private estates to make land available to all. Govan noted that figures often described as "agrarian" in our current expansive usage, including Thomas Jefferson himself, never embraced the term because of its strident anti-property implications.

Govan worried that the term *agrarian* had lost specificity. It had devolved, he said, to the point where it carried a "penumbra of undefined meaning, an implication of hostility to industry, commerce, and finance." He didn't place all the blame on the Southern Agrarians, but did hold them responsible for fostering that misuse by the public. Yet in the twentieth century the widespread adoption of the term with its new connotations suggests that it has met a need, and it has passed into common usage. Thomas Jefferson may not have been an agrarian in his own time, but he is one now. It is useful for neoagrarians to remember the original meaning of the word, however, and to continue to ask themselves how citizens in an industrial society are to gain adequate access to the land. How widespread an ownership of land, and in what forms, do today's agrarians propose?

I'll Take My Stand generated more substantial controversies as well. One charge is that the Southern Agrarians egregiously misrepresented—"whitewashed" may be an apt term—the often grim reality of rural life in the South and the degree to which it ever resembled an ideal agrarian society of small yeomen farmers. A second charge, following from that, is that they (and by extension all agrarians) ignored the extent to which industrial development delivers humanity from the miseries that accompany agricultural life.

Indeed, there was a contradiction near the heart of the Southern Agrarian project. The Agrarians insisted that culture and economy were intimately linked. Yet the culture they defended rested on an economy markedly less in line with their agrarian ideals than they let on. In the early twentieth century, the South, particularly the Deep South, remained largely devoted to the socially and environmentally rapacious extraction of industrial commodities, especially lumber and cotton—and Appalachia was beginning to learn all about coal. The South may have been more agricultural and rural than other regions, but it was not Jeffersonian—except in the painful sense that Jefferson himself embodied many of the same contradictions about land ownership, race relations, and economic viability.

The main instruments of industrial extraction in the South were not machines, of course, but small tenant farmers, and especially black sharecroppers. An insight into the African American experience in the Southern countryside during the first half of the twentieth century is provided by *All God's Dangers: The Life of Nate Shaw*. In this book, a Harvard scholar recorded the life history of "Nate Shaw," whose real name was Ned Cobb, using his story to capture the hardship and sheer terror of life for many in the rural South. But it also conveyed something else: a remarkable set of agrarian values. In their love for and knowledge of the land, their dedication to hard work, their self-reliance, their aspiration to independent freehold ownership, and their willingness to stand up to oppression to defend that aspiration, Cobb and his family exemplified the yeoman ideal. Something vital was lost when this African American agrarian culture fled the rural South, but it was something that had never been granted the freedom to flourish.

In addition to racism, there was a second problem in aligning agrarian values with the backwardness of the rural South, as critics would charge after the publication of *I'll Take My Stand*. Resisting industrial progress appeared to be an endorsement of squalor. Gerald Johnson, a journalist who reviewed the book, pointed out that the South had already tried an agrarian program of sorts during the latter decades of the nineteenth century by failing to embrace industrialization. "And what did she get out of it? The South of 1900 is your answer—a hookworm-infested, pellagra-smitten, poverty-stricken, demagogue-ridden 'shotgun civilization.'" Johnson added that the lives of too many Southern farmers "bore a remarkable resemblance to the lot of the Russian serf prior to the imperial ukase of 1861." A similar critique leveled by H. L. Mencken is included in this volume.

In defense of the Southern Agrarians, they wrote at the beginning of the Great Depression, when the ability of industrial society to deliver the goods to wage laborers and keep them from poverty had also collapsed, leaving many without the alternative of independent self-reliance. And if the alternative of agrarian yeoman society was not what really prevailed across the rural South, it is perhaps what *should* have prevailed, had the Agrarians' values been followed. Furthermore, the Agrarians did not oppose all industrial development—they opposed the unchecked power of industrialism to govern human affairs. They also wrote at a time when it appeared that the rise of industrialism would lead everywhere to totalitarian states. They were prescient enough to see that even if the state did not end by centralizing all authority, industrial corporatism might arrive at a similar end. State power might be of some use in restraining corporate power, but it was as likely to end up serving corporate power more often than not. The Agrarians sought a countervailing force in the retention of a large population of independent farmers, and their culture, on the land. Neoagrarians continue to ask not how we can do without cities and industry, but what sort of agriculture and rural society is best for the countryside, and what set of values can best restrain and guide industrial power. The Southern Agrarians still make for provocative reading on that score.

Twelve Southerners

When plans for the I'll Take My Stand *symposium took shape, contributors began writing their essays before any agreement on general principles had been formulated. Sensing the need to provide a set of organizing ideas, key leaders of the Agrarian group, especially John Crowe Ransom and Donald Davidson, tried to hammer out a document that would tie all the planned essays together. Originally, the document presented a brief preamble followed by a list of seventeen specific claims and proposals. The first article stated, "The good life must be lived much closer to the land than the ruling American ideal permits." After asserting the "spiritual poverty" of the industrial age, this early draft offered a remedy that was "simple but radical: The cancellation of the Ideal of Industrial Progress."*

The final introduction differed in form from the first draft, but held to the central premises of the original articles. This final version of the introduction was written largely by John Crowe Ransom, in close consultation with Donald Davidson and Andrew Lytle.

From *I'll Take My Stand* (1930)

Introduction: A Statement of Principles

The authors contributing to this book are Southerners, well acquainted with one another and of similar tastes, though not necessarily living in the same physical community, and perhaps only at this moment aware of themselves as a single group of men. By conversation and exchange of letters over a number of years it had developed that they entertained many convictions in common, and it was decided to make a volume in which each one should furnish his views upon a chosen topic. This was the general background. But background and consultation as to the various topics were enough; there was to be no further collaboration. And so no single author is responsible for any view outside his own article. It was through the good fortune

Fugitives reunion, 1956 (from left to right, Allen Tate, Merrill Moore,
Robert Penn Warren, John Crowe Ransom, Donald Davidson).
Reproduced by permission of Vanderbilt University Special Collections
and University Archives, Nashville, TN.

of some deeper agreement that the book was expected to achieve its unity. All the articles bear in the same sense upon the book's title-subject: all tend to support a Southern way of life against what may be called the American or prevailing way; and all as much as agree that the best terms in which to represent the distinction are contained in the phrase, Agrarian *versus* Industrial.

But after the book was under way it seemed a pity if the contributors, limited as they were within their special subjects, should stop short of showing how close their agreements really were. On the contrary, it seemed that they ought to go on and make themselves known as a group already consolidated by a set of principles which could be stated with a good deal of particularity. This might prove useful for the sake of future reference, if they should undertake any further joint publication. It was then decided to prepare a general introduction for the book which would state briefly the common convictions of the group. This is the statement. To it every one of the contributors in this book has subscribed.

Nobody now proposes for the South, or for any other community in this country, an independent political destiny. That idea is thought to have been finished in 1865. But how far shall the South surrender its moral, social, and economic autonomy to the victorious principle of Union? That question remains open. The South is a minority section that has hitherto been jealous of its minority right to live its own kind of life. The South scarcely hopes to determine the other sections, but it does propose to determine itself, within the utmost limits of legal action. Of late, however, there is the melancholy fact that the South itself has wavered a little and shown signs of wanting to join up behind the common or American industrial ideal. It is against that tendency that this book is written. The younger Southerners, who are being converted frequently to the industrial gospel, must come back to the support of the Southern tradition. They must be persuaded to look very critically at the advantages of becoming a "new South" which will be only an undistinguished replica of the usual industrial community.

But there are many other minority communities opposed to industrialism, and wanting a much simpler economy to live by. The communities and private persons sharing the agrarian tastes are to be found widely within the Union. Proper living is a matter of the intelligence and the will, does not depend on the local climate or geography, and is capable of a definition which is general and not Southern at all. Southerners have a filial duty to discharge to their own section. But their cause is precarious and they must seek alliances with sympathetic communities everywhere. The members of the present group would be happy to be counted as members of a national agrarian movement.

Industrialism is the economic organization of the collective American society. It means the decision of society to invest its economic resources in the applied sciences. But the word science has acquired a certain sanctitude. It is out of order to quarrel with science in the abstract, or even with the applied sciences when their applications are made subject to criticism and intelligence. The capitalization of the applied sciences has now become extravagant and uncritical; it has enslaved our human energies to a degree now clearly felt to be burdensome. The apologists of industrialism do not like to meet this charge directly; so they often take refuge in saying that they are devoted simply to science! They are really devoted to the applied sciences and to practical production. Therefore it is necessary to employ a certain skepticism even at the expense of the Cult of Science, and to say, It is an Americanism, which looks innocent and disinterested, but really is not either.

The contribution that science can make to a labor is to render it easier by the help of a tool or a process, and to assure the laborer of his perfect economic security while he is engaged upon it. Then it can be performed with leisure and enjoyment. But the modern laborer has not exactly received this benefit under the industrial regime. His labor is hard, its tempo is fierce, and his employment is insecure. The first principle of a good labor is that it must be effective, but the second principle is that it must be enjoyed. Labor is one of the largest items in the human career; it is a modest demand to ask that it may partake of happiness.

The regular act of applied science is to introduce into labor a labor-saving device or a machine. Whether this is a benefit depends on how far it is advisable to save the labor. The philosophy of applied science is generally quite sure that the saving of labor is a pure gain, and that the more of it the better. This is to assume that labor is an evil, that only the end of labor or the material product is good. On this assumption labor becomes mercenary and servile, and it is no wonder if many forms of modern labor are accepted without resentment though they are evidently brutalizing. The act of labor as one of the happy functions of human life has been in effect abandoned, and is practiced solely for its rewards.

Even the apologists of industrialism have been obliged to admit that some economic evils follow in the wake of the machines. These are such as overproduction, unemployment, and a growing inequality in the distribution of wealth. But the remedies proposed by the apologists are always homeopathic. They expect the evils to disappear when we have bigger and better machines, and more of them. Their remedial programs, therefore, look forward to more industrialism. Sometimes they see the system righting itself spontaneously and without direction: they are Optimists. Sometimes they rely on the benevolence of capital, or the militancy of labor, to bring about a fairer division of the spoils: they are Coöperationists or Socialists. And sometimes they expect to find super-engineers, in the shape of Boards of Control, who will adapt production to consumption and regulate prices and guarantee business against fluctuations: they are Sovietists. With respect to these last it must be insisted that the true Sovietists or Communists—if the term may be used here in the European sense—are the Industrialists themselves. They would have the government set up an economic super-organization, which in turn would become the government. We therefore look upon the Communist menace as a menace indeed, but not as a Red one; because it is simply according to the blind drift of our industrial development to expect in America at last much the same economic system as that imposed by violence upon Russia in 1917.

Turning to consumption, as the grand end which justifies the evil of modern labor, we find that we have been deceived. We have more time in which to consume, and many more products to be consumed. But the tempo of our labors communicates itself to our satisfactions, and these also become brutal and hurried. The constitution of the natural man probably does not permit him to shorten his labor-time and enlarge his consuming-time indefinitely. He has to pay the penalty in satiety and aimlessness. The modern man has lost his sense of vocation.

Religion can hardly expect to flourish in an industrial society. Religion is our submission to the general intention of a nature that is fairly inscrutable; it is the sense of our rôle as creatures within it. But nature industrialized, transformed into cities and artificial habitations, manufactured into commodities, is no longer nature but a highly simplified picture of nature. We receive the illusion of having power over nature, and lose the sense of nature as something mysterious and contingent. The God of nature under these conditions is merely an amiable expression, a superfluity, and the philosophical understanding ordinarily carried in the religious experience is not there for us to have.

Nor do the arts have a proper life under industrialism, with the general decay of sensibility which attends it. Art depends, in general, like religion, on a right attitude to nature; and in particular on a free and disinterested observation of nature that occurs only in leisure. Neither the creation nor the understanding of works of art is possible in an industrial age except by some local and unlikely suspension of the industrial drive.

The amenities of life also suffer under the curse of a strictly-business or industrial civilization. They consist in such practices as manners, conversation, hospitality, sympathy, family life, romantic love—in the social exchanges which reveal and develop sensibility in human affairs. If religion and the arts are founded on right relations of man-to-nature, these are founded on right relations of man-to-man.

Apologists of industrialism are even inclined to admit that its actual processes may have upon its victims the spiritual effects just described. But they think that all can be made right by extraordinary educational efforts, by all sorts of cultural institutions and endowments. They would cure the poverty of the contemporary spirit by hiring experts to instruct it in spite of itself in the historic culture. But salvation is hardly to be encountered on that road. The trouble with the life-pattern is to be located at its economic base, and we cannot rebuild it by pouring soft materials from the top. The young men and women in colleges, for example, if they are already placed in a false way of life, cannot make more than an inconsequential acquaintance with the arts and humanities transmitted to them. Or else the understanding of these arts and humanities will but make them the more wretched in their own destitution.

The "Humanists" are too abstract. Humanism, properly speaking, is not an abstract system, but a culture, the whole way in which we live, act, think, and feel. It is a kind of imaginatively balanced life lived out in a definite social tradition. And, in the concrete, we believe that this, the genuine humanism, was rooted in the agrarian life of the older South and of other parts of the country that shared in such a tradition. It was not an abstract moral "check" derived from the classics—it was not soft material poured in from the top. It was deeply founded in the way of life itself—in its tables, chairs, portraits, festivals, laws, marriage customs. We cannot recover our native humanism by adopting some standard of taste that is critical enough to question the contemporary arts but not critical enough to question the social and economic life which is their ground.

The tempo of the industrial life is fast, but that is not the worst of it; it is accelerating. The ideal is not merely some set form of industrialism, with so many stable industries, but industrial progress, or an incessant extension of industrialization. It never proposes a specific goal; it initiates the infinite series. We have not merely capitalized certain industries; we have capitalized the laboratories and inventors, and undertaken to employ all the labor-saving devices that come out of them. But a fresh labor-saving device introduced into an industry does not emancipate the laborers in that industry so much as it evicts them. Applied at the expense of agriculture, for example, the new processes have reduced the part of the population supporting itself upon the soil to a smaller and smaller fraction. Of course no single labor-saving process is fatal; it brings on a period of unemployed labor and unemployed capital, but soon a new industry is devised which will put them both to work again, and a new commodity is thrown upon the market. The laborers were sufficiently embarrassed in the meantime, but, according to the theory, they will eventually be taken care of. It is now the public which is embarrassed; it feels obligated to purchase a commodity for which it had expressed no desire, but it is invited to make its budget equal to the strain. All might yet be well, and stability and comfort might again obtain, but for this: partly because of industrial ambitions and partly because the repressed creative impulse must break out somewhere, there will be a stream of further labor-saving devices in all industries, and the cycle will have to be repeated over and over. The result is an increasing disadjustment and instability.

It is an inevitable consequence of industrial progress that production greatly outruns the rate of natural consumption. To overcome the disparity, the producers, disguised as the pure idealists of progress, must coerce and wheedle the public into being loyal and steady consumers, in order to keep the machines running. So the rise of modern advertising—along with its twin, personal salesmanship—is the most significant development of our industrialism. Advertising means to persuade the consumers to want exactly what the applied sciences are able to furnish them. It consults the happiness of the consumer no more than it consulted the happiness of the laborer. It is the great effort of a false economy of life to approve itself. But its task grows more difficult every day.

It is strange, of course, that a majority of men anywhere could ever as with one mind become enamored of industrialism: a system that has so little regard for individual wants. There is evidently a kind of thinking that rejoices in setting up a social objective which has no relation to the individual. Men are prepared to sacrifice their private dignity and happiness to an abstract social ideal, and without asking whether the social ideal produces the welfare of any individual man whatsoever. But this is absurd. The responsibility of men is for their own welfare and that of their neighbors; not for the hypothetical welfare of some fabulous creature called society.

Opposed to the industrial society is the agrarian, which does not stand in particular need of definition. An agrarian society is hardly one that has no use at all for industries, for professional vocations, for scholars and artists, and for the life of cities. Technically, perhaps, an agrarian society is one in which agriculture is the leading vocation, whether for wealth, for pleasure, or for prestige—a form of labor that is pursued with intelligence and leisure, and that becomes the model to which the other forms approach as well as they may. But an agrarian regime will be secured readily enough where the superfluous industries are not allowed to rise against it. The theory of agrarianism is that the culture of the soil is the best and most sensitive of vocations, and that therefore it should have the economic preference and enlist the maximum number of workers.

These principles do not intend to be very specific in proposing any practical measures. How may the little agrarian community resist the Chamber of Commerce of its county seat, which is always trying to import some foreign industry that cannot be assimilated to the life-pattern of the community? Just what must the Southern leaders do to defend the traditional Southern life? How may the Southern and Western agrarians unite for effective action? Should the agrarian forces try to capture the Democratic party, which historically is so closely affiliated with the defense of individualism, the small community, the state, the South? Or must the agrarians—even the Southern ones—abandon the Democratic party to its fate and try a new one? What legislation could most profitably be championed by the powerful agrarians in the Senate of the United States? What anti-industrial measures might promise to stop the advances of industrialism, or even undo some of them, with the least harm to those concerned? What policy should be pursued by the educators who have a tradition at heart? These and many other questions are of the greatest importance, but they cannot be answered here.

For, in conclusion, this much is clear: If a community, or a section, or a race, or an age, is groaning under industrialism, and well aware that it is an evil dispensation, it must find the way to throw it off. To think that this cannot be done is pusillanimous. And if the whole community, section, race, or age thinks it cannot be done, then it has simply lost its political genius and doomed itself to impotence.

ANDREW NELSON LYTLE

Like several of the most prominent contributors to I'll Take My Stand, *Andrew Lytle was a child of the upper South, having been born in Murfreesboro, Tennessee, in 1902. He was educated at the Sewanee Military Academy before entering Vanderbilt University in the early 1920s. He arrived at Vanderbilt in time to take part in the tail end of the Fugitive gatherings, contributing one poem to the group's journal. His participation initiated long relationships with Ransom, Tate, Warren, and other Fugitives.*

Lytle had a long and varied career. After Vanderbilt, he attended the Yale School of Drama and, returning to the South, helped manage his father's farm, Cornsilk, where he tried to combine writing and agriculture. Eventually he focused on teaching and on his literary pursuits. He published his first book, a biography of Confederate cavalry leader Nathan Bedford Forrest, in 1931, and his last, a meditation on Sigrid Undset's Kristin Lavransdatter, *in 1992. In between, he published several novels, short stories, and volumes of literary criticism. He edited the* Sewanee Review *in the mid-1940s, then again from 1961 to 1972. He also taught creative writing and literature at the University of the South for many years, and at various other colleges and universities as a visiting scholar.*

His brief sojourn in the Northeast during the late 1920s convinced Lytle of the virtues of his native region and the disorders of the industrial society of the North. He was also infuriated by the way the South was portrayed during the Scopes trial, an anger he shared with other Fugitive friends. When Tate, Davidson, and Ransom raised the possibility of responding to the perceived insults, Lytle was enthusiastic. Other organizers were also glad to get Lytle's perspective, especially because of his deeper direct experience with farming and his knowledge of the South's rural culture.

The essay he wrote for the collection, "The Hind Tit," explored his concerns over the fate of the South's yeoman farmers and the folk culture they embodied. Lytle's local focus evolved over time to take in a much broader perspective. He came to see the disruptions brought on by modern technology as an affliction for the whole Western world, which he thought had entered a "satanic phase." Where industry and the modern economy dominated, as symbolized by the spread of the automobile, Lytle argued that communities dissolved and a spiritual dimension of life faded: "I can't believe that any society is strong which holds physical comfort as its quest."

From "The Hind Tit" (1930)

I

When we remember the high expectations held universally by the founders of the American Union for a more perfect order of society, and then consider the state of life in this country today, it is bound to appear to reasonable people that somehow the experiment has proved abortive, and that in some way a great commonwealth has gone wrong.

There are those among us who defend and rejoice in this miscarriage, saying we are more prosperous. They tell us—and we are ready to believe—that collectively we are possessed of enormous wealth and that this in itself is compensation for whatever has been lost. But when we, as individuals, set out to find and enjoy this wealth, it becomes elusive and its goods escape us. We then reflect, no matter how great it may be collectively, if individually we do not profit by it, we have lost by the exchange. This becomes more apparent with the realization that, as its benefits elude us, the labors and pains of its acquisition multiply.

To be caught unwittingly in this unhappy condition is calamitous; but to make obeisance before it, after learning how barren is its rule, is to be eunuched. For those who are Southern farmers this is a particularly bitter fact to consider. We have been taught by Jefferson's struggles with Hamilton, by Calhoun's with Webster, and in the woods at Shiloh or along the ravines of Fort Donelson where the long hunter's rifle spoke defiance to the more accelerated Springfields, that the triumph of industry, commerce, trade, brings misfortune to those who live on the land.

Since 1865 an agrarian Union has been changed into an industrial empire bent on conquest of the earth's goods and ports to sell them in. This means warfare, a struggle over markets, leading, in the end, to actual military conflict between nations. But, in the meantime, the terrific effort to manufacture ammunition—that is, wealth—so that imperialism may prevail, has brought upon the social body a more deadly conflict, one which promises to deprive it, not of life, but of living; take the concept of liberty from the political consciousness; and turn the pursuit of happiness into a nervous running-around which is without the logic, even, of a dog chasing its tail.

This conflict is between the unnatural progeny of inventive genius and men. It is a war to the death between technology and the ordinary human functions of living. The rights to these human functions are the natural rights of man, and they are threatened now, in the twentieth, not in the eighteenth, century for the first time. Unless man asserts and defends them he is doomed, to use a chemical analogy, to hop about like sodium on water, burning up in his own energy.

But since a power machine is ultimately dependent upon human control, the issue presents an awful spectacle: men, run mad by their inventions, supplanting themselves with inanimate objects. This is, to follow the matter to its conclusion, a moral and spiritual suicide, foretelling an actual physical destruction.

The escape is not in socialism, in communism, or in sovietism—the three final stages industrialism must take. These change merely the manner and speed of the suicide; they do not alter its nature. Indeed, even now the Republican government and the Russian Soviet Council pursue identical policies toward the farmer. The Council arbitrarily raises the value of its currency and forces the peasant to take it in exchange for his wheat. This is a slightly legalized confiscation, and the peasants have met it by refusing to grow surplus wheat. The Republicans take a more indirect way—they raise the tariff. Of the two policies, that of the Russian Soviet is the more admirable. It frankly proposes to make of its farmers a race of helots.

We have been slobbered upon by those who have chewed the mad root's poison, a poison which penetrates to the spirit and rots the soul. And the time is not far off when the citizens of this one-time Republic will be crying, "What can I do to be saved?" If the farmers have been completely enslaved by that time, the echo to their question will be their only answer. If they have managed to remain independent, the answer lies in a return to a society where agriculture is practiced by most of the people. It is in fact impossible for any culture to be sound and healthy without a proper respect and proper regard for the soil, no matter how many urban dwellers think that their victuals come from groceries and delicatessens and their milk from tin cans. This ignorance does not release them from a final dependence upon the farm and that most incorrigible of beings, the farmer. Nor is this ignorance made any more secure by Mr. Haldane's prognostication that the farm's ancient life will become extinct as soon as science rubs the bottle a few more times. The trouble is that already science has rubbed the bottle too many times. Forgetting in its hasty greed to put the stopper in, it has let the genius out.

But the resumption by the farmer of his place of power in the present order is considered remote. Just what political pressure he will be able to bring upon the Republicans to better his lot is, at the moment, unknown. Accepting the most pessimistic view, the continued supremacy of this imperialism and his continued dependency upon it, his natural enemy, the wealth-warrior who stands upon the bridge of high tariff and demands tribute, he is left to decide upon immediate private tactics. How is the man who is still living on the land, and who lives there because he prefers its life to any other, going to defend himself against this industrial imperialism and its destructive technology?

One common answer is heard on every hand: Industrialize the farm; be progressive; drop old-fashioned ways and adopt scientific methods. These slogans are powerfully persuasive and should be, but are not, regarded with the most deliberate circumspection, for under the guise of strengthening the farmer in his way of life they are advising him to abandon it and become absorbed. Such admonition coming from the quarters of the enemy is encouraging to the land-owner in one sense only: it assures him he has something left to steal. Through its philosophy of Progress it is committing a mortal sin to persuade farmers that they can grow wealthy by adopting its methods. A farm is not a place to grow wealthy; it is a place to grow corn.

It is telling him that he can bring the city way of living to the country and that he will like it when it gets there. His sons and daughters, thoroughly indoctrinated with these ideas at state normals, return and further upset his equilibrium by demanding the things they grew to like in town. They urge him to make the experiment, with threats of an early departure from his hearth and board. Under such pressure it is no wonder that the distraught countryman, pulled at from all sides, contemplates a thing he by nature is loath to attempt . . . experimentation.

If it were an idle experiment, there would be no harm in such an indulgence; but it is not idle. It has a price and, like everything else in the industrial world, the price is too dear. In exchange for the bric-à-brac culture of progress he stands to lose his land, and losing that, his independence, for the vagaries of its idealism assume concrete form in urging him to over-produce his money crop, mortgage his land, and send his daughters to town to clerk in ten-cent stores, that he may buy the products of the Power Age and keep its machines turning. That is the nigger in the woodpile . . . keep the machines turning!

How impossible it is for him to keep pace with the procession is seen in the mounting mortgages taken by banks, insurance companies, and the hydra-headed loan companies which have sprung up since the World War. In spite of these acknowledged facts, the Bureau of Agriculture, the State Experimental Stations, farm papers, and county agents, all with the best possible intentions, advise him to get a little more progressive, that is, a little more productive. After advising this, they turn around and tell him he must curtail his planting. They also tell him that he (meaning his family) deserves motor-cars, picture shows, chain-store dresses for the women-folks, and all the articles in Sears-Roebuck catalogues. By telling him how great is his deserving, they prepare the way to deprive him of his natural deserts.

He must close his ears to these heresies that accumulate about his head, for they roll from the tongues of false prophets. He should know that prophets do not come from cities, promising riches and store clothes. They have always come from the wilderness, stinking of goats and running with lice and telling of a different sort of treasure, one a corporation head would not understand. Until such a one comes, it is best for him to keep to his ancient ways and leave the homilies of the tumble-bellied prophets to the city man who understands such things, for on the day when he attempts to follow the whitewash metaphysics of Progress, he will be worse off than the craftsman got to be when he threw his tools away. If that day ever comes, and there are strong indications that it may, the world will see a new Lazarus, but one so miserable that no dog will lend sympathy enough to lick the fly dung from his sores. Lazarus at least groveled at the foot of the rich man's table, but the new Lazarus will not have this distinction. One cannot sit at the board of an insurance company, nor hear the workings of its gargantuan appetite whetting itself on its own digestive processes. . . .

II

On a certain Saturday, a group of countrymen squatted and lay about the Rutherford County court-house yard. . . . One remarked to the others that "as soon as a farmer begins to keep books, he'll go broke shore as hell."

Let us take him as a type and consider the life of his household before and after he made an effort to industrialize it. Let us set his holdings at two hundred acres, more or less—a hundred in cultivation, sixty in woods and pasture, and forty in waste land, too rocky for cultivation but offering some pasturage. A smaller acreage would scarcely justify a tractor. And that is a very grave consideration for a man who lives on thirty or fifty acres. If the pressure becomes too great, he will be forced to sell out and leave, or remain as a tenant or hand on the large farm made up of units such as his. This example is taken, of course, with the knowledge that the problem on any two hundred acres is never the same: the richness of the soil, its qualities, the neighborhood, the distance from market, the climate, water, and a thousand such things make life on every farm distinctly individual.

The house is a dog-run with an ell running to the rear, the kitchen and dining-room being in the ell, if the family does not eat in the kitchen; and the sleeping-rooms in the main part of the house. . . .

Over the front doorway is a horseshoe, turned the right way to bring luck to all who may pass beneath its lintel. The hall is almost bare, but scrubbed clean. At the

Thomas Anshutz, *The Way They Live*, 1879. Image copyright © The Metropolitan Museum of Art/Art Resource, New York.

back is a small stairway leading to the half-story. This is where the boys sleep, in their bachelorhood definitely removed from the girls. To the left is the principal room of the house. The farmer and his wife sleep there in a four-poster, badly in need of doing over; and here the youngest chillurn sleep on pallets made up on the floor.

The large rock fireplace is the center of the room. The home-made hickory chairs are gathered in a semicircle about it, while on the extreme left of the arc is a rough hand-made rocker with a sheep-skin bottom, shiny from use, and its arms smooth from the polishing of flesh, reserved always for "mammy," the tough leather-skinned mother of the farmer. Here she sets and rocks and smokes near enough for the draught to draw the smoke up the chimney. On the mantel, at one end, is dry leaf tobacco, filling the room with its sharp, pungent odor. A pair of dog-irons rests on the hearth, pushed against the back log and holding up the ends of the sticks which have burnt in two and fallen among the hot ashes. The fire is kept burning through the month of May to insure good crops, no matter how mild and warm its days turn out to be. The top rock slab is smoked in the middle where for generations the wind has blown suddenly down the chimney, driving heavy gusts to flatten against the mantel and spread out into the room. A quilting-frame is drawn into the ceiling, ready to be lowered into the laps of the women-folks when the occasion demands, although it is gradually falling into disuse. Beneath it, spreading out from the center of the floor, a rag rug covers the wide pine boards which, in turn, cover the rough-hewn puncheons that sufficed during the pioneer days. From this room, or rather from the hearth of this room, the life of the dwelling moves.

If this is the heart of the house, the kitchen is its busiest part. The old, open fireplace has been closed in since the war, and an iron range has taken its place. This much machinery has added to the order of the establishment's life without disrupting it. Here all the food is prepared, and the canning and preserving neces-sary to sustain the family during the winter is done. . . .

Before dawn the roosters and the farmer feel the tremendous silence, chilling and filling the gap between night and day. He gets up, makes the fires, and rings the rising bell. He could arouse the family with his voice, but it has been the custom to ring the bell; so every morning it sounds out, taking its place among the other bells in the neighborhood. Each, according to his nature, gets up and prepares for the day: the wife has long been in the kitchen when the boys go to the barn; some of the girls help her, while the farmer plans the morning work and calls out directions.

One or two of the girls set out with their milk-pails to the barn, where the cows have been kept overnight. There is a very elaborate process to go through with in milking. First the cow must be fed to occupy her attention; next, the milker kneels or sits on a bucket and washes the bag which will have gotten manure on it during the night (she kneels to the right, as this is the strategic side; the cow's foot is some-how freer on the left). After the bag is clean, the milking begins. There is always a variation to this ritual. When the calf is young, the cow holds back her milk for it; so the calf is allowed to suck a little at first, some from each teat, loosening the milk with uniformity, and then is pulled off and put in a stall until his time comes. There is one way to pull a calf off, and only one. He must be held by the ears and the tail at the same time, for only in this manner is he easily controlled. The ears alone, or the tail alone, is not enough.

This done, the milking begins. The left hand holds the pail, while the right does the work, or it may be the reverse. The hand hits the bag tenderly, grabs the teat, and closes the fingers about it, not altogether, but in echelon. The calf is then let out for his share. If he is young and there are several cows, it will be all that is left, for careful milkers do not strip the cow until the calf is weaned. The strippings are those short little squirts which announce the end, and they are all cream.

The milk is next brought back to the house, strained, and put in the well to cool. This requires a very careful hand, because if it happens to spill, the well is ruined. The next step is to pour up the old milk and let it turn—that is, sour—for churning. Some will be set aside to clabber for the mammy whose teeth are no longer equal to tougher nourishment. What she does not eat is given to the young chickens or to the pigs.

After breakfast the farmer's wife, or one of the girls, does the churning. This process takes a variable length of time. If the milk is kept a long time before it is poured up, the butter is long in coming. Sometimes witches get in the churn and throw a spell over it. In that case a nickel is dropped in to break the charm. The butter, when it does come, collects in small, yellow clods on top. These clods are separated from the butter-milk and put in a bowl where the rest of the water is worked out. It is then salted, molded, and stamped with some pretty little design. After this is done, it is set in the well or the spring to cool for the table. The process has been long, to some extent tedious, but profitable, because insomuch as it has taken time and care and intelligence, by that much does it have a meaning.

Industrialism gives an electric refrigerator, bottled milk, and dairy butter. It takes a few minutes to remove it from the ice to the table, while the agrarian process has taken several hours and is spread out over two or three days. Industrialism saves time, but what is to be done with this time? The milkmaid can't go to the movies, read the sign-boards, and go play bridge all the time. In the moderate circumstances of this family, deprived of her place in the home economy, she will be exiled to the town to clerk all day. If the income of the family can afford it, she remains idle, and therefore miserable.

The whole process has been given in detail as an example of what goes on in every part of an agrarian life. The boys, coming in to breakfast, have performed in the same way. Every morning the stock must be fed, but there is always variety. They never shuck the same ears of corn, nor do they find the mules in the small part of the stall, nor the hogs in the same attitudes, waiting to be slopped. The buckets of milk did not move regularly from cow to consumer as raw material moves through a factory. The routine was broken by other phenomena. Breakfast intervened. One morning the cow might kick the pail over, or the milkmaid might stumble over a dog, or the cow come up with a torn udder. It is not the only task she performs, just as feeding the stock is not the only task done by the boys. The day of each member of the family is filled with a mighty variety. . . .

After the midday meal is over the family takes a rest; then the men go back to the fields and the women to those things yet to be done, mending clothes, darning, knitting, canning, preserving, washing or ironing or sewing. By sundown they are gathered about the supper table, and afterward set before the fire if it is winter, or upon the porch in warmer weather. One of the boys will get out his guitar and play "ballets" handed down from father to son, some which have originated in the new country, some which have been brought over from the Old World and changed to fit the new locale. Boys from the neighborhood drop in to court, and they will jine in, or drive away with the gals in hug-back buggies. If they are from another neighborhood, they are sure to be rocked or shot at on the way over or on the way home.

If the gathering is large enough, as it is likely to be when crops are laid by, it will turn into a play-party. Most of these games practiced by the plain people have maintained the traditions brought from England and Scotland, while the townsmen lost their knowledge of them in a generation. For example, "The Hog Drovers" is a version of the English folk-game, "The Three Sailors." The Southern country, being largely inland, could only speculate upon the habits of sailors, but they knew all about the hog drovers. Every year droves of razorbacks, with their eyelids sewed together to hinder them from wandering off into the woods, were driven ten or eleven miles a day toward the Eastern markets. They would be stopped at private farms along the route, where pens had been put up to receive them, to feed. The drovers, nomadic and as careless as sailors, could not be made to keep promises. Parents, therefore, were careful of their daughters.

The game comes from, and is a copy of, the life of the people. A boy seats himself upon a chair in the middle of the room with a gal in his lap. He is the head of the house, and she is his daughter. The other gals are seated around the walls, waiting their turns; while the boys, representing the hog drovers, enter two abreast in a sort of a jig, singing the first stanza:

> "Hog drovers, hog drovers, hog drovers we air,
> A-courtin yore darter so sweet and so fair,
>> Can we git lodgin' here, oh, here,
>> Can we git er-lodgin' here?"

They stop in front of the old man, and he answers:

> "Oh, this is my darter that sets by my lap,
> And none o' you pig-stealers can git her from pap,
>> And you can't git lodgin' here, oh, here,
>> And you can't git er-lodgin' here."

The boys then jig about the chair, singing:

"A good-lookin' darter, but ugly yoreself—
We'll travel on further and sit on the shelf,
And we don't want lodgin' here, oh, here,
    And we don't want er-lodgin' here."

They jig around the room, then return. The old man relents. Possibly it has as its genesis a struggle between greed and the safety of his daughter's virtue:

"Oh, this is my darter that sets by my lap,
And Mr. *So-and-so* can git her from pap
    If he'll put another one here, oh, here,
    If he'll put another one here."

The boy who is named jigs to one of the gals, brings her to the old man, takes his darter to the rear of the line, and the game starts over. After every couple has been paired off, they promenade all and seek buggies or any quiet place suitable for courting. . . .

Besides these play-parties people pleasured themselves in other ways. There were ice-cream socials, old-time singings, like the Sacred Harp gatherings, political picnics and barbecues, and barn dances. All of these gatherings which bring the neighborhood together in a social way are unlike the "society" of industrialism. Behind it some ulterior purpose always lurks. It becomes another province of Big Business and is invaded by hordes of people who, unable to sell themselves in the sterner marts, hope to catch their prey in his relaxed moments and over the tea tables make connections which properly belong to the office. This practice prostitutes society, for individuals can mingle socially from no motive except to enjoy one another's company.

JOHN CROWE RANSOM

John Crowe Ransom was born in Pulaski, Tennessee, in 1888, the son and grandson of Methodist ministers. He was the third of five children in a close-knit family with a strong interest in books and learning. Throughout all the years of his schooling, Ransom discussed his growing interests in literature and philosophy with his father, who provided a sympathetic sounding board, even as his thoughts strayed from Christian orthodoxy.

Ransom's intellectual inclinations led him to undergraduate studies at Vanderbilt, then on to Oxford as a Rhodes scholar from 1910 to 1913. At Oxford, he studied in the Greats program, which focused on classical literature and philosophy. He served in the army during the First World War, and by the time of his service he had begun writing poetry, sharing the poems with fellow Southerner and soldier Donald Davidson. Poems

About God, *the first of Ransom's published volumes, appeared in 1919. Over the next ten years, he published several more books of poetry, building a strong reputation and attracting admirers including Robert Frost and Robert Graves.*

By the late 1920s, Ransom's impulse to write poetry was waning, but his interest in political and social matters was burning bright. In an important departure from his previous work, he wrote God without Thunder, *an "unorthodox defense" of religious orthodoxy. In the book, Ransom argued that science was of limited value and that myth and literature could capture and communicate aspects of reality that eluded the abstractions of the scientific method.*

In God without Thunder, *Ransom defended the reality and importance of aesthetic and cultural dimensions of human experience, and he continued to do so in his agrarian writings. His agrarianism grew out of a desire to protect the integrity of the still-agricultural South and its culture against the encroachments of modern industry and commerce, which, in Ransom's eyes, entailed an economic reductionism that was deeply entwined with scientific reductionism.*

As a corollary to his agrarianism, Ransom, along with Davidson and others, also defended regionalism as an organizing principle for addressing the political, economic, and cultural dangers facing the United States in the 1930s. A firm believer in the physical realities of given locations, Ransom suggested that regional integrity, necessarily dependent on agriculture, should be acknowledged and protected for the fullest development of the nation's people. "The Aesthetic of Regionalism," excerpted here, appeared in the American Review.

From "The Aesthetic of Regionalism" (1934)

The eastbound train out of Albuquerque, climbing into the mountains, winds through dry and scrubby country which has a certain fascination for green visitors from the green regions and looks incapable of supporting human life. This visitor was going to pull down his window shade and try simply to keep cool, when he was surprised by the sight of human habitation after all, and on a rather large scale: a populous Indian pueblo. A second appeared presently, and then another. One displayed a very good church, but all were worth passing that particular day for this reason: it was threshing time. On the outskirts of each town were the threshing-floors, evidently of home-made concrete and belonging each to a family or unit of the tribal economy. On the floors Indians were beating out the grain; on some the work was nearly done, the grain had been separated from the chaff and lay in a golden pile. The threshers were old and young, of both sexes, and beautifully arrayed. They laughed, and must have felt pleased with their deities, because the harvest was a success, and bread was assured them for the winter.

So this was regionalism; flourishing on the meanest capital, surviving stubbornly, and brilliant. In the face of the efforts of the insidious white missions and the aggressive government schools to "enlighten" these Indian people, their culture persists, though for the most part it goes back to the Stone Age, and they live as they always have lived. It may be supposed that they find their way of living satisfactory, and are so far from minding it that they prefer it above others, receiving from it the two benefits which a culture can afford. First, the economic benefit; for they live where white men could scarcely live, they have sufficient means, and they are without that special insecurity which white men continually talk about, and which has to do with such mysterious things as the price of wheat; they thresh, bake, and eat their own grain and do not have to suffer if they cannot sell it. And second, a subtler but scarcely less important benefit in that their way of living is pleasant; it "feels" right, it has aesthetic quality. As a matter of fact, the Indian life in that one animated scene appeared to the philosophical regionalist one to be envied by the pale-faces who rode with him in painful dignity on the steel train, reflecting upon private histories and futures, but neither remembering nor expecting anything so bright and charming as this. . . .

Regionalism is as reasonable as non-regionalism, whatever the latter may be called: cosmopolitanism, progressivism, industrialism, free trade, interregionalism, internationalism, eclecticism, liberal education, the federation of the world, or simple rootlessness; so far as the anti-regional philosophy is crystallized in such doctrines. Regionalism is really more reasonable, for it is more natural, and whatever is natural is persistent and must be rationalized.

The reasonableness of regionalism refers first to its economic, and second to its aesthetic.

A regional economy is good in the sense that it has always worked and never broken down. That is more than can be said for the modern, or the interregional and industrial economy. Regionalism is not exactly the prevalent economy today; it has no particular status in Adam Smith's approach to economic theory, which contemplates free trade, and which has proved very congenial to the vast expansions of the nineteenth and twentieth centuries; therefore regionalism suffers a disability. Yet just now, by reason of the crash of our non-regional economy, it tends to have its revival. Of the two economies, the regional is the realistic one. The industry is in sight of the natural resources of the region and of its population. The farmers support themselves and support their cities; and the city merchants and manufacturers have their eyes on a local market and are not ambitious to build up trade with the distant regions; perhaps it occurs to them that an interregional or world trade cannot be controlled. The quantity and quality of world trade which a given community carried on even as late as 1900 are probably changed beyond recognition now, for a great variety of reasons, of which some were predictable and others were not; but at any rate a community can be badly hurt by the storm, if it chooses to fish in the ocean. Regionalism offers an economy as safe as it is modest.

Now it must be great fun to produce on a grand scale, so long as there is consumption for what you produce. The philosophical regionalist is quite disposed to grant that, and to concede the importance of the producer's having fun. But too much fun runs to mischief. It is agreed now that producers' fun must be curtailed, and producers regulated, as if they were irresponsible boys, unable to be trusted with their freedom, and with their grandiose concept of trade as something which will always love them and take care of their production. The interregional business men of the future will not look like joyous producers so much as communistic ants and wasps; and as between the economy of big business and interregionalism, with its privations, and the old regional economy in which producers had every reason to be realistic, and could be left to their own discretion, there is indeed some show of reason for the latter.

The aesthetic of regionalism is less abstract, and harder to argue. Preferably it is a thing to try, and to feel, and that is what it is actually for some Europeans, and for the Indians of our Southwest. They do not have to formulate the philosophy of regionalism. But unfortunately regionalism for white America is so little an experience that it is often obliged to be a theory.

Coming to the theory, the first thing to observe is that nature itself is intensely localized, or regional; and it is not difficult to imagine that the life people lead in one of the highly differentiated areas of the earth's surface is going to have its differences also. Some persons, with a sociological bias, suppose that the local peculiarities of life and custom, for example in the Southern highlands, are due to the fact that the population is old and deeply inbred, and has developed a kind of set because it has been out of communication with the world. Other persons, who are economists, think at once of the natural resources of the region, and the sort of subsistence it affords to its population, and find there the key to the cultural pattern. Both must be right; regionalism is a compound effect with two causes. But the primary cause is the physical nature of the region. A region which is physically distinct supports an economic unit of society; but its population will have much more of "domestic" trade than of foreign, and it will develop special ways and be confirmed in them.

As the community slowly adapts its life to the geography of the region, a thing happens which is almost miraculous; being no necessity of the economic system, but a work of grace perhaps, a tribute to the goodness of the human heart, and an event of momentous consequence to what we call the genius of human "culture." As the economic patterns become perfected and easy, they cease to be merely economic and become gradually aesthetic. They were meant for efficiency, but they survive for enjoyment, and men who were only prosperous become also happy.

The first settlers in a region are occupied with its conquest, and driven by a pure economic motive. Human nature at this stage is chiefly biological, and raw; physical nature, being harried and torn up by violence, looks raw too. But physical nature is perfectly willing to yield to man's solicitations if they are intelligent. Eventually the economic pattern becomes realistic, or nicely adapted to the bounty which nature is prepared in this region to bestow. It is as if man and nature had declared a truce and written a peace; and now nature not only yields up her routine concessions, but luxuriates and displays her charm; and men, secured in their economic tenure, delight in this charm and begin to represent it lovingly in their arts. More accurately, their economic actions become also their arts. It is the birth of natural piety: a transformation which may be ascribed to man's intuitive philosophy; by religious persons, such as Mary Austin, to the operation of transcendental spirit in nature, which is God. It is certainly the best gift that is bestowed upon the human species. The arts make their appearance in some ascending order, perhaps indicated like this: labour, craft, and business insist upon being transacted under patterns which permit the enjoyment of natural background; houses, tools, manufactured things do not seem good enough if they are only effective but must also be ornamental, which in a subtle sense means natural; and the fine arts arise, superficially pure or non-useful, yet faithful to the regional nature and to the economic and moral patterns to which the community is committed. It is in this stage that we delight to find a tribe of Navajos, or some provincial population hidden away in Europe.

For now the expert travelers come through, saying, Here is a region with a regionalism, and this is a characteristic bona fide manifestation of human genius. The region is now "made" in the vulgar sense (useless to a philosophical regionalist) that the curious and eclectic populations of far-away capitals will mark it on their maps, collect its exhibits for their museums, and discuss it in their literary essays. But for the regionalists who live in the region it is made already, because they have taken it into themselves by assimilation.

The regionalists receive the benefit of regionalism, not the distant eclectics; it is they who have the piety, and for whom the objects and activities have their real or pious meaning. This piety is directed first towards the physical region, the nature who has always given them sustenance and now gives them the manifold of her sensibilia. It is also directed towards the historic community which has dwelt in this region all these generations and developed these patterns. It is their region and their community, and their double attachment might well seem too powerful, and too natural, and also too harmless, to excite the wrath of any reputed philosophers; or it may be the envy, if the philosophers are so abstract and intellectual that they have never sufficiently felt such attachments; yet, whatever the motive be, some philosophers do actually represent themselves as aggrieved by it. . . .

The machine economy, carried to the limit with the object of "maximum efficiency," is the enemy of regionalism. It always has been; not only at the present stage of affairs has the issue between them become really acute, and been raised specifically and publicly in many places; for example, in Southern communities, now agitated as to their proper alignment between the Southern "tradition" and the "new" industrialism. The industrialism is not new, but the awakening of the Southern communities to its menace is new.

The machine economy was bad enough in coming to America, where the regionalisms were at many different periods of growth, but it came to the perfected cultures of Europe with the disruptive force of a barbarian conquest, turning the clock back, cancelling the gains of many mellowing centuries. (Such strong terms will apply of course to those regions which sooner or later allowed the machine economy to take charge of things.) It is no wonder that a good many pious European thinkers have been appalled by a sort of havoc which was much less visible on this side, and which the pious American thinkers, if any, have therefore been at much less pains to think about.

The new economy restored to the act of labour the tension from which it had delivered itself so hardly and so slowly. It returned the labouring population, and in some degree the whole business population, to a strictly economic status; a status with which the Europeans were fairly unfamiliar, and which their history recorded only putatively as the possible status of serfs, or the possible status of the original savages fighting for subsistence; and a status in all respects more ignominious than that of pioneers and frontiersmen in America. For under this economy the labourer is simply preoccupied with tending his abstract machine, and there is no opportunity for aesthetic attitudes. And not much material for them, either, since it is now more and more the machine which makes the contact with nature and not the man. But most of the machines are concerned with processing the materials taken out of the land, and they are housed in factories, while the factories are housed in cities. Therefore the landed population tends to lose its virtue, and the population as a whole becomes more and more urbanized. Now a city of any sort removes men from direct contact with nature, and cannot quite constitute the staple or normal form of life for the citizens, so that city life is always something less than regional. But the cities of a machine age are peculiarly debased. They spring up almost overnight, a Detroit, an Akron, a Los Angeles. They are without a history, and they are without a region, since the population is imported from any sources whatever; and therefore they are without a character.

Allen Tate

Allen Tate, one of the central figures among the Southern Agrarians, was born in Winchester, Kentucky, in 1899. His childhood was troubled by discord between his parents and by declining financial stability. The family moved frequently and Tate suffered from poor health, making an education difficult to pursue at times. Yet he was a precocious reader and excelled in his studies when he enrolled at Vanderbilt.

While at Vanderbilt, Tate developed a deep interest in literature, influenced in part by his teacher John Crowe Ransom. The friendship between the two, though rocky at times, lasted the rest of their lives. Tate entered into other long friendships at Vanderbilt that provided a foundation for his personal and professional life: with Donald Davidson, onetime roommate Robert Penn Warren, Andrew Lytle, and others. Tate began his distinguished career in poetry and literary criticism when Ransom invited him to join the Fugitive circle in Nashville.

His contribution to I'll Take My Stand *was the essay "Remarks on the Southern Religion," in which he explored themes that would occupy him throughout his life. He defended the religious imagination, as he would the poetic imagination, against the threats he believed were posed by industrialism, utilitarianism, and scientific reductionism. "Since there is, in the Western mind, a radical division between the religious, the contemplative, the qualitative, on the one hand, and the scientific, the natural, the practical on the other, the scientific mind always plays havoc with the spiritual life when it is not powerfully enlisted in its cause; it cannot be permitted to operate alone." For Tate, an agrarian society—one more stable, more rooted in the past, and less afflicted by industrial giantism—could nourish a culture in which the contemplative and the practical could flourish in their proper balance.*

After I'll Take My Stand *was released, Tate led the effort to publish a collection of essays that would build on the previous effort. The result was* Who Owns America? *Tate's essay for the latter book focused on the nature of property and its relation to individuals and the body politic. The essay, published in a slightly different form by the* American Review *in 1936, is excerpted here.*

From "Notes on Liberty and Property" (1936)

A LITERARY man is likely to think that property is quite simple: it is something you *own*. A second glance dispels the illusion. For property rights even in the simplest society are not absolute, but relative. And only by thinking of them as relative—subject to obligation, limitation, and even confiscation—is it possible to understand any kind of property, particularly the modern corporate variety. The simplicity of mere ownership does not bear analysis, even by a literary man.

If property is a relative term, so is liberty, and in exactly the same way. For to the extent to which a man controls the property by which his welfare is insured, is he possessed of liberty. It is impossible to think of liberty apart from property, property apart from liberty.

But liberty since the time of Marx has ceased to mean merely individual liberty. Here, then, the crucial issue between property and collectivism is whether any meaning that the word liberty has can be attributed to a group or is strictly the attribute of an individual who enjoys a certain control of the means of production. Can a group own property? If it can, may it be said that a group as large as a whole state can own it? This question, to be answered in any way that makes sense, must be looked at practically. For legal ownership does not always mean effective ownership. There is a point at which effective ownership ceases, although the legal fictions sustaining "property" may hold that beyond that point ownership endures. Effective ownership ceases at the point where a certain kind of effective control ceases. So a defender of the institution of private property will question not only the collectivist state but also large corporate property.

When the means of production are "owned" by the people, the control passes to the state. When a large part of the means of production, say one of the heavy industries, is owned by thirty thousand stockholders, the control of their ownership passes to a small group of men. In each case, collectivist ownership or corporate ownership, the property rights are legal. A large group then may legally own property. But is its ownership in any sense *effective*? A man owns a hundred thousand dollars' worth of stock in the United States Steel Corporation. His property rights in that corporation entitle him, apart from the largely fictitious "privileges" of such ownership, to a certain cash dividend. He may also sell his stock. The dividend and the privilege of selling the stock are his sole property rights. He cannot effectively question the amount of the dividend, nor can he dictate the policy of the corporation. He has no control over the portion of the means of production that he owns: he has no effective ownership.

In a collectivist state, in which private accumulations of capital are severely limited or forbidden, a man would not have a hundred thousand dollars to "invest." He would not be permitted to "save" the surplus income of his labor so that he could apply it to further production—he could not "let his money work for him." The collectivist state itself would accumulate the capital for future production: the individual would "own" that capital only in the sense that it would be there for him to apply his labor to. And if the units of production are properly balanced—corn with wheat, wheat with steel, steel with cosmetics, all with one another—he may expect a certain security. But he is not free. For it cannot be said that he in any sense controls the means of production. Control, the power to direct production and to command markets, is freedom under finance-capitalism.

The history of property in the United States is a struggle, from 1787 on, of one kind of property against another. Small ownership, typified by agriculture, has been worsted by big, dispersed ownership—the corporation. This must be kept steadily in mind. Without this fact it is easy to fall into the trap of the Big Business interests today, who are trying to convince people that there is *one* kind of property—just *property*, whether it be a thirty-acre farm in Kentucky or a stock certificate in the United States Steel Corporation. For if there is a contest merely between property and non-property—between real private property, as the average American understands it, and collectivism—the small owner will come to the support of the big corporation. And this is what the big corporation is using every means to make the small owner do.

The owner of the small farm, of the small factory, of the village store, owns a distinct kind of property. It is the familiar, historical kind. The reason why the "little man" confidently identifies his interests with the big interests is that he cannot imagine another kind of property than his own. He thinks that there is just "property," and that he has been less successful in accumulating it than Mr. Mellon. Of course the corporations know better. And they take advantage of the innocent rectitude of the owner of genuine property. There could not be a more grotesque proof of this intention of Big Business than the Liberty League, which uses liberty and property as slogans of what little liberty, what little property, they still have.

A movement to restore property to the citizens of this country must be based upon a broad distinction. The people must be shown the fundamental difference between private property, which means effective control by the owner, and corporate property, which usually means control by a clique of the many owners. The people must learn that corporate property is no less hostile to their interests than state, or collective, ownership—that the corporation is socially less responsible and perhaps eventually less efficient than collectivism.

The joint-stock corporation is the enemy of private property in the same sense as communism is. The collectivist state is the logical development of corporate owner-ship and, if it comes, it will signalize the final triumph of Big Business. "All the arts," said Walter Pater, "strive toward the condition of music." Corporate structure strives toward the condition of Moscow.

It will have reached that condition when the integration of the big monopolies requires still further concentration of control, in the hands of the state, and when ownership is so dispersed that it will be co-extensive with society as a whole.

What is effective ownership? It is not a metaphysical essence. Unlike liberty it is not a thing of the spirit. Common sense can recognize it. The effective ownership of property entails personal responsibility for the action of a given portion of the means of production. A true property system will be composed of a large proportion of owners whose property is not to be expressed solely in terms of exchange-value, but retains, for the owner, the possibility of use-value. Liberty is the power of the owner to choose between selling and using; not absolute power of choice, but choice relative to "conditions." As the freedom to "use" disappears, liberty begins to dis-appear. There has never been a society in which use-value has been the exclusive kind of value; no such society is being recommended now. But it must remain the basis of liberty.

A farmer owns a hog. It has two values—use-value and exchange- or market-value. The farmer's ownership is effective because he has the relatively free choice between killing the hog for his smoke-house and selling it on the market.

No such choice is open to the stockholder in the giant corporation. He holds a certificate of rights and expectations. In order to make good the rights and to fulfill the expectations of the "owners" the corporation has got to sell its commodity. Its concern is wholly with exchange-value. The "liberty" available to the corporation consists in the degree of *power* it has from time to time over the market. If it lacks this power it has no liberty whatever. The farmer, if he is protected by system of prices and distribution favorable to agriculture, enjoys a kind of liberty, the real kind, that can function apart from power over others. Now suppose a corporation makes tires. The market for tires in a given year is bad. It cannot eat the tires, nor can it operate enough cars of its own to consume them. Neither can the stockholder consume tires to the amount of "expectations" (dividends) due him. He may look at the pretty pictures on his stock certificate, and starve—or he may sell the stock at a price that he cannot influence in his favor.

It is not suggested that everybody make his own tires in a system that requires by law universal production for use. It is rather that finance-capitalism has become so top-heavy with a crazy jig-saw network of exchange-value that the individual citizen is wholly at the mercy of the shifting pieces of the puzzle at remote points where he cannot possibly assert his own needs and rights. This was not originally the American system. We began with the belief that society should be supported by agriculture, the most stable basis of society because it is relatively less dependent upon the market than any other kind of production.

Now this is elementary, and that is why Big Business does not include it in its propaganda today. Nor is Big Business interested in the responsibility of property, an attribute of ownership no less important than legal title itself. Responsibility is a function of control, and is necessary to effective ownership. A stock certificate is a symbol of a certain amount of capital working somewhere to produce a certain amount of exchange-value from which the "owner" hopes to derive a certain amount of profit. But dispersed ownership guided by concentrated control deprives the stockholder of the onerous privileges of responsibility. For control alone makes responsibility possible. It doesn't make it inevitable. The history of the big corporation shows that the men in control, having a remote, symbolic, paper connection with the owners, violate their responsibility in two ways—by milking the stockholders and by stealing from new capital issues. . . .

The struggle is not new. It is the meaning of American history. Hamilton and Jefferson are the symbols of the struggle. Its story is too well known to need re-telling. The next phase of the contest is doubtless near, but how the lines will be drawn it is impossible to predict. There are two general possibilities. We shall drift with the corporate structure of emasculated ownership until all trace of widespread control vanishes: that would be the tyrant state where corporations would be bigger than now and the two thousand men [who control the wealth of the country] reduced, say, to twenty. Or we shall return to real politics, resume our political character, and reassert the rights of effective ownership.

I am not suggesting that the American Telephone and Telegraph Company break up into jealous units, one for each county. But I do suggest, if the institution of property, corporate or private, is to survive at all, that we keep only enough centralization—of production as well as control—to prevent gross economic losses and the sudden demoralization of large classes of workers. Our objective has been the big corporation. We must change it. Our objective should be the private business. Corporations are not yet big enough to satisfy the corporations. Nor doubtless will property ever be widely enough distributed to please the absolute distributist. Distributed property should nevertheless be the aim.

Or put it this way: we have been mere economists, and now we have got to be political economists as well. Economics is the study of wealth, and it points ways to greater production of wealth. But political economy is the study of human welfare.

We have tried to produce as much wealth as possible. It cannot be denied that technology and corporate ownership have combined to increase staggeringly the aggregate wealth of modern states. But it is an equivocal wealth. The aggregate wealth of a nation may be stupendous, and the people remain impoverished. Let us assume what need not be true, that the total wealth of the property state would not be so great as the total wealth of the tyrant state. Yet the well-being of the people would be greater all round. So, if we are to achieve so desirable an end, we have got to add politics to economics in order to get a sum that we may, perhaps, call free citizens. For politics is—or should be—concerned with the welfare of persons, which is not always the same as their capacity to produce the maximum of goods.

The skeptics about the property state, and even some of its friends who misunderstand it, assume that we are advocating something like this: Every man must live on a farm, hew his own logs for his cabin, make his own clothing—after tending the sheep and growing the cotton—raise all his food, and refuse to have electric lights. I should like to use this derisive idyll as a boomerang. Even though production for use throughout society is now neither possible nor desirable, it should not be forgotten that the nearer a society is to production for use, the freer it is. We are not, therefore, crying for absolute liberty; we do want a little of it—as much as can be got when the majority of men own small units of production, whether factories or farms.

We do not ask everybody to live on a farm, nor—since we are allowing ourselves a little exchange-value in the property state—do we ask everybody to rush out as soon as he has read this essay and buy a small store, a small factory, a small automobile, or a small football team.

At present the buyer of a farm would probably, in a year, be glad to run from his debts, and give it to the insurance company; or should he not be glad to run, he had better try to be. A farm now is not necessarily property. We want to make it property again. A small grocery store may represent certain paper property rights, but in view of the six chain stores surrounding it, it does not represent the same property rights as it did a hundred years ago. We want the store to be property again. Altogether it does seem to be a modest wish. For it is not only necessary to buy the farm or the factory, it is necessary to keep it. It can be kept if we can restore property rights that unite again ownership and control.

H. L. Mencken

Several contributors to I'll Take My Stand *cited the Scopes trial and the negative publicity the South received at the time as a critical factor in turning their minds toward a defense of their native region. Nobody better exemplified the sensibility of the South's critics than the journalist and editor H. L. Mencken.*

Mencken was a native of Baltimore, born in 1880. He began his career in journalism in 1899 when he took a job at the Baltimore Morning Herald. *He moved on to the* Baltimore Sun, *becoming managing editor in 1906. He developed a range of interests that went far beyond the daily news, and over the next forty years he would become one of the nation's leading critical voices. As editor of* Smart Set *magazine and the* American Mercury, *and as the author of several books, Mencken commented on aspects of contemporary life ranging from politics to literature to music. Through his writings, Mencken crusaded, as his biographer Fred Hobson put it, "against American provincialism, Puritanism, and prudery— all of which he believed he found, to a degree larger than elsewhere, in the states below the Potomac and Ohio." In his criticism, he proved a reliable iconoclast.*

Mencken's notoriety among Southern readers was cemented in 1917 by the publication of his essay "The Sahara of the Bozart," a broadside against Southern culture. The essay infuriated many in the region, but some Southerners identified with Mencken's opposition to Southern parochialism and generally critical sensibility. At Vanderbilt University, student Allen Tate carried a copy around campus. Within a few years, he would be sending poems to Mencken for publication.

Less publicly, Mencken cultivated a critical movement within the South to do battle with Southern philistinism and fundamentalist religion. Through his correspondence, he encouraged novelists such as James Branch Cabell and offered the American Mercury *as a venue for writers whose views he found congenial, including journalist W. J. Cash, best known today for his book* The Southern Mind.

But it was for his very public scourging of the South during the spectacle over the teaching of evolution that took place in Tennessee in the summer of 1925, popularly known as the Scopes Monkey Trial, that Mencken is best remembered, and that scourging helped galvanize John Crowe Ransom, Allen Tate, and Donald Davidson to organize the effort to produce I'll Take My Stand. *After the book came out, Mencken wrote a brief, and not entirely unsympathetic, review of the volume in the* American Mercury. *Then, in 1935, he wrote a more extended consideration of the Southern Agrarians' movement in the* Virginia Quarterly Review, *from which this excerpt is taken.*

From "The South Astir" (1935)

Between 1857, when Hinton R. Helper published "The Impending Crisis," and 1909, when Edgar Gardner Murphy published "The Basis of Ascendancy," there was hardly any free and rational discussion, in the South itself, of the fundamental Southern problems. To be sure, a touch of realism got into an occasional newspaper editorial, and once in a great while some unsuccessful candidate for office, his withers wrung, broke out with an uncomfortable truth, but these were only minor eddies in a torrential current, the flow of which was all the other way. During that fateful half-century the Southern people remained immersed in the psychological mists that had arisen from Appomattox. A few simple dogmas sufficed for their thinking, such as it was. Beyond the borders of those dogmas it was an indecorum to go, and even a kind of treason. . . .

Unfortunately, liquidating a dogma is always a very tedious and hazardous business. All the fools holler for it loudly, for it saves them the trouble of thinking, and the minority of prudent men find it difficult to think in the din. Such a hollering of fools, with no counter uproar by the prudent, gives the noisiest of the former the appearance of being leaders of opinion, and if they go on long enough that appearance is converted into reality. This is what happened in the South. If there had been only one dogma to liquidate, the prudent might have summoned up their wits and speech, combined against it, and so disposed of it. But the South had a long series of them, ranging all the way from the completely plausible to the downright insane, and it was an almost impossible task to grapple with the whole lot. So no gladiator attempted it, and for long years the fools had everything their own way. Unimpeded, they gradually converted Southern politics into a hollow jousting of mountebanks, reduced Southern economics to the barber-shop level, filled Southern literature with the scents of musk, magnolia, and Jockey Club, and brought Southern theology to such an estate that it began to be almost indistinguishable from voodooism. The nadir was probably reached toward the turn of the century. This was the dismal time which saw the emergence of such mephitic shapes as Vardaman and Blease, the stampede to Bryan, the triumph of sentimentality in Southern letters, and the heyday of the evangelist.

How and why such pestilences end is always a bit mysterious. All one can say with any assurance is that there seems to be a law of diminishing returns for imbecility, just as there is for taxes. Over a long stretch of years the general appetite for buncombe and banality looks to be quite without limit, and then of a sudden the byways begin to rustle with whispers of doubt. Such uprisings, of course, seldom if ever originate among the actual folk; they are the artifacts of minorities in the vague regions above. . . .

Unfortunately, free thought is not necessarily wise thought, and something of the sort, I fear, must be said against some of the ideas which now float about the South. Their proponents, having declared war to the death upon nonsense, begin to fire so wildly that in many cases they also do execution upon sense, and in consequence the strategic effect of their bombardment is much less than it ought to be. I point, for a sufficient example, to the case of the so-called Agrarians, a band of earnest young revolutionaries chiefly resident in Tennessee. There can be no doubt whatever of their good faith, nor of their possession of a certain kind of intelligence. They are patriots in the best sense, and they tackle some of the fundamental Southern problems with eager and adventurous minds, and state their conclusions with great assurance. But only too often, alas, what they have to offer is only a little less absurd than the old balderdash that they seek to supplant.

Their primary error is that of social reformers at all times and everywhere: they conjure up a beautiful Utopia, prove that life in it would be pleasant, and then propose that everyone begin to move in tomorrow. Carried away by their ardor, they overlook the massive detail that it really doesn't exist, and cannot be imagined as existing in the actual world. How, indeed, could the South, even if it would, go back to the bucolic economy of the days before the Civil War, or set up any other economy of comparable outlines? It is tied to the industrial system so tightly that any cutting loose would have the effect, not of a mere revolution, but of a cataclysm, and out of that cataclysm nothing could emerge save chaos. Certainly it is impossible to imagine an orderly and self-sufficient society emerging. Did it do so in Russia? Far from it. In Russia the attempt to upset the industrial system produced so vast a congeries of evils that it had to be revived at once, and today it dominates the whole of Russian life in a way that would not be tolerated in any part of the Western World. It is childish to deny the existence of the thing itself because its name has been changed. The Russian who slaves away on those melodramatic power dams or in those endless foundries and rolling-mills is ten times the serf that any Southern linthead has ever been, and, since he has no voice in his own affairs, and may not even groan without risking his neck, he seems doomed to sweat in his chains until the megalomaniacal libido of his masters is satisfied, or another cataclysm somehow delivers him. He went into Utopia dreaming of forty acres and a mule, but what he has found there is only a forced job at meager wages, without any intervals for political campaigns, revivals, or strikes. Such is industrialism after it has been reënacted with Marxian amendments.

Certainly it would be silly to argue that the variety prevailing in the South is so bad, or even half so bad. That it needs an occasional overhauling is plain enough, and that it should be watched pretty sharply at all times is also evident, but that it is incurably inimical to the common weal and ought to be abolished forthwith is surely at least doubtful. Strike a balance between its costs and its benefits, and you will find that the latter overtop the former immensely, estimated by any rational standard. It may be, as the brethren say, that it tends to accumulate large and perhaps dangerous fortunes, to give money a preponderant influence in politics, and to foster the growth of a disinherited proletariat; but at the same time it must be manifest that it also tends, in spite of occasional depressions, to increase the general wealth, to set up salutary impediments to demagogy, and to offer a new hope to an already disinherited peasantry. Wherever it has got the firmest lodgment, as in the North Carolina Piedmont, you will not only find a higher level of physical well-being than in the agrarian areas, but also a higher tolerance of ideas. Even the Agrarian Habakkuks themselves are the clients of industrialism, which supplies them generously with the canned-goods, haberdashery, and library facilities that are so necessary to the free ebullition of the human intellect. Left to the farmers of Tennessee, they would be clad in linsey-woolsey and fed on sidemeat, and the only books they could read would be excessively orthodox.

Thus a note of falsetto gets into their revelations of the New Jerusalem. If they argued for the multiplication of subsistence farms in the South, and let it go at that, no one would challenge them, for it is obvious to all that tenant-farming is an unmitigated curse there, as it is elsewhere; but when they add their banal borrowings from the *New Masses* and the *New Republic* they greatly enfeeble their case. The New South can no more do without industrialism than the New Russia has been able to do without it. The only question before the house is whether it can be held under such restraints that it will stop short of its cancerous development in, say, New England, Pennsylvania, and the English Midlands, and along the great Russian rivers. It seems to me that there is no impediment to so policing it, given a reasonable revival of the old Southern skill at government. But if, as a part of some new and extra-fantastic New Deal, it is abolished altogether, then the subsistence farming that the young professors pine for will go with it, and the ideal planter of their dreams, sitting comfortably on his own land, will become a *kulak* hunted down by secret agents and reduced, when taken, to a kind of slavery that no white man in America has ever suffered.

This tendency to overlook obvious consequences and implications is characteristic of a great deal of the current discussion of public problems in the South, as it is of such discussions elsewhere, and it is especially characteristic of the writings of the younger and more radical publicists. They have made the fatal discovery that it is much easier to concoct Utopias than it is to polish the car or wipe the baby's nose, and so they take a great many leaps into the empyrean, forgetting that what goes up must come down. Worse, they not only run to preposterous conclusions; they also have traffic with very dubious premises, and are not above pulling and hauling good ones to fit their uses. Here the advocates of Regionalism are especially peccant. That they are right when they argue that the South should grapple resolutely with its own problems, and try to solve them in accord with its own best interests and its own private taste, is something that no one will deny. But when they go on to argue, as Mr. Donald Davidson seems to do in a recent article, that it should cut itself off from the rest of the country altogether, then they come close to uttering rubbish. It can, in point of fact, no more cut itself off from the rest of the country than it can cut itself off from the industrial organization of Christendom. Its best interests are bound to be colored and conditioned, not only by the best interests of the North and

West, but also by their notions as to what would be good for it, and what it deserves to have. And its canons of taste can no more be formulated in a vacuum than its principles of politics can be so formulated. As Haeckel—a foreigner, and hence a scoundrel—long ago pointed out, the cell does not act, it *reacts*, and that is quite as true of the cells in the human cortex as it is of the amoebae in a test-tube. When the flow of ideas from without is cut off, or hampered by filters and barriers, then the bubbling of ideas within slows down. That, in brief, was what was the matter with the South during the long half-century after the war. Too many cultural Tibets were set up, and too many survive to this day. Certainly it would be folly to try to get rid of them by surrounding the whole region with new Himalayas.

NED COBB (NATE SHAW)

"Nate Shaw" was the pseudonym used for Ned Cobb, an Alabama farmer, in the 1974 book All God's Dangers: The Life of Nate Shaw. *The book was based on Cobb's relation of his life story to historian and writer Theodore Rosengarten.*

In December of 1968, Rosengarten traveled to the rural South with a friend, Dale Rosen, a senior at Harvard University who was gathering information for a thesis about the Alabama Sharecroppers Union. The union had been organized by the Communist Party in 1931 to protect the interests of black and white farmers struggling to weather the Depression. Rosen had learned from poet John Beecher that an early member of the union who had been involved in some of the group's most notable activities still lived in Tallapoosa County, Alabama. Rosen and Rosengarten visited eighty-three-year-old Ned Cobb to talk about his involvement with the Sharecroppers Union. Two and a half years later, Ted Rosengarten returned to Alabama to record Cobb's recollections. All God's Dangers, *which touched on all aspects of Cobb's life, was drawn from this oral history.*

Cobb was born in 1885, the son of former slaves. As an adult, he became an energetic and successful farmer. By the 1920s, he was beginning to buy land of his own. Unfortunately, his exertions came at a time of falling cotton prices that bottomed out with the rest of the nation's economy in the Great Depression. Cobb joined the Alabama Sharecroppers Union in order to resist the foreclosures that he and his neighbors faced. In 1932, Cobb and several other union members were involved in a shootout with local police while defending a farmer whose livestock was in danger of repossession. Several participants were killed or wounded, including Cobb, who survived a load of buckshot in the back. He was arrested and spent twelve years in prison. These events formed the basis of John Beecher's 1940 poem "In Egypt Land."

As Ned Cobb's story unfolds in All God's Dangers, *readers encounter an exceptionally determined and resilient man. In his own words, Cobb had "no get-back." He gave little ground in his battles with the landlords and merchants who tried to take advantage*

Ned Cobb, pictured with his wife, Viola,
and oldest son, Andrew, taken after church, 1907. Courtesy
Theodore Rosengarten.

*of him, nor with the police, nor with the boll weevils that devastated cotton crops across
the South. It was Cobb's remark about the weevil that inspired the title of the book: "All
God's dangers ain't a white man." Ned Cobb's story also brings to light an era that may be
removed in time but remains familiar in some of its circumstances and challenges. These
excerpts convey Cobb's perception of the importance of animal husbandry, as expressed
through the care of his mules, and his descriptions of cultivating cotton and food crops.
While Cobb understood and deeply resented the impact of racism on his career, these
passages also demonstrate his faith in hard work and divine providence.*

From *All God's Dangers* (1974)

First thing I'd do, on the average, every day of my life, after I got my clothes on,
dash out and feed them mules. Next thing, if I had to draw water or anything else
around the house needed to be done, I'd do it. My wife would get breakfast done and

the whole family would set down at the table and eat. Then I was ready to go back to the lot and turn them mules out where they could get water themselves or either I'd draw water at the well. Done it myself and I was very shrewd on them jobs. If my boys fed and watered my mules—maybe some mornin I had some other little thing I wanted to get done real early, first thing. My boys would go to the lot—I could trust em; I always told my boys whenever they'd feed em, "Them mules aint goin to eat no more than they want. Just give em a good feed, plenty."

Sweet feed, oats, sometimes sweet feed and oats mixed, just a certain amount all put in one vessel to make up their feed. Sometimes it was ear corn—never did shell no corn and give em; a mule that's kept in good shape will eat that corn off of them cobs, sometimes he'll leave a few nubs at the end of the ears—mighty seldom my mules would ever eat up the cobs. You get a mule hungry, don't give him enough ear corn, he'll bite the nub end of that corn and eat it. The most of them cobs, my mules left em right there in that trough for me to pick up and throw out when I put out another feed of corn. Fed em three times a day just like I et—them animals, my mules or my horses, I considered the next thing to my family. Fed em mornins early, day dinner when I'd carry em to the barn, nights when I'd put em in the lot. Very seldom that I'd fasten my mules in the stable; unless it was bad weather, I'd leave the door open so when they got done eatin they could walk out in the lot. A mule won't never overeat hisself if he's used to getting plenty to eat. But you'd better not fool with no horses thataway. A horse will overeat his stomach, a horse will outeat a mule—it's nature for em to do it.

You can't trust a horse like you can a mule. A horse is not like a mule and a mule is not like a horse. A horse and a mule has got, as a rule, different ways. When it comes to heavy duties, little mule or big mule, he's kinder and more willin to work than a horse. Of course, any animal that aint willin he's goin to show it to you. That'll be a sorry mule, but it's more in the nature of a horse to act that way. It's known: better to have you a mule out there than a horse because that mule goin to bow when you call him and your business will pick up. If it's a thoroughgoin mule, aint no trouble to get work out of him, he'll bow to anything. Some horses when you hitch em up, they got a lot of fidgetin to do, and some of em just aint goin to work for you to save your life—maybe he'll travel, maybe he'll pull a buggy, but when you put him to plowin, or put him to a two-horse wagon, right there's where you goin to have your trouble with a horse.

Twice a day if I'm plowin, mornin and night, I'd brush and curry my mules. Keep my mules in thrifty condition, keep em lookin like they belonged to somebody and somebody was carin for em. Mule love for you to curry him; he'll stand there just as pretty and enjoy it so good—run that comb over him first one way then the other, backwards and forwards, scrub him good, all over, the fur under his stomach if necessary, and from his head back. Brush the head—mule don't love for you to hit his head much with a curry comb.

When I'd get him curried, hang my curry comb back up on the barn wall—my curry comb had just about gived out at the time I sold that last mule—reach back and get that brush, brush him all over far as I could reach that brush under him; brush his legs down as low as his hocks. Clean that mule up so he'll feel good, then go get his gear and put it on, take him and hitch him to the plow, go to plowin. He feels good to be clean just like a person do. . . .

If I was goin to a wagon, many a time I'd use strop harness on my mules, with a crupper under their tails, to haul logs. I started off usin britchin on em, them big leather strops behind the back legs and hips, but that wouldn't do to keep them animals cool for haulin lumber. Best thing to do, averagely, for you to haul lumber with your mules, run them leather strops across them mules' hips down to them traces—and that's what you call strop harness. I used em for haulin lumber so my mules wasn't covered up in britchin and leathers, keep em hot. And put brakes on that wagon to hold that wagon off of em goin down hills. I'd load anywhere from a thousand to—wouldn't put on a foot over a thousand if I was under any reasonable hill at all; buck it down, pick up my lines, call up there, "All right, babies, let's get it." You'd see them big heifers fall out then. O, my mules just granted me all the pleasure I needed, to see what I had and how they moved. . . .

I had my business cut and dried for farmin; I couldn't do that sawmill job and my farmin both successfully. I'd just as well to sell both my mules outright at that point and take what I could get for em. I had good money tied up in them mules and I couldn't let em stand in the lot and not plow em. And I absolutely had my heart in farmin. I knowed that what little lookin out I had done on the farm brought me up to them high-priced mules and I didn't know definitely at that time bout what the sawmills would bring me. I had it planned at home to make a livin and that's where I decided was my best hope. And so I told him [Cobb's employer, George Pike] I was goin back to my farm. And he told me to my head—I knowed it was a joke; I had sense enough to know jokes when I heard em—"Nate, you think because you got a damn good pair of mules you goin to come back and haul lumber when the lumber gets dry enough to haul."

I looked at him and laughed.

He said, "Well, I aint goin to let you haul a bit."

I knowed that if he didn't let me haul I was goin to live right on. Mighta been tight or mighta been good, I knowed I was goin to live. Sawmill work weren't goin to keep me eatin. But he was just jokin with me—and besides, he weren't the boss. They done found out what there was to me. So I went on and quit, went on back to my farm. . . .

I plowed, in plowin my mules, I'd take my two-horse plow, jump in there right after Christmas, take my stalk cutter, hitch my mules to it, and cut my stalks just as soon as they was dry enough that they'd break good. When that was done, I'd go on and get my wood all in for my year's cookin; then get my land broke with my two-horse plow—I've tried it all sorts of ways. Sometimes I'd break my land maybe two or three weeks, a month before I got ready to plant. I'd go back, that land was just like it never been plowed. It showed it'd been plowed, but doggone it, it was packed again tight as wax. Some years I broke my land early—that gived the vegetation and the cotton stalks a chance to rot up some; and the land would pack back or not pack back, accordin to conditions I couldn't control.

I don't say breakin your land and gettin it ready in January is the only way. You got plenty of time, you on time if you have it prepared by the first of April. But if you don't, you better hustle like the diggers to get it ready or you'll make a feeble crop. Jump in there the first of April and you might have a little you want to rebreak or even a patch that aint been broke, but the majority is ready for plantin. The best way, if you don't want to lose a corn crop, regardless to the cotton, if you want to make a cotton crop and a corn crop too, successfully, you get out there and hustle and watch the time, be on time the first time because sometimes you might have to plant that crop over; you don't know. If you can't be on time—you must use your time, a heap of times, when God gives it to you. God's a man, you can't start His time, you can't stop it. Some folks don't use the time God gives em; that's why they're liable to come up defeated. . . .

If the weather's just right, you can plant the first days of the week and by the middle of the next week, you got a pretty stand of cotton all over your field. Cotton and weeds all come up together and when they come up the weeds beats cotton growin. Just as soon as that cotton will bear your cultivators, when it's just bustin the ground, drop your fenders on your mule plow and work that cotton. That fender bolted to your plow beam—it won't allow the dirt to fall in them drills and stop that cotton comin up.

Just as soon as that cotton gets up to a stand, best time to chop it out provided you feel that the cold won't get what you leave there. But you don't always know about that. If you chop it too early, decidin wrong about what the weather's goin to do, you go out there and thin out your crop, that bad weather come and get the balance of it. Or you might chop it wrong if you don't have the experience of it. You'll be diggin out too deep, guano and all, and you'll bruise the plants. But if you chop it out right and the weather don't change on you, it stays mild and in favor of growin, what you leave there will make a good crop.

Dry weather can cut the growth of that cotton to an extent. Too much rain can cause it to overgrow itself. Cotton's a sun weed, cotton's a sun weed. Too much water and it'll grow too fast, sap runs too heavy in the stalk and it'll make more stalk than cotton. Cotton is a kind of sunflower; it just takes so much rain. You can make a good cotton crop a heap of times in a dry summer, owin to how much rain you had in the spring. Anyway, you just runnin a risk farmin; you don't know whether you goin to win or lose. You can bet on it ever so much, but sometimes you lose your bet and if you a poor farmer, you in a hole then. It's a heap like gamblin—you don't know what you goin to win, you don't know what you goin to lose. You take a farmin man out there, he dependin on his cotton and his corn. He don't know till he gather it what he goin to get, if he goin to get enough to come up even. I believe a man can put too much trust in his crop; he'll bet too heavy on it and he's subject to lose. He's takin a desperate risk. How can you get out there and plant any kind of seed—cotton, corn, peas, tomatoes, vegetable seed of any kind, of the least and the most, what's about *you* to make it sprout and come up? What's about you? You got no power. God got the power. But God has got a part for you to do—He aint goin to come down here and run nary a seed of no sort for you. God aint goin to come down here and run nary a furrow out there in that field, plow that ground up—He aint goin to do it. He give you wisdom and knowledge to do it. But if you set around and don't do nothing, won't work, you burnt up, you hear? God fixes it so, He's so merciful and kind—you can't sprout no seed, but go to work and tear up that land and plant that seed yourself. God requires you to do it and if you don't, He aint goin to come down here and do it. Still and all, think about it: the power's in the Almighty. You can't make your seed grow! It takes sunshine and water to make vegetables come up—sunshine when He sees fit to send it, water when He sees fit to send it. You get out here and do your part, what He put you here to do, because God aint goin to do your labor, He aint goin to do it. . . .

I wasn't raisin under five and six bales of cotton every year that I stayed on the Bannister place. Raised corn—I kept my corn to feed my stock, met all my expenses with cotton and what I was makin off my lumber haulin job. I didn't never want for no vegetable, what I had I growed em. Okra, anything from okra up and down— collards, tomatoes, red cabbages, hard-headed cabbages, squash, beans, turnips, sweet potatoes, ice potatoes, onions, radishes, cucumbers—anything for vegetables. And fruits, fruits for eatin purposes and cookin, pies, preserves—apples, peaches, plums, watermelons, cantaloupes, muskmelons. I quit growin muskmelons for one reason: they got to where some years the worms would take to em. They is a differ- ent melon to a watermelon. The inside of a muskmelon is yellow like the inside of a cantaloupe—they're good tastin, sweet. Cut em open and scrape the seed out of em, sprinkle if you like a little salt over em. Sometimes I've seen people sprinkle a little black pepper over em too.

I raised many a crop of ice potatoes but you had to keep them ice potatoes sprin- kled for bugs, anointed. Just planted enough for family use as long as they'd last. Same with sweet potatoes—I'd store em, put em in a bank. Sometimes I'd have four banks of sweet potatoes, and the seed ones in a kiln to theirselves and the eatin ones to theirselves. Clean out a wide space for whatever number of potatoes I had to bank, accordin to the amount I raised. Cut me out a big circle, made it flatter in the center than at the outer edges, and put me a layer of pine straw in there, then salt my potatoes and pour em in there until I had a big pile. Then I'd cover em sufficient with pine straw and if I thought they needed it, I'd use somethin like pasteboard, apply it around that bank to keep the dirt from runnin on the potatoes, take my shovel and ditch all around them banks, pile the dirt up to the top—just leave a hole enough at the top of the bank for the potatoes to get air. Sometimes I'd even put a tin top, bought tin, over it to keep out the rain. I'd have potatoes then for table purposes way up in the spring of the year. Eventually they'd lose their sweetness and get pethy. Chop em up then and feed em to my stock, mules if it was necessary; but I always had plenty of corn without havin to feed my mules sweet potatoes. But hogs, I'd feed my hogs on em a plenty.

I had my own cows to milk—my children started to milkin on the Bannister place, my oldest boy and oldest daughter and still their mother wasn't milkin. They milked when I couldn't milk. After them children got to be nine years old they helped me regular. I went to the lot and showed em how to milk and they was glad to learn. My wife didn't have to milk no cow; she didn't have to go to the field, I'd drive her out the field.

And I killed all the meat we could use until I killed meat again—from winter to winter. I had a white man walk through my yard—two of em, Mr. Albert Clay and Mr. Craven. I don't know Mr. Craven's given name but that was Mr. Clay's brother-in-law. Come through my yard one Saturday evenin and I had killed three big hogs, me and my little boys, and had em stretched out over the yard after I cut em up.

They walked up to my back yard on the north side of the house—that old house I was livin in was built east and west—and they come up from towards my barn. I was surprised in a way but I didn't let it worry me, people go where they want to and walk anywhere they want to. Mr. Albert Clay and Mr. Craven come up from towards my barn. My barn set west of the house and back behind the barn was my pasture. Well, they come right up cross the back yard—that yard was covered with meat from three big hogs I'd just killed and had the meat put out; it was all of nine hundred pounds of meat. They looked there in that yard and stools, boxes, tables, benches, and everything had planks across em and them planks was lined with meat, just killed and cut out. Dressed, gutted, and cut open but not fully cut up, layin out, ready for salt.

Mr. Craven made a big moderation. "Where'd you get all this meat, Shaw? What are you goin to do with all this meat?"

It was all over the yard, coverin everything in sight. Three great big hogs weighed over three hundred pounds apiece. Had more meat there than you could shake a stick at.

"That's more meat than ever I seed any nigger"—that's the way he said it—"I aint never seed that much meat that no one nigger owned it."

They looked hard, didn't stop lookin. After a while they crept on out of there, still stretchin their eyes at that meat. They didn't like to see a nigger with too much; they didn't like it one bit and it caused em to throw a slang word about a "nigger" havin all this, that, and the other. I didn't make no noise about it. I didn't like that word, but then that word didn't hurt me; it was some action had to be taken to hurt me. I just rested quiet and went on preparin that meat. . . .

Well, I was a Negro of this type: regardless to what people said, regardless to how much I knowed that they was a enemy to me, I just pulled myself along anyhow to the best of my abilities and knowledge. Didn't hold myself back like they wanted me to. So that man that spoke that word about my meat, then I went on up the road and done his son a favor, picked him up and carried him home, cut him out of some of that walkin, and he jumped up and admired my mule and buggy in a way that made it clear he didn't like to see no nigger have a outfit like that—and it would have made him spiteful to see a *white* man with it—happened all one evenin. . . .

1923, I got what the boll weevil let me have—six bales. Boll weevil et up the best part of my crop. Didn't use no poison at that time, just pickin up squares. All you could do was keep them boll weevils from hatchin out and goin back up on that cotton. Couldn't kill em.

The boll weevil come into this country in the teens, between 1910 and 1920. Didn't know about a boll weevil when I was a boy comin up. They blowed in here from the western countries. People was bothered with the boll weevil way out there in the state of Texas and other states out there before we was here. And when the boll weevil hit this country, people was fully ignorant of their ways and what to do for em. Many white employers, when they discovered them boll weevils here, they'd tell their hands out on their plantations—some of em didn't have plantations, had land rented in their possession and put a farmin man out there; he was goin to gain that way by rentin land and puttin a man out there to work it; he goin to beat the nigger out of enough to more than pay the rent on it. And the white man didn't mind rentin land for a good farmer. That rent weren't enough to hurt him; he'd sub-rent it to the fellow that goin to work it or put him out there on halves. Didn't matter how a nigger workin a crop, if he worked it it's called his until it was picked out and ginned and then it was the white man's crop. Nigger delivered that cotton baled up to the white man—so they'd tell you, come out to the field to tell you or ask you when you'd go to the store, "How's your crop gettin along?" knowin the boll weevil's eatin away as he's talking. Somebody totin news to him every day bout which of his farmers is pickin up squares and which ones aint.

"You seen any squares fallin on the ground?"

Sometimes you'd say, "Yes sir, my crop's losin squares."

He'd tell you what it was. Well, maybe you done found out. He'd tell you, "Pick them squares up off the ground, keep em picked up; boll weevil's in them squares. If you don't, I can't furnish you, if you aint goin to keep them squares up off the ground."

Boss man worryin bout his farmers heavy in debt, if he ever goin to see that money. Mr. Lemuel Tucker, when I was livin down there on Sitimachas Creek, he come to me, "You better pick them squares up, Nate, or you won't be able to pay me this year."

Don't he know that I'm goin to fight the boll weevil? But fight him for my benefit. He goin to reap the reward of my labor too, but it aint for him that I'm laborin. All the time it's for myself. Any man under God's sun that's got anything industrious about him, you don't have to make him work—he goin to work. But Tucker didn't trust me to that. If a white man had anything booked against you, well, you could just expect him to ride up and hang around you to see that you worked, especially when the boll weevil come into this country. To a great extent, I was given about as little trouble about such as that as any man. I didn't sit down and wait till the boss man seed my sorry acts in his field. I worked. I worked.

Me and my children picked up squares sometimes by the bucketsful. They'd go out to the field with little sacks or just anything to hold them squares and when they'd come in they'd have enough squares to fill up two baskets. I was industrious enough to do somethin about the boll weevil without bein driven to it. Picked up them squares and destroyed em, destroyed the weevil eggs. Sometimes, fool around there and see a old weevil himself.

I've gived my children many pennies and nickels for pickin up squares. But the fact of the business, pickin up squares and burnin em—it weren't worth nothin. Boll weevil'd eat as much as he pleased. Consequently, they come to find out pickin up them squares weren't worth a dime. It was impossible to get all them squares and the ones you couldn't get was enough to ruin your crop. Say like today your cotton is illuminated with squares; come up a big rain maybe tonight, washin them squares out of the fields. Them boll weevils hatches in the woods, gets up and come right back in the field. You couldn't keep your fields clean—boll weevil scheming to eat your crop faster than you worked to get him out.

My daddy didn't know what a boll weevil was in his day. The boll weevil come in this country after I was grown and married and had three or four children. I was scared of him to an extent. I soon learned he'd destroy a cotton crop. Yes, all God's dangers aint a white man. . . .

Who sent the corn weevil here? And who sent the boll weevil and all sorts of pesters, who put em here? Who created the heaven and earth and everything therein? God put all these pesters and insects here. As bad a old thing as a snake, God put him here; and He put them things here—maybe, I wouldn't accuse God of nothin wrong—to trouble people. Folks in this world needs pesterin to wake em up to their limit. And to my best opinion, God put the different weevils here and the weevils does their duty. Some things may do more than God put em here to do—that's the human, he do more things than God put em here to do. But God thought so much of this human race He created humans in His image and His own likeness, and still they're the worst things God got on this earth to one another. God knew you'd do it but He gave you a chance, He put you here, He put His holy righteous words here for you to read and look over; and if you can't read God's words you still can believe some things. He thought so much of you He gived you knowledge He didn't give the other animals. And He gived you a soul to save, He made you responsible to that knowledge: a man is responsible, a woman is responsible, for the acts of their flesh and blood and the thoughts on their minds.

I can't hate God's pesters, definitely, because they doin what God put em here to do. The boll weevil, he's a smart bird, sure as you born. And he's here for a purpose. Who knows that purpose? And who is it human that can say for sure he knows his own purpose? He got all the wisdom and knowledge God give him and God even sufferin him to get a book learnin and like that—and what the boll weevil can do to me aint half so bad to what a man might do. I can go to my field and shake a poison dust on my crop and the boll weevil will sail away. But how can I sling a man off my back? . . .

It approaches my mind like this: what is labor? It's a trait of man. God put us all here to work. I know people who use to would work—always there was some of em wouldn't work at all—but since the government been givin em a hand-down, they wouldn't mind the flies off their faces. They'll tell you quick, "I'm drawin, I aint goin to hit a lick with a snake."

You used to could have a field full of hands, but now you can't hardly get one or two in there; it's just out of the question. Aint available no more—they quit it, you can't hire em. Some of em run off and get em a public job; some of em will sit right in the house on their tail or loaf the road, and aint goin to do nothin much nowhere. I've heard em say, my color talking it: "I wouldn't tell a mule to get up if he was sittin in my lap." But I never heard em say they wouldn't eat the fruits of the earth. He won't plow, he won't chop cotton, he won't pick cotton, but still he's goin to wear these clothes made out of cotton. Everything that's used to sustain life and nurture your well-bein, it comes off a farm of some kind—if it's a little farm if it aint no more than just for family use.

I hear em talk, who I know used to plow and seed em plow; because they has once in days past made crops under the white man's administration and didn't get nothin out of it, he don't want to farm today regardless to what he could make out there; he don't want to plow no mule—that was his bondage and he turnin away from it. He huntin him a public job, leavin the possession and use of the earth to the white man.

You take these public companies, they keeps em laid off these jobs to an extent; the ones that's got em hired turns em loose, out, until way after a while, maybe they call em back, mess em around that way. But a man needs regular work—if he wants to work, it's right, it's honorable for him to work and try to help hisself. But he aint goin to want to work if they liable to turn him loose anytime they please. Work em to death or don't work em enough, that's how they do around here. Cut off that regular work. If I got you hired out there to work on any sort of a public job, I'd be figurin like this: I'm already makin—I'm the boss man, understand—I'm already makin money on that job, if I give you regular work to do and pay you, why, that'd cut me down some. I'm already gettin good anyhow and I'll just lay you off for a few days—they doin that right here, at the cotton mill. I don't know how the people gets along that's laid off thataway. It's like a farmer don't have no land to work or it aint his land he's workin, he just gettin paid by the day, and he don't get the benefit of the crop; just give him enough to keep him goin today till tomorrow. . . .

I'm yet a laborin man, I makes baskets. White folks comes from all parts to get my baskets. One family come, by the name of Hooker, man and his wife, out of West Point, Georgia. Why, if it wasn't for makin baskets and sellin baskets, I wouldn't hardly have no business with white folks at all. . . .

There's a Mr. Lorne Ray, man that bought out my tools and kitchen material, he buys every basket he can get his hands on. I can't make em fast enough for the public. I go to keep it cleaned off around here to make it look like somebody lives here. And then, too, I got a garden needs hoein and a hog to look after. The Lord keeps me strong but if I take any more on me than I do, I wouldn't make it.

Terry Evans, "Carl with Twin Calf," 1994, Matfield Green, Kansas. Courtesy Terry Evans.

7. Back to the Land Again, 1940–Present

A child could not grow up in a better place than a farm; for at the heart
of civilization is the byre, the barn, and the midden.

EDWIN MUIR, *An Autobiography*

The trajectory of Scottish poet Edwin Muir's life, which began on a remote Or-
kney farm in 1887 and lasted well into the atomic age, illustrates how compressed
recent history has become. "I was born before the Industrial Revolution," he wrote
in a diary, "and am now about two hundred years old. But I have skipped a hundred
and fifty of them." The rapid succession from an agricultural to an industrial to a
postindustrial economy casts Muir's claim that the farm is at the heart of civilization into sharp relief, in America as elsewhere. For the second half of the twentieth
century saw not only a steep decline in the number of American farmers, who are
now a tiny minority of the population, but the almost complete collapse of our rural
economy and of traditional agrarian culture.

If the farm has moved from the center of civilization to the periphery, what can
we say about the state of our society? One possibility is that we are in a period of irresistible cultural foundering, set adrift from an agricultural (or any other) anchor.
Another possibility is that Progress has brought us something unprecedented: an
improved civilization in which only a relative handful of people are directly involved
in the basic tasks of raising the food and fiber society needs, and thus large majorities
are happily divorced from the land. That is, industrial civilization assumes that agriculture is essential only for providing caloric fuel for human culture, not for defin-
ing it. Agrarians believe that assumption to be dangerously shortsighted. Modern
agrarianism amounts to a determined effort to restore some core agrarian sensibility,
some meaningful connection with the land, to the center of industrial civilization.

From the nation's birth, agrarians of various stripes have looked to farm life as the antithesis of an emerging urban, industrial existence. The farm has been seen as a place of economic independence, where a man could be his own boss and perform work that, while difficult and demanding, was also diverse and stimulating. Farm work would not alienate him (or, more recently, her) from his own spirit, family, community, or nature. That has been the agrarian ideal, and many—perhaps most—American farmers have struggled to achieve it, and still do. But, as we have seen, they have done so within a system that has come increasingly to exemplify industrial production. Agriculture has become radically specialized and streamlined, relying upon large machines and other technological inputs. Farming has become enormously productive, but also in many ways as isolating and alienating an industry as the American economy has yet devised.

How did this come to pass? At the time of World War II there were still about 6 million farms in America, as there had been for decades; and farmers comprised almost 20 percent of the workforce. America was no longer an agrarian nation by any stretch, but it still had a robust, if beleaguered, agricultural sector consisting mostly of rather small family farms. Following the farm crisis of the 1920s and the ecological and economic disasters of the 1930s, a strong movement to promote a permanent, ecological agriculture had taken shape. In 1940, a group of progressive agricultural thinkers including Hugh Hammond Bennett, Louis Bromfield, Rexford Tugwell, Paul Sears, Henry A. Wallace, Morris Cooke, and Russell Lord formed a national organization called the Friends of the Land. This was no bantamweight or gadfly group—it was stocked with leading agricultural policymakers and writers who reached a wide audience. They envisaged a stable agricultural and rural sector founded upon soil conservation and the application of ecological principles. In this powerfully agrarian vision, the practitioners of permanent agriculture were to be millions of educated small, diversified farmers living in harmony with the land. A prominent spokesman for the Friends of the Land was Aldo Leopold, whose now-famous "land ethic" was grounded in this new ecological agrarianism that aimed to reconcile farming and conservation, as Liberty Hyde Bailey had proposed a generation before.

Needless to say, this was not the vision of American agriculture that actually prevailed. Instead, the postwar decades saw a great boom in farm production, accompanied by a rapid decline in the number of farmers, engineered once again by those who saw the central purpose of farming to be the provision of abundant cheap commodities. "Get big or get out!" declared Secretary of Agriculture Ezra Taft Benson in the 1950s, and his advice was echoed by his successor Earl Butz in the 1970s, who told farmers to "adapt or die." In the years following World War II, the number of farms fell rapidly to about 2 million, and by the end of the twentieth century farmers comprised less than 2 percent of the workforce. This was accomplished through the massive application of technologies designed to increase the productivity of both land and labor, including large tractors and combines, high-yielding seeds, chemical fertilizers and pesticides, and a large expansion in irrigation. While much of the nation's farming is still done by independent operators, many of whom still espouse agrarian values, they are typically running large, specialized farms in an increasingly empty countryside.

Some people, such as Secretary Butz, consider this development an overwhelming success. "Agricultural efficiency—food-producing efficiency—is the first key to making the Great American Dream come true," said Butz. In countries whose agricultural productivity failed to match that of the United States, "there just isn't enough manpower, or womanpower, left to produce things to bring their every-day living standards up to the American level." Butz noted that by spending historically small proportions of their income on food, Americans had more to spend on "all the good things that make existence more satisfying." The point made by Butz and others of similar stamp was that an economy with such a degree of specialization, able to provide the basic commodity of food very cheaply, opens opportunities for people to develop a wide variety of skills and products. In agriculture itself, however, this formula keeps most farmers perpetually on the brink of ruin and in half a century has pushed two-thirds of them over the edge. Such creative destruction may be regrettable for those whose livelihoods are destroyed in any industry, but in agriculture, where the farm is at once a job, a way of life, a family home, and the basis of a community, perhaps more is at stake.

Victor Davis Hanson tells one such story of the struggle to preserve a family farm in California's San Joaquin Valley during the 1980s. The land had been in his family for generations. They raised grapes for raisins as well as orchard crops in the fertile land near Fresno. High production had been depressing raisin prices over many years, culminating in a crash in 1983. Hanson left the academic world to join his brother and extended family in an effort to reverse their fortunes on the farm. One gamble they made was to plant the Royal seedless grape, a new variety bred especially for the winter holiday market when other grapes were no longer available. The bet was that this particular variety could be sold at high prices, with little or no competition. Hanson later learned that the Royal seedless was just about the most finicky, labor-intensive grape ever foisted by agricultural science upon honest farmers. But what really lost the bet was the implementation of the free-trade policies of Ronald Reagan's administration. Chilean grapes were beginning to flood into the United States during the winter months, just when the Hansons' Royals were due to go to market. Agricultural specialization was proving to be an increasingly global phenomenon, and whatever the benefits to American consumers, the increased competition brought the Hanson family farm to disaster. Nor was Hanson alone. He notes that around two thousand family farms per week were lost through most of the 1980s, many farmers having followed Butz's advice and taken on overleveraged mortgages to buy more land and equipment during the boom years of the 1970s. The boom-and-bust cycle in American agriculture had repeated itself once again.

In describing the plight he faced, Hanson draws attention to a kind of cultural

contradiction within the agriculture practiced in the San Joaquin Valley and throughout the nation: the very qualities that made people successful farmers also made
them business failures. Agriculture, in Hanson's eyes, has always been a tragic
struggle of men and women to wrest food from an intractable natural world, as
punitive as it is nourishing. Victory is never complete; it is more a kind of temporary
stay against dissolution. Catastrophe—the result of bad weather, pests, rapacious
invaders, or unlucky judgment—is virtually inevitable. The pressure to survive
and overcome these difficult odds drives farmers toward excellence, and the true
agrarian stoically embraces the challenge. Hanson, a classical scholar, points out
that farming for the Greeks was "the best tester of good and bad men." There is
little place in Hanson's hard-bitten agrarian philosophy for Liberty Hyde Bailey's
holy earth, or Aldo Leopold's land ethic.

This may have been an apt formula for breeding farmers capable of withstanding
the rigors of nature, but what about the added rigors of the market? In the global
economy of the late twentieth century, this ancient formula for agrarian survival
seemed to dissolve. Farmers had not only to contend with the natural forces they
had always faced but also to compete with equally tenacious agriculturalists right
next door and half a world away. Their very productivity and industriousness threatened their livelihood through overproduction. Although Hanson points to parallels
between the farming culture he grew up with and Greek traditions dating back
more than twenty-five hundred years, the desperate struggles of his San Joaquin
neighbors are more akin, perhaps, to the ideas of Adam Smith. In any given year,
Hanson relates, success in raisin farming "hinges largely on how many of one's
kindred growers pick too late and see their crops putrefy, diminishing the always
present oversupply. . . . Your profit often depends on someone else's perdition. 'It's
not enough that I get by,'" one neighbor told Hanson. "'You guys, you have to fail.'"

But American farmers have not always been driven solely by such a narrow competitive ethos. Saying this runs the risk of suggesting that the opposite might have
been characteristic of rural America: that it was a communal Arcadia, untainted by
commercial and personal ambitions. Of course it never was that, either. But the effort
to avoid sentimentalizing America's agrarian past has its own pitfalls. It assumes
that the dominant modern reality of trial by market has always had as natural and
inevitable a place in American farming as trial by weeds and weather, and that other
agrarian values have always been subordinate to that reality and must always remain
so. Most agrarian writing has vigorously challenged that assumption, continuing
to champion values that preceded the market revolution, values that have endured
in spite of it. In preindustrial agrarian America, a large measure of household self-
reliance generally came first, followed by complex networks of local exchange of
both work and products among neighbors, along with some surplus or specialty
marketed for cash. This limited the opportunity for large profits but also insulated
against financial ruin. And while Hanson's hard-driving California fruit growers
appear to have little room in their lives for anything but relentless work and lost

sleep (at least in his telling), rural people in other times and places have cultivated distinctive musical traditions, recreations, customs for marriage and child-rearing, and all the other aspects of coherent communal life. The loss of what remained of such culture was a theme of John Crowe Ransom and the other Southerners who contributed in 1930 to *I'll Take My Stand*. The long century of decline in rural New England has also been movingly evoked by writers such as Robert Frost, Donald Hall, and Hayden Carruth.

The determination to move beyond simply mourning the past to attempt to keep American agrarian culture alive has been taken up by writers such as Gene Logsdon, Wendell Berry, and Wes Jackson, though their work may prove even more valuable in preparing the ground for a whole new breed of agrarian. In *What Are People For?* Berry calls the late twentieth-century mass movement of people from the farm to the city, celebrated by the likes of Earl Butz, one of the most dismaying migrations in human history. In a favorite metaphor, Berry likens the sustaining of culture to the building of soil. Like fertile soil, culture is husbanded through slow accumulation and cultivation, always under threat of erosion. People come to understand what ought to be valued only gradually, through their own efforts but also through contact with their community. "There is no one to teach young people but older people, and so older people must do it," Berry writes. "That they do not know enough to do it, that they have never been smart enough or experienced enough or good enough to do it, does not matter." Teaching ought not to be the work solely of professionals in schools. Berry particularly regrets the separation of children in the industrial economy from the work life of adults. Far better, as in the agrarian economy, for children to be present while adults work, learning how the work is done as they watch and play, learning about their elders as people. This richness of human connections in a healthy community mirrors the rich biological life of soil. Breaking connections between people robs individuals of irreplaceable understandings, perspectives, and memories. The movement of talented young people to cities is just such a deprivation. Berry points out that one apparent goal of the contemporary economy is for the young to supersede their parents, not to succeed them. With bonds between the generations broken, parents and children alike abide in a "dimensionless" present, where shared memory is diminished and the future promises only further loss. Berry also fears that the quality of thought among leaders in government, business, and academia is spoiled by their distance from local realities. They lack the practical experience and necessary affection for real places that might give them pause when considering the possible costs of their actions. In a similar vein, Wes Jackson calls upon us to deliberately rebuild local connections and "become native to our places," and to replace university majors in "upward mobility" with majors in "homecoming."

Like Victor Davis Hanson, Berry is a frequent critic of consumerism and the shallowness of contemporary culture. But in certain fundamental attitudes, they diverge sharply. Where Hanson's agrarianism is harsh and tragic, Berry's thinking tends toward the providential. For Berry, our world is ordered so that people can and should, in the ordinary course of things, satisfy their most profound longings through work, family, community, and engagement with nature. But this satisfaction is bounded and conditional. One must not degrade the natural world through one's work or play, and one's appetites must be appropriate and seemly. The current economy fails on both counts, through exploitative use of nature on one side, and the fostering of unlimited wants on the other.

In an essay entitled "Economy and Pleasure" from the book *What Are People For?* Berry argues that work itself should be pleasing. "In the right sort of economy, our pleasure would not be merely an addition or by-product or reward; it would be both an empowerment of our work and its indispensable measure." He acknowledges the pains of hard physical labor in making this claim, as he recalls many years of picking tobacco. Unusually resistant to mechanization, the tobacco harvest is handled through teamwork, often providing a convivial experience. "None of us would say that we take pleasure in all of it all of the time, but we do take pleasure in it, and sometimes the pleasure can be intense and clear. . . . [B]ecause of the unrelenting difficulty of the work, everything funny or amusing is relished." Of the swapping of stories about harvests past, reminiscences about former coworkers, and chatter about the weather, Berry speculates: "The conversation, one feels, is ancient. Such talk in barns and at row ends must go back without interruption to the first farmers."

Neoagrarians such as Berry wonder if the late twentieth century marks a watershed, the point at which the heretofore unbroken chain of connection to farmers who have gone before is almost completely lost. They often point to the Amish as the best—and perhaps only—example of traditional agrarian culture being carried on in a way that is coherent and thriving. Yet the agrarian ideal continues to attract newcomers. The lure of agricultural life has been a major motive behind the late twentieth-century surge in people moving "back to the land," as well as behind the sustainable farming and "locavore" movements of today. The distance between these latter-day agrarianisms and older traditions such as that of the Amish is obvious enough. But however romantic and naïve may have been the original urge to discover a "simpler" life, one that integrates work, community, and nature, the survivors from the 1960s and 1970s have been at it for long enough now to be taken seriously. Many of these agrarian writers of the modern era have come to terms in their own lives with the challenges of maintaining a healthy farm within an industrial civilization, and their passion has been tempered by experience.

Helen and Scott Nearing embodied this ideal during their homesteading years in Vermont and Maine. Their labors in providing food and shelter on their land, largely by hand, would have been accomplishment enough, but through it all they managed to write, travel, and remain politically active. Their achievements inspired many others to attempt to follow a similar course. The agrarian ideal is carried forward in good part by their heirs, some of whom are now reaching a similar maturity. Their numbers remain small, perhaps, but are not diminishing. Whether this new agrarian line can substitute for the disappearing connections to traditional agrarian culture mourned by Jackson and Berry is an open question. But the great providential irony of their work may be that in writing about what is being lost, they have inspired a new agrarianism with intellectual roots in Henry Thoreau and Helen Nearing as much as in J. Hector St. John de Crèvecoeur and Thomas Jefferson.

ALDO LEOPOLD

Aldo Leopold is commemorated as a pioneering wildlife ecologist and conservationist, but his interest in agriculture was a constant throughout his career. Born in Burlington, Iowa, in 1887, Leopold knew the farms and fields of his hometown as well as the forests and riverbanks. He was a keen observer of ecosystem dynamics, and acutely aware of the impact of farms on the natural habitats with which they were intertwined.

Leopold received his MA from the Yale School of Forestry in 1909 and entered the U.S. Forest Service, moving to the American Southwest. There he served as supervisor of Carson National Forest and explored the landscapes of Arizona and New Mexico, drafting land- and game-management plans. Leopold was largely responsible for the first Forest Service wilderness designation, in the Gila National Forest in 1924. He finished his Forest Service career with the Forest Products Laboratory in Wisconsin, leaving in 1928 to continue his work on game management. In 1933 he joined the University of Wisconsin's Department of Agricultural Economics with a chair in Game Management, and he is credited with moving the field toward the broader ecological approach of wildlife management. In 1935 Leopold was instrumental in the founding of the Wilderness Society, and in 1937 he helped to found the Wildlife Society.

Aldo Leopold's greatest contribution, however, may be his essays about conservation. With a simple and direct prose style, he conveyed complex ideas about the ethical implications of people's relationship to the land. His most famous publication, A Sand County Almanac, *was published posthumously in 1949, after his untimely death from a heart attack while fighting a grass fire in 1948, near the rundown Baraboo, Wisconsin, farm where he and his family experimented with ecological restoration.*

Leopold presented "The Farmer as a Conservationist" during Farm and Home Week at the University of Wisconsin in early 1939. It displays his evolving vision of ecological ethics and his hope that the American farm might develop into its own rich biotic community, with elements of the wild pre–European settlement landscape allowed to coexist alongside the cultivated land.

From "The Farmer as a Conservationist" (1939)

Conservation means harmony between men and land.

When land does well for its owner, and the owner does well by his land; when both end up better by reason of their partnership, we have conservation. When one or the other grows poorer, we do not.

Few acres in North America have escaped impoverishment through human use. If someone were to map the continent for gains and losses in soil fertility, waterflow, flora, and fauna, it would be difficult to find spots where less than three of these four basic resources have retrograded; easy to find spots where all four are poorer than when we took them over from the Indians.

As for the owners, it would be a fair assertion to say that land depletion has broken as many as it has enriched.

It is customary to fudge the record by regarding the depletion of flora and fauna as inevitable, and hence leaving them out of the account. The fertile productive farm is regarded as a success, even though it has lost most of its native plants and animals. Conservation protests such a biased accounting. It was necessary, to be sure, to eliminate a few species, and to change radically the distribution of many. But it remains a fact that the average American township has lost a score of plants and animals through indifference for every one it has lost through necessity.

What is the nature of the process by which men destroy land? What kind of events made it possible for that much-quoted old-timer to say: "You can't tell me about farming; I've worn out three farms already and this is my fourth"?

Most thinkers have pictured a process of gradual exhaustion. Land, they say, is like a bank account: if you draw out more than the interest, the principal dwindles. When Van Hise said "Conservation is wise use," he meant, I think, restrained use.

Certainly conservation means restraint, but there is something else that needs to be said. It seems to me that many land resources, when they are used, get out of order and disappear or deteriorate before anyone has a chance to exhaust them.

Look, for example, at the eroding farms of the cornbelt. When our grandfathers first broke this land, did it melt away with every rain that happened to fall on a thawed frost-pan? Or in a furrow not exactly on contour? It did not; the newly broken soil was tough, resistant, elastic to strain. Soil treatments which were safe in 1840 would be suicidal in 1940. Fertility in 1840 did not go down river faster than up into crops. Something has got out of order. We might almost say that the soil bank is tottering, and this is more important than whether we have overdrawn or underdrawn our interest.

Look at the northern forests: did we build barns out of all the pineries which once covered the lake states? No. As soon as we had opened some big slashings we made a path for fires to invade the woods. Fires cut off growth and reproduction. They outran the lumberman and they mopped up behind him, destroying not only the timber but also the soil and the seed. If we could have kept the soil and the seed, we should be harvesting a new crop of pines now, regardless of whether the virgin crop was cut too fast or too slow. The real damage was not so much the overcutting, it was the run on the soil-timber bank.

A still clearer example is found in farm woodlots. By pasturing their woodlots, and thus preventing all new growth, cornbelt farmers are gradually eliminating woods from the farm landscape. The wildflowers and wildlife are of course lost long before the woodlot itself disappears. Overdrawing the interest from the woodlot bank is perhaps serious, but it is a bagatelle compared with destroying the capacity of the woodlot to yield interest. Here again we see awkward use, rather than over-use, disordering the resource. . . .

Conservation, then, is keeping the resource in working order, as well as preventing over-use. Resources may get out of order before they are exhausted, sometimes while they are still abundant. Conservation, therefore, is a positive exercise of skill and insight, not merely a negative exercise of abstinence or caution. . . .

Can a farmer afford to devote land to woods, marsh, pond, wind-breaks? These are semi-economic land uses—that is, they have utility but they also yield non-economic benefits.

Can a farmer afford to devote land to fencerows for the birds, to snag-trees for the coons and flying squirrels? Here the utility shrinks to what the chemist calls "a trace."

Can a farmer afford to devote land to fencerows for a patch of ladyslippers, a remnant of prairie, or just scenery? Here the utility shrinks to zero.

Yet conservation is any or all of these things.

Many labored arguments are in print proving that conservation pays economic dividends. I can add nothing to these arguments. It seems to me, though, that something has gone unsaid. It seems to me that the pattern of the rural landscape, like the configuration of our own bodies, has in it (or should have in it) a certain wholeness. No one censures a man who loses his leg in an accident, or who was born with only four fingers, but we should look askance at a man who amputated a natural part on the grounds that some other is more profitable. The comparison is exaggerated; we had to amputate many marshes, ponds and woods to make the land habitable, but to remove any natural feature from representation in the rural landscape seems to me a defacement which the calm verdict of history will not approve, either as good conservation, good taste, or good farming. . . .

Sometimes I think that ideas, like men, can become dictators. We Americans have so far escaped regimentation by our rulers, but have we escaped regimentation by our own ideas? I doubt if there exists today a more complete regimentation of the human mind than that accomplished by our self-imposed doctrine of ruthless utilitarianism. The saving grace of democracy is that we fastened this yoke on our own necks, and we can cast it off when we want to, without severing the neck. Conservation is perhaps one of the many squirmings which foreshadow this act of self-liberation.

The principle of wholeness in the farm landscape involves, I think, something more than indulgence in land-use luxuries. Try to send your mind up in an airplane; try to see the *trend* of our tinkerings with fields and forests, waters and soils. We have gone in for governmental conservation on a huge scale. Government is slowly but surely pushing the cutovers back into forest; the peat and sand districts back into marsh and scrub. This, I think, is as it should be. But the cow in the woodlot, ably assisted by the ax, the depression, the June beetle, and the drouth, is just as surely making southern Wisconsin a treeless agricultural steppe. There was a time when the cessation of prairie fires added trees to southern Wisconsin faster than the settlers subtracted them. That time is now past. In another generation many southern counties will look, as far as trees are concerned, like the Ukraine, or the Canadian wheatlands. A similar tendency to create *monotypes*, to block up huge regions to a single land-use, is visible in many other states. It is the result of delegating conservation to government. Government cannot own and operate small parcels of land, and it cannot own and operate good land at all.

Stated in acres or in board feet, the crowding of all the timber into one place may be a forestry program, but is it conservation? How shall we use forests to protect vulnerable hillsides and riverbanks from erosion when the bulk of the timber is up north on the sands where there is no erosion? To shelter wildlife when all the food is in one county and all the cover in another? To break the wind when the forest country has no wind, the farm country nothing but wind? For recreation when it takes a week, rather than an hour, to get under a pine tree? Doesn't conservation imply a certain interspersion of land-uses, a certain pepper-and-salt pattern in the warp and woof of the land-use fabric? If so, can government alone do the weaving? I think not.

It is the individual farmer who must weave the greater part of the rug on which America stands. Shall he weave into it only the sober yarns which warm the feet, or also some of the colors which warm the eye and the heart? Granted that there may be a question which returns him the most profit as an individual, can there be *any* question which is best for his community? This raises the question: is the individual farmer capable of dedicating private land to uses which profit the community, even though they may not so clearly profit him? We may be over-hasty in assuming that he is not.

I am thinking, for example, of the windbreaks, the evergreen snow-fences, hundreds of which are peeping up this winter out of the drifted snows of the sandy counties. Part of these plantings are subsidized by highway funds, but in many others the only subsidy is the nursery stock. Here then is a dedication of private land to a community purpose, a private labor for a public gain. These windbreaks do little good until many land-owners install them; much good after they dot the whole countryside. But this "much good" is an undivided surplus, payable not in dollars, but rather in fertility, peace, comfort, in the sense of something alive and growing. It pleases me that farmers should do this new thing. It foreshadows conservation.

It may be remarked, in passing, that this planting of windbreaks is a direct reversal of the attitude which uprooted the hedges, and thus the wildlife, from the entire cornbelt. Both moves were fathered by the agricultural colleges. Have the colleges changed their mind? Or is an Osage windbreak governed by a different kind of economics than a red pine windbreak? . . .

The landscape of any farm is the owner's portrait of himself.

Conservation implies self-expression in that landscape, rather than blind compliance with economic dogma. What kinds of self-expression will one day be possible in the landscape of a cornbelt farm? What will conservation look like when transplanted from the convention hall to the fields and woods?

Begin with the creek: it will be unstraightened. The future farmer would no more mutilate his creek than his own face. If he has inherited a straightened creek, it will be "explained" to visitors, like a pock-mark or a wooden leg.

The creek banks are wooded and ungrazed. In the woods, young straight timber-bearing trees predominate, but there is also a sprinkling of hollow-limbed veterans left for the owls and squirrels, and of down logs left for the coons and fur-bearers. On the edge of the woods are a few wide-spreading hickories and walnuts for nutting. Many things are expected of this creek and its woods: cordwood, posts, and sawlogs; flood-control, fishing and swimming; nuts and wildflowers; fur and feather. Should it fail to yield an owl-hoot or a mess of quail on demand, or a bunch of sweet william or a coon-hunt in season, the matter will be cause for injured pride and family scrutiny, like a check marked "no funds."

Visitors when taken to the woods often ask, "Don't the owls eat your chickens?" Our farmer knows this is coming. For answer, he walks over to a leafy white oak and picks up one of the pellets dropped by the roosting owls. He shows the visitor how to tear apart the matted felt of mouse and rabbit fur, how to find inside the whitened skulls and teeth of the bird's prey. "See any chickens?" he asks. Then he explains that his owls are valuable to him, not only for killing mice, but for excluding other owls which *might* eat chickens. His owls get a few quail and many rabbits, but these, he thinks, can be spared.

The fields and pastures of this farm, like its sons and daughters, are a mixture of wild and tame attributes, all built on a foundation of good health. The health of the fields is their fertility. On the parlor wall, where the embroidered "God Bless Our Home" used to hang in exploitation days, hangs a chart of the farm's soil analyses. The farmer is proud that all his soil graphs point upward, that he has no check dams or terraces, and needs none. He speaks sympathetically of his neighbor who has the misfortune of harboring a gully, and who was forced to call in the CCC [Civilian Conservation Corps]. The neighbor's check dams are a regrettable badge of awkward conduct, like a crutch.

Separating the fields are fencerows which represent a happy balance between gain in wildlife and loss in plowland. The fencerows are not cleared yearly, neither are they allowed to grow indefinitely. In addition to bird song and scenery, quail and pheasants, they yield prairie flowers, wild grapes, raspberries, plums, hazelnuts, and here and there a hickory beyond the reach of the woodlot squirrels. It is a point of pride to use electric fences only for temporary enclosures.

Around the farmstead are historic oaks which are cherished with both pride and skill. That the June beetles did get one is remembered as a slip in pasture management not to be repeated. The farmer has opinions about the age of his oaks, and their relation to local history. It is a matter of neighborhood debate whose oaks are most clearly relics of oak-opening days, whether the healed scar on the base of one tree is the result of a prairie fire or a pioneer's trash pile.

Martin house and feeding station, wildflower bed and old orchard go with the farmstead as a matter of course. The old orchard yields some apples but mostly birds. The bird list for the farm is 161 species. One neighbor claims 165, but there is reason to suspect he is fudging. He drained his pond; how could he possibly have 165?

His pond is our farmer's special badge of distinction. Stock is allowed to water at one end only; the rest of the shore is fenced off for the ducks, rails, redwings, gallinules, and muskrats. Last spring, by judicious baiting and decoys, two hundred ducks were induced to rest there a full month. In August, yellow-legs use the bare mud of the water-gap. In September the pond yields an armful of waterlilies. In the winter there is skating for the youngsters, and a neat dozen of rat-pelts for the boys' pin-money. The farmer remembers a contractor who once tried to talk drainage. Pondless farms, he says, were the fashion in those days; even the Agricultural College fell for the idea of making land by wasting water. But in the drouths of the thirties, when the wells went dry, everybody learned that water, like roads and schools, is community property. You can't hurry water down the creek without hurting the creek, the neighbors, and yourself.

The roadside fronting the farm is regarded as a refuge for the prairie flora: the educational museum where the soils and plants of pre-settlement days are preserved. When the professors from the college want a sample of virgin prairie soil, they know they can get it here. To keep this roadside in prairie, it is cleaned annually, always by burning, never by mowing or cutting. The farmer tells a funny story of a highway engineer who once started to grade the cutbanks all the way back to the fence. It developed that the poor engineer, despite his college education, had never learned the difference between a silphium and a sunflower. He knew his sines and cosines, but he had never heard of the plant succession. He couldn't understand that to tear out all the prairie sod would convert the whole roadside into an eyesore of quack and thistle.

In the clover field fronting the road is a huge glacial erratic of pink granite. Every year, when the geology teacher brings her class out to look at it, our farmer tells how once, on a vacation trip, he matched a chip of the boulder to its parent ledge, two hundred miles to the north. This starts him on a little oration on glaciers; how the ice gave him not only the rock, but also the pond, and the gravel pit where the kingfisher and the bank swallows nest. He tells how a powder salesman once asked for permission to blow up the old rock "as a demonstration in modern methods." He does not have to explain his little joke to the children.

He is a reminiscent fellow, this farmer. Get him wound up and you will hear many a curious tidbit of rural history. He will tell you of the mad decade when they taught economics in the rural kindergarten, but the college president couldn't tell a bluebird from a blue cohosh. Everybody worried about getting his share; nobody worried about doing his bit. One farm washed down the river, to be dredged out of the Mississippi at another farmer's expense. Tame crops were over-produced, but nobody had room for wild crops. "It's a wonder this farm came out of it without a concrete creek and a Chinese elm on the lawn." This is his whimsical way of describing the early fumblings for "conservation."

Helen Nearing and Scott Nearing

Contemporaries of back-to-the-landers Ralph Borsodi and Louis Bromfield, Helen and Scott Nearing did not publish their account of their experiments on the land until 1954. By then, after more than two decades of simple living, their message had earned them a following among those seeking a sustainable alternative lifestyle. The Nearings found many new disciples among those who returned to the land during the 1960s and 1970s.

In 1932 the Nearings moved from New York City to the hills of northern New England, looking to establish self-sufficiency on their land. A critical component of their subsistence plan was that their work cycle would include time to engage in contemplation and their avocations. This permission to focus on the life of the mind meant that they continued to write and lecture extensively, and by the early 1970s, the Nearings had written over fifty books between the two of them. They sought greater freedom on the farm, rather than less, and so they were willing to trade some material abundance for that leisure time. A maple syrup operation provided most of their cash income, along with the material for another widely read book.

Like many contemporary agrarians, the Nearings eschewed modern technologies and chemical inputs, and chose not to buy expensive tools—borrowing or trading if they needed heavy equipment. The coherency of the Nearings' vision won them a readership that remains strong even today. As one commentator observed, "By trying to get back to a simpler life they find themselves ahead of their time. That is why they are fast becoming culture heroes, models for a generation of young people." The Nearings left Vermont in 1952 after the unbridled development from the boom in skiing at nearby Stratton Mountain changed the culture of their community. They moved to the Maine coast, where they established a self-sufficient homestead from the ground up all over again.

The impact of the Nearings' experiment in country living reached its apotheosis during the early 1970s, at the height of that era's back-to-the-land movement. Between 1970 and

Helen Nearing picking berries, no date. From the
Scott and Helen Nearing Papers at the Thoreau Institute
at Walden Woods, Lincoln, MA. Courtesy of the
Walden Woods Project and the Good Life Center.

1972, Living the Good Life *sold over sixty-five thousand new copies, and hundreds of*
visitors traveled to the Nearings' Maine farm to witness their ongoing work. The following
excerpts from Living the Good Life *describe the vision that motivated their experiments*
in living close to the land.

From *Living the Good Life: How to Live Sanely and Simply in a Troubled World* (1954)

PREFACE

This is a book about a twentieth century pioneering venture in a New England
community. Most of the subject matter is derived from twenty years of living in the
backwoods of Vermont. The book aims to present a technical, economic, sociological and psychological report on what we tried to do, how we did it, and how well or
ill we succeeded in achieving our purposes.

Scott Nearing hoeing, no date. From the Scott and
Helen Nearing Papers at the Thoreau Institute
at Walden Woods, Lincoln, MA. Courtesy of the
Walden Woods Project and the Good Life Center.

During the deepest part of the Great Depression, in 1932, we moved from New York City to a farm in the Green Mountains. At the outset we thought of the venture as a personal search for a simple, satisfying life on the land, to be devoted to mutual aid and harmlessness, with an ample margin of leisure in which to do personally constructive and creative work. With the passage of time and the accumulation of experience we came to regard our valley in Vermont as a laboratory in which we were testing out certain principles and procedures of more general application and concern.

It was, of course, an individual experience, meeting a special need, at a particular time. When we moved to Vermont we left a society gripped by depression and unemployment, falling a prey to fascism, and on the verge of another world-wide military free-for-all; and entered a pre-industrial, rural community. The society from which we moved had rejected in practice and in principle our pacifism, our vegetarianism and our collectivism. So thorough was this rejection that, holding such views, we could not teach in the schools, write in the press or speak over the radio, and were thus denied our part in public education. Under these circumstances, where could outcasts from a dying social order live frugally and decently, and at the same time have sufficient leisure and energy to assist in the speedy liquidation of the disintegrating society and to help replace it with a more workable social system? . . .

We had tried living in several cities, at home and abroad. In varying degrees we met the same obstacles to a simple, quiet life,—complexity, tension, strain, artificiality, and heavy overhead costs. These costs were payable only in cash, which had to be earned under conditions imposed upon one by the city,—for its benefit and advantage. Even if cash income had been of no concern to us, we were convinced that it was virtually impossible to counter city pressures and preserve physical health, mental balance and social sanity through long periods of city dwelling. After careful consideration we decided that we could live a saner, quieter, more worthwhile life in the country than in any urban or suburban center.

We left the city with three objectives in mind. *The first was economic.* We sought to make a depression-free living, as independent as possible of the commodity and labor markets, which could not be interfered with by employers, whether businessmen, politicians or educational administrators. *Our second aim was hygienic.* We wanted to maintain and improve our health. We knew that the pressures of city life were exacting, and we sought a simple basis of well-being where contact with the earth, and home-grown organic food, would play a large part. *Our third objective was social and ethical.* We desired to liberate and dissociate ourselves, as much as possible, from the cruder forms of exploitation: the plunder of the planet; the slavery of man and beast; the slaughter of men in war, and of animals for food.

We were against the accumulation of profit and unearned income by non-producers, and we wanted to make our living with our own hands, yet with time and leisure for avocational pursuits. We wanted to replace regimentation and coercion with respect for life. Instead of exploitation, we wanted a use economy. Simplicity should take the place of multiplicity, complexity and confusion. Instead of the hectic mad rush of busyness we intended a quiet pace, with time to wonder, ponder and observe. We hoped to replace worry, fear and hate with serenity, purpose and at-one-ness.

After twenty years of experience, some of it satisfactory and some of it quite the reverse, we are able to report that:

1. A piece of eroded, depleted mountain land was restored to fertility, and produced fine crops of high quality vegetables, fruits and flowers.
2. A farm economy was conducted successfully without the use of animals or animal products or chemicalized fertilizers.
3. A subsistence homestead was established, paying its own way and yielding a modest but adequate surplus. About three-quarters of the goods and services we consumed were the direct result of our own efforts. Thus we made ourselves independent of the labor market and largely independent of the commodity markets. In short, we had an economic unit which depression could affect but little and which could survive the gradual dissolution of the United States economy.

4. A successful small-scale business enterprise was organized and operated, from which wagery was virtually eliminated.
5. Health was maintained at a level upon which we neither saw nor needed a doctor for the two decades.
6. The complexities of city existence were replaced by a fairly simple life pattern.
7. We were able to organize our work time so that six months of bread labor each year gave us six months of leisure, for research, travelling, writing, speaking and teaching.
8. In addition, we kept open house, fed, lodged, and visited with hundreds of people, who stayed with us for days or weeks, or much longer.

We have not solved the problem of living. Far from it. But our experience convinces us that no family group possessing a normal share of vigor, energy, purpose, imagination and determination need continue to wear the yoke of a competitive, acquisitive, predatory culture. Unless vigilante mobs or the police interfere, the family can live with nature, make themselves a living that will preserve and enhance their efficiency, and give them leisure in which they can do their bit to make the world a better place. . . .

Chapter 1. We Search for the Good Life

Many a modern worker, dependent on wage or salary, lodged in city flat or closely built-up suburb and held to the daily grind by family demands or other complicating circumstances, has watched for a chance to escape the cramping limitations of his surroundings, to take his life into his own hands and live it in the country, in a decent, simple, kindly way. Caution, consideration for relatives or fear of the unknown have proved formidable obstacles, however. After years of indecision he still hesitates. Can he cope with country life? Can he make a living from the land? Has he the physical strength? Must one be young to start? Where can he learn what he needs to know? Can he build his own house? Can he feed his family from the garden? Must he keep animals? How much will a farm tie him down? Will it be but a new kind of drudgery all over again? These and a thousand other questions flood the mind of the person who considers a break with city living.

This book is written for just such people. We maintain that a couple, of any age from twenty to fifty, with a minimum of health, intelligence and capital, can adapt themselves to country living, learn its crafts, overcome its difficulties, and build up a life pattern rich in simple values and productive of personal and social good.

Changing social conditions during the twenty years that began in 1910 cost us our professional status and deprived us of our means of livelihood. Whether we liked it or not we were compelled to adjust to the new situation which war, revolution and depression had forced upon the western world. Our advancing age (we were approaching fifty) certainly played some part in shifting our viewpoint, but of far greater consequence were the world developments.

Beyond these social pressures our choices were in our own hands, and their consequences would descend upon our own heads. We might have stayed on in the city, enduring and regretting what we regarded as essentially unsatisfactory living conditions, or we might strike out in some other direction, perhaps along a little-used path.

After a careful first-hand survey of developments in Europe and Asia, as well as in North America, we decided that western civilization would be unable hence-forth to provide an adequate, stable and secure life even for those who attempted to follow its directives. If profit accumulation in the hands of the rich and powerful continued to push the economy toward ever more catastrophic depressions; if the alternative to depression, under the existing social system, was the elimination of the unmarketable surplus through the construction and uses of ever more deadly war equipment, it was only a question of time before those who depended upon the system for livelihood and security would find themselves out in the cold or among the missing. In theory we disapproved of a social order activated by greed and functioning through exploitation, acquisition and accumulation. In practice, the outlook for such a social pattern seemed particularly unpromising because of the growing nationalistic sentiments among colonial peoples and the expanding collectivist areas. Added to this, the troubles which increasingly bedeviled western man were most acute at the centers of civilization and were multiplying as the years passed. Under these conditions we decided that we could not remain in the West and live a good life unless we were able to find an alternative to western civilization and its outmoded culture pattern.

Was there an alternative? We looked in three directions for an answer. First we considered and rejected the possibility of living abroad as refugees from what was for us a revolting and increasingly intolerable social situation. Even two decades ago, in the early 1930s, movement was far easier than it is today. In a very real sense, the world lay open before us. Where should we go in search of the good life? We were not seeking to escape. Quite the contrary, we wanted to find a way in which we could put more into life and get more out of it. We were not shirking obligations but look-ing for an opportunity to take on more worthwhile responsibilities. The chance to help, improve and rebuild was more than an opportunity. As citizens, we regarded it as an assignment. Therefore, we decided not to migrate.

As a second alternative to staying in the urban culture pattern of the West, we checked over the possibilities of life in a cooperative or an intentional community. In the late 1920s the chances of such a solution were few, far between and unprom-ising. We would have preferred the cooperative or communal alternative, but our experience, inquiries and investigations convinced us that there were none available or functioning into which we could happily and effectively fit.

Finally, we decided on the third alternative, a self-sufficient household economy, in the country, and in the United States, which we would try to make solvent, efficient and satisfying. Having made this decision, our next task was to define our purposes and adjust them to the possibilities of our situation.

We were seeking an affirmation,—a way of conducting ourselves, of looking at the world and taking part in its activities that would provide at least a minimum of those values which we considered essential to the good life. As we saw it, such values must include: simplicity, freedom from anxiety or tension, an opportunity to be useful and to live harmoniously. Simplicity, serenity, utility and harmony are not the only values in life, but they are among the important ideals, objectives and concepts which a seeker after the good life might reasonably expect to develop in a satisfactory natural and social environment. As things stand today, it is not this combination of values, but rather their opposite (that is, complexity, anxiety, waste, ugliness and uproar) which men associate with the urban centers of western civilization.

Our second purpose was to make a living under conditions that would preserve and enlarge joy in workmanship, would give a sense of achievement, thereby promoting integrity and self-respect; would assure a large measure of self-sufficiency and thus make it more difficult for civilization to impose restrictive and coercive economic pressures, and make it easier to guarantee the solvency of the enterprise.

Our third aim was leisure during a considerable portion of each day, month or year, which might be devoted to avocational pursuits free from the exacting demands of bread labor, to satisfying and fruitful association with one's fellows, and to individual and group efforts directed toward social improvement.

Our search for the good life brought us face to face with several immediate questions: Where to live the good life? How to finance the enterprise? And finally there was the central problem of how to live the good life once we had found the place and the economic means. . . .

We took our time, and during many months looked through the north-eastern states. Finally we settled on Vermont. We liked the thickly forested hills which formed the Green Mountains. The valleys were cosy, the people unpretentious. Most of the state was open and wild, with little of the suburban or summer vacation atmosphere.

We also picked Vermont for economy's sake. In New York, New Jersey and Eastern Pennsylvania, where we first inquired, land values were high, even in the depression years. By comparison, the prices and costs in Vermont were reasonable.

Where should we go in Vermont? On the map it is a small state compared with some of its big neighbors. From a distance it seemed an easy matter to take a run through the area and check its possibilities, but when we reached the Green Mountains with their steep, curving highways and began to thread our way through the endless mazes of back roads or went on foot from valley to valley, along logging roads and trails which lost themselves in thickets of underbrush which choke the hillsides, Vermont looked big and baffling. We decided that we needed help. We read the farm ads, gratefully accepted suggestions from friends, and finally fell into the amiable clutches of ex-farmer and present real-estate salesman, L. P. Martin of Newfane, Vermont. . . .

After escorting us around the southern part of the state for three consecutive days, they sold us a farm in the town of Winhall. Actually, we bought the first farm they showed us, but between that first view and the purchase we looked at dozens of others. None appealed to us as much as the old Ellonen place, in the Pikes Falls valley which covers part of three townships—Stratton, Winhall, and Jamaica. So we went back there on a chill day in the autumn of 1932, and signed an agreement to buy the place.

Its setting and view are lovely. Nestled against a northern slope, the Ellonen place looks up at Stratton Mountain and "The Wilderness," a name applied to the 25,000 acre pulp reserves owned by paper companies. Stratton is a wild, lonely, heavily wooded 4,000 foot mountain, inhabited by 50 or 60 people, where in Daniel Webster's time there had been 1500. "A few score abandoned farms, started in a lean land, held fiercely so long as there was any one to work them, and then left on the hill-sides. Beyond this desolation are woods where the bear and the deer still find peace, and sometimes even the beaver forgets that he is persecuted and dares to build his lodge" [from Rudyard Kipling's *Letters of Travel*].

Our new place was a typical run-down farm, with a wooden house in poor repair, a good-sized barn with bad sills and a leaky roof, a Finnish bath house, and 65 acres of land from which the timber had been cut. "Conveniences" consisted of a pump and a black iron sink in the kitchen and a shovel-out backhouse at one end of the woodshed. The place had a plenteous spring of excellent water, a meadow, a swamp or two, and some rough land facing south and stretching perhaps a third of a mile up Pinnacle Mountain, which lay to the east of Stratton [Mountain]. The farm was located on a dirt road seven miles from the Jamaica Post Office and two miles from the hamlet of Bondville. Both villages together had under 600 people in them, and along our ten mile stretch of back-road there were not more than a dozen families. . . .

CHAPTER 2. OUR DESIGN FOR LIVING

. . . We were in the country. We had land. We had all the wood we could use, for the cutting. We had an adequate supply of food from the gardens. We had time, a purpose, energy, enough ingenuity and imagination, a tiny cash income from maple and a little cash money on hand.

We were on a run-down, run-out farm. We were living in a poorly built wooden house through which the winter winds swept like water through a sieve. We owned a timber tract that would come into its own only in twenty to thirty years. We owned the place next door, another run-down farm, equipped with wretched buildings. Our soil was swampy, rough and rocky, mostly covered with second growth, but there was a small amount of good timber left on it. Our gardens were promising, but the main garden was too low and wet to be really productive.

We were in good health. We were solvent in that we had no debts. We were fairly hopeful of the future, but inexperienced in the ways of subsistence living and somewhat uncertain as to how we should proceed. After due consideration and the spirit of the times, we drew up a ten year plan.

This plan was not made out of whole cloth, all at once. It was modified by experience, as we went along. It was flexible, but in principle and usually in practice we stuck to it. Suppose we set down the main points which the plan covered when we outlined it in the middle 1930's.

1. *We wish to set up a semi-self-contained household unit, based largely on a use economy, and, as far as possible, independent of the price-profit economy which surrounds us.*

The Great Depression had brought millions of bread-winners face to face with the perils which lurked for those who, in a commodity economy based on wage-paid labor, purchase their livelihood in the open market. The wage and salary workers did not own their own jobs, nor did they have any part in deciding economic policy nor in selecting those who carried policy into effect. The many unemployed in 1932 did not lose their jobs through any fault of their own, yet they found themselves workless, in an economy based on cash payment for the necessaries and decencies. Though their incomes had ceased, their outgo for food, shelter and clothing ate up their accumulated savings and threw them into debt. Since we were proposing to go on living in this profit-price economy, we had to accept its dread implications or find a workable alternative. We saw that alternative in a semi-subsistence livelihood.

We would attempt to carry on this self-subsistent economy by the following steps: (1) Raising as much of our own food as local soil and climatic conditions would permit. (2) Bartering our products for those which we could not or did not produce. (3) Using wood for fuel and cutting it ourselves. (4) Putting up our own buildings with stone and wood from the place, doing the work ourselves. (5) Making such implements as sleds, drays, stone-boats, gravel screens, ladders. (6) Holding down to the barest minimum the number of implements, tools, gadgets and machines which we might buy from the assembly lines of big business. (7) If we had to have such machines for a few hours or days in a year (plough, tractor, rototiller, bull-dozer, chain-saw), we would rent or trade them from local people instead of buying and owning them.

2. *We have no intention of making money, nor do we seek wages or profits. Rather we aim to earn a livelihood, as far as possible on a use economy basis. When enough bread labor has been performed to secure the year's living, we will stop earning until the next crop season.*

Ideas of "making money" or "getting rich" have given people a perverted view of economic principles. The object of economic effort is not money, but livelihood. Money cannot feed, clothe or shelter. Money is a medium of exchange,—a means of securing the items that make up livelihood. It is the necessaries and decencies which are important, not the money which may be exchanged for them. And money must be paid for, like anything else. Robert Louis Stevenson wrote in *Men and Books*, "Money is a commodity to be bought or not to be bought, a luxury in which we may either indulge or stint ourselves, like any other. And there are many luxuries that we may legitimately prefer to it, such as a grateful conscience, a country life, or the woman of our inclination."

People brought up in a money economy are taught to believe in the importance of getting and keeping money. Time and again folk told us, "You can't afford to make syrup. You won't make any money that way." One year a neighbor, Harold Field, kept a careful record of the labor he put in during the syrup season and of the sale price of his product, and figured that he got only 67 cents an hour for his time. In view of these figures, the next year he did not tap out because sugaring paid less than wage labor. But, during that syrup season he found no chance to work for wages, so he didn't even make the 67 cents an hour.

Our attitude was quite different. We kept careful cost figures, but we never used them to determine whether we should or should not make syrup. We tapped our trees as each sap season came along. Our figures showed us what the syrup had cost. When the season was over and the syrup on hand, we wrote to various correspondents in California or Florida, told them what our syrup had cost, and exchanged our product for equal value of their citrus, walnuts, olive oil or raisins. As a result of these transactions, we laid in a supply of items at no cash outlay, which we could not ourselves produce. Our livelihood base was broadened as the result of our efforts in the sugar bush and the sap house.

We also sold our syrup and sugar on the open market. In selling anything, we tried to determine exact costs and set our prices not in terms of what the traffic would bear but in terms of the costs,—figuring in our own time at going day wages.

Just as each year we estimated the amount of garden produce needed for our food, so we tried to foresee the money required to meet our cash obligations. When we had the estimated needs, we raised no more crops and made no more money for that period. In a word, we were trying to make a livelihood, and once our needs in this direction were covered, we turned our efforts in other directions,—toward social activities, toward avocations such as reading, writing, music making, toward repairs or replacements of our equipment.

3. *All of our operations will be kept on a cash and carry basis. No bank loans. No slavery to interest on mortgages, notes and I.O.U.'s.*

Under any economy, people who rent out money live on easy street. Whether as individuals or banking establishments, they lend money, take security and live on a rich harvest of interest and the proceeds of forced sales. The money lenders are able to enjoy comfort and luxury, without doing any productive labor. It is the borrowing producers who pay the interest or lose their property. Farmers and home owners by the thousands lost everything they had during the Great Depression because they could not meet interest payments. We decided to buy for cash or not at all. . . .

6. *So long as the income from the sale of maple syrup and sugar covers our needs we will not sell anything else from the place. Any garden or other surpluses will be shared with neighbors and friends in terms of their needs.*

This latter practice was carried out generally in the valley. Rix Knight had extra pear trees. In a good season he distributed bushels to any of us who had no pears. Jack Lightfoot let us pick his spare apples and let others cut Christmas greens, free of charge. We brought firewood to those who needed it, and many garden products. Our chief delight was growing, picking and giving away sweet peas. We grew these in profusion,—double rows 60 to 100 feet long, each year. Whenever taking a trip to town in blooming season (July to frost of late September) we filled baskets and basins with dozens of bunches and gave them out during the day to friends and strangers alike. Grocers, dentist friends, gas station attendants, utter strangers on the street,—all were the delighted recipients of the fragrant blossoms. One woman, after endeavoring to pay for a large bunch, was heard to go off muttering, "I've lived too near New York too long to understand such practices."

7. *We will keep no animals.* Almost without exception, Vermont farmers have animals, often in considerable variety. We do not eat animals, or their products, and do not exploit them. We thus escape the servitude and dependence which tie both farmer and animal together. The old proverb "No man is free who has a servant" could well read "No man is free who has an animal."

Animal husbandry on a New England farm involves building and maintaining not only sheds but barns and the necessary fences, and also the cutting or buying of hay. Into this enterprise goes a large slice of the farmer's time. Farm draft animals work occasionally but eat regularly. Many of them eat more than they produce and thus are involuntary parasites. All animals stray at times, even with the best of fences, and like all runaway slaves, must be followed and brought back to servitude. The owners of horses, cattle, pigs and chickens wait on them regularly, as agrarian chamber maids, feeding, tending them and cleaning up after them. Bernard Shaw has said: "Millions of men, from the shepherd to the butcher, become mere valets of animals while the animals live, and their executioners afterwards."

We believe that all life is to be respected—non-human as well as human. Therefore, for sport we neither hunt nor fish, nor do we feed on animals. Furthermore, we prefer, in our respect for life, not to enslave or exploit our fellow creatures. Widespread and unwarranted exploitation of domestic animals includes robbing them of their milk or their eggs as well as harnessing them to labor for man. Domestic animals, whether cows, horses, goats, chickens, dogs or cats are slaves. Humans have the power of life or death over them. Men buy them, own them, sell them, work them, abuse and torture them and have no compunctions against killing and eating them. They compel animals to serve them in multitudinous ways. If the animals resist, rebel or grow old, they are sent to the butcher or else are shot out of hand.

Cats and dogs live dependent subservient lives under the table tops of humans. Domestic pets kill and drive away wild creatures, whose independent, self-respecting lives seem far more admirable than those of docile, dish-fed retainers. We enjoy the wild creatures, and on the whole think they are more lithe, beautiful and healthy than the run of cats and dogs, although some of our best friends in Vermont have been canine and feline neighbors.

While remaining friends with all kinds of animals, we preferred to be free from dependents and dependence. Many a farmer, grown accustomed to his animal-tending chores and to raising food for animals instead of for himself, could thus find his worktime cut in half.

EARL L. BUTZ

Earl Butz was born in 1909 on a 160-acre livestock farm in the northeastern Indiana community of Albion. He studied at Purdue University, received Purdue's first PhD in agricultural economics in 1937, and worked for the rest of his career as a farm economist and federal policymaker.

Butz is best known for his work in government; he served as assistant secretary of agriculture from 1954 to 1957 under Ezra Taft Benson in the Eisenhower administration, and as secretary of agriculture under Presidents Nixon and Ford from 1971 to 1976. Early in his tenure as secretary, in 1972, Butz engineered a $1 billion grain sale to the Soviet Union, providing most of the American wheat reserve to the USSR—just on the brink of a Midwestern drought that dropped American yields and led to domestic shortages. Consequently, grain prices rose, and in response Butz encouraged farmers to plant "fencerow to fencerow." This launched another phase of expansion and consolidation in American agriculture, characterized by Butz's famous catchphrases, like (echoing Benson) "Get big or get out" or "Get bigger, get better, or get out."

The expansion of agriculture during this period helped to solidify the U.S. government's cheap food policy that has dominated ever since, a policy rooted in the thinking of Edwin Nourse and other economists earlier in the century. Butz's response to rising food prices in American supermarkets in 1973 was to encourage huge yields and use government payments to farmers to subsidize low prices for processors and consumers. As Butz once said, "That's the basis of our affluence now, the fact that we spend less on food. It's America's best-kept secret." This superabundance of cheap grain underpins factory meat production, corn syrup, and other key elements of the modern industrial food economy.

Through speeches and policy prescriptions, the Department of Agriculture under Butz encouraged agricultural consolidation and efficiency, as the following excerpt demonstrates. Ultimately, the high prices of the 1970s and Butz's encouragement drove many farmers into debt in order to expand their operations, which paved the way for the farm crisis of the 1980s, when inflation and declining prices squeezed the farmers who had borrowed heavily the decade before.

From "Agribusiness in the Machine Age" (1960)

Modern agriculture is much broader than the narrow dictionary definition—"the art or science of cultivating the ground." It is the whole business of supplying food and fiber for a growing population at home and abroad.

The art or science of cultivating the ground is but one link in the long chain of feeding and clothing people. The chain begins many jobs before we reach the farm and continues several processes after our newly produced food and fiber leave the farm gate. For this whole complex of agricultural production and distribution functions some persons use the term "agribusiness."

The United States has shifted in the past century from a predominantly agricultural economy to an industrial and commercial economy. The industrial sector is now beginning to give way, at least relatively, to a growing variety of personal services. This is possible primarily because we have been able to specialize in our production and processing of food and fiber by the increased use of science, technology, and mechanization and an increasing output per worker in farming and in agribusiness. We have transferred many farm jobs off the farm, such as the shift of farm power from horses and oats to tractors and gasoline, but we do the entire job with much less manpower than formerly.

Agriculture is now in the middle of its third great revolution. Agricultural engineers have had a big part in all three revolutions.

The first revolution came in the middle of the 19th century, when we began to substitute animal energy for human energy. The invention of the reaper is the best known event associated with it. This and other developments called for considerable retooling in agriculture. They increased output per worker on farms and started us on the path of feeding our growing Nation with a constantly shrinking proportion of our total population in the field. Agriculture began to take on some characteristics of a commercial enterprise, although sometimes the change was almost imperceptible.

The second great revolution began in the 1920's, with the substitution of mechanical energy for animal energy. It likewise increased the commercialization of agriculture, shifted a number of production functions off the farm, increased output per worker substantially, and resulted in a further reduction in the proportion of our total working population on farms.

The third revolution is the undergirding of agricultural production and marketing with vast amounts of science, technology, and business management. This revolution has been in progress for a decade or two, but at an accelerated pace during the past few years. This revolution is transferring still additional production and marketing functions off the farm and continues to underscore the importance of specialization at all levels of the agribusiness complex.

Agricultural historians a generation hence may characterize the decade of the 1950's as the decade of the scientific breakthrough. In this decade we experienced an unprecedented number of discoveries, which have changed agriculture from stem to stern.

The decade of the 1960's opens with the march of agricultural science in full stride. Agriculture is changing from a way of living to a way of making a living. It is changing from a business of arts and crafts to a business undergirded with large amounts of science and technology.

It is wrong to think of agriculture as a declining industry.

American agriculture is an expanding industry in every important respect except one—the number of people required to run our farms. Our agricultural plant each year uses more capital, more science and technology, more managerial capacity, more purchased production inputs, more specialized marketing facilities, and more research than the year before.

We do not think of air transportation as a declining industry just because a pilot in a jet airliner can now take 100 passengers from coast to coast in half a day, compared with 20 passengers in a day and a half two decades ago. This, like agriculture, is a strong and growing industry. . . .

The declining trend in farm population is itself a sign of a strong agriculture. Brainpower has replaced horsepower as the essential ingredient on our farms. The total United States agricultural output increased by two-thirds in the past two decades, while the number of farmworkers declined some 3 million. This means that production per worker on our farms has doubled in the past 20 years. This remarkable increase in production efficiency can be matched by no other major sector of the American economy.

Progress of this kind can be continued only if we have capable and well-informed men on our farms. We will need fewer farmers in the future, but they must be better. They will be operating on a fast track, and the race will go to the swift.

We MUST broaden our thinking about agriculture to include the businesses that supply our farmers with items used in production, as well as the processing and distributing concerns that handle the food and fiber produced on farms. . . .

The agricultural world and the industrial world are not two separate communities with merely a buyer-seller relationship. They are so bound together and so interrelated that we must think of them jointly if we are to reach sound conclusions about either one.

The modern commercial farm resembles a manufacturing plant in many respects. The large amount of equipment in use on the farm represents a substitution of capital and machinery for labor. . . .

The modern farm operator is much less self-sufficient than his father was. He buys many goods and services needed in his production that father produced on the farm. In a very real sense, he assembles "packages of technology" that have been put together by others on a custom basis. For example, he buys his tractors and petroleum, whereas his father produced horses and oats. Think for a moment of the technology that goes into the modern feedbag, with its careful blending of proteins, antibiotics, minerals, and hormones, as contrasted with the ear corn and a little tankage put out for the hogs in his grandfather's day.

This development obviously calls for a high level of managerial capacity. The manager of the modern commercial family farm must make more managerial decisions each week covering a much wider range of subject matter than does his counterpart in the city. He has more capital invested, takes greater risks, faces stiffer competition, and has more opportunity for reward if he does a good job. . . .

Today's farm production is a synthesis of several scientific disciplines.

The earning capacity of the average farmer used to be limited primarily by his physical strength and the amount of work he could do. He substituted some animal muscle for human muscle, but not a great deal. He substituted very little mechanical energy for muscle power. Agriculture was primarily a means of converting muscle energy into farm produce.

Human energy is much less important in today's farm operation. Energy can be purchased so much more cheaply than it can be provided by man. Today's farm operator is a combination manager-applicator of the life sciences, the physical sciences, and the social sciences. The research undergirding modern agriculture ranges all the way from physics to physiology, from biology to business. It is just as complex and just as far on the periphery of knowledge as is the research done in the laboratories of the nuclear scientist, for example.

The first claim of any society upon its total population resources is to get enough food to keep the population alive and well. This is true in primitive societies, in semi-developed societies and in highly developed societies. We do this so efficiently in this country that almost nine-tenths of our population is available to produce the wide variety of goods and services that make up the American standard of living. . . .

Inasmuch as the typical food production cycle is annual, and the human hunger cycle is daily, it became necessary to devise means of preserving and storing an annual food supply to meet daily food needs in locations far removed from areas of production. The result was the development of a commercial food processing and distributing industry which today feeds our vast urban population much better than their farmer ancestors fared. The national diet has improved materially in terms of quality and variety.

The interdependence of the various segments of the agribusiness chain is obvious. When these functions were mostly performed on the farm, there was a high degree of integration among them. The individual family farm saw to this, for failure to do so would mean loss of income and perhaps hunger.

In recent times, especially in some kinds of commodities, there has been a pronounced tendency to integrate the various functions of production and marketing through contractual arrangements of one kind or another.

This process has come to be known as vertical integration. Although contractual integration arrangements are controversial and are viewed with suspicion by some people, they are a manifest effort on the part of the industry to seek such economies as can be attained by careful coordination of the entire chain of production, processing, and marketing.

Vertical integration is essentially an attempt to combine the advantages of specialization in modern society with the good features of a system in which all the steps were fully coordinated, as they had to be on grandfather's farm. A certain amount of vertical integration is inevitable—and beneficial—in the kind of agribusiness we have today. . . .

With such large amounts of capital and technology involved, management has become the key factor in successful farm operation. This is in sharp contrast to a generation or two ago, when the farm unit was much more self-sufficient than now, with much less capital involved, with much less science applied, and with many fewer critical managerial decisions to be made.

The very cornerstone of our high standard of life is our ever-increasing efficiency in the production and marketing of food and fiber, made possible by the specialized functions that characterize agribusiness. Increased efficiency in production and distribution of food and fiber and the subsequent release of manpower for other work are the first prerequisites for an industrialized society. The first claim of any organized society on its total production resources is food. The cry for food has echoed through the ages. Food remains man's first physical need.

VICTOR DAVIS HANSON

Victor Davis Hanson, born in 1953, was raised in California's San Joaquin Valley. His mother was a state judge and his father a school administrator, but extended family members were farmers. They raised fruit trees and grape vines in Selma, near Fresno. Hanson attended Santa Cruz State University as an undergraduate, then received his PhD in classics from Stanford in 1980.

He has led a varied career that included full-time farming as well as teaching and writing both scholarly and polemical works. Hanson has written extensively on military history and its intersection with agricultural life and practices in rural Greece. In his 1995 book The Other Greeks, *he argues that the signal achievements of classical Greece were rooted in the countryside and its independent farmers, rather than in Athens or other cities. From the farmer, Greece got not only its food but also its democratic bent and its regard for the individual.*

To Hanson, the agrarian roots of Greek culture issued in a legacy that continues, barely, to the present day. It has been renewed through contact with the land and the struggle of individual farmers to produce food from an intractable, capricious natural world. Hanson's agrarianism is that not of Virgil and his Georgics, *but of hard-bitten Hesiod. Hanson described the latter's* Works and Days *as "a more melancholy, more angry account of the necessary pain and sacrifice needed to survive on the land, [which] did not find, and has not found, Virgil's popular audience. No wonder: Hesiod's soil* is not kind, but unforgiving, and so must be mastered if it is not to master the farmer himself."

What is true of the land is no less true of human nature, according to Hanson: "Is not the vanishing agrarian the true heir of Western culture? In the spirit of the Greeks, he, nearly alone now in this country, believes that man, like wild species of trees and vines, is feral. As orchards and vineyards are tamed by agriculture, so too culture—law, statute, tradition, and custom—domesticates man, teaches him to become productive, and so forces us all to repress and abandon our innate savagery."

In the following excerpt from his book The Land Was Everything, Hanson reflects on his return to farming after leaving academia in the 1980s. He reflects as well on the place of the agrarian in a society whose connection to the land is increasingly tenuous.

From *The Land Was Everything: Letters from an American Farmer* (2000)

How I hate to dwell in these accumulated and crowded cities! They are but the confined theatre of cupidity; they exhibit nothing but the action and reaction of a variety of passions which, being confined within narrower channels, impel one another with the greatest vigour. The same passions are more rare in the country; and, from their greater extent and expansion, they are but necessary gales. I always delight in the country. Have you never felt at the returning of spring a glow of general pleasure, an indiscernible something that pervades our whole frame, an inward involuntary admiration of everything which surrounds us?
J. HECTOR ST. JOHN DE CRÈVECOEUR, *Sketches of Eighteenth-Century America*

I. COUNTRY VERSUS CITY

Most professionals in America always suspect that a farmer and his ilk are either ignorant or crass, and surely not professionals. The farmer, in turn, believes that you would have to be crazy to live in town, crazier still to wear such garb and endure such reproach at the office, where the sleek and stupid can as easily excel as the more real and intelligent fail. Each is the absolute antithesis of the other, the agrarian requiring action above all, the professional anything but force and audacity. Absolutes versus nuance; natural ill manners at odds with studied refinement. The former, of the outdoors, values independence and commitments that are ironclad, the latter, inside, often sees those very ideals as sheer recklessness and obstinacy. The latter likes—no, must have—associations and committees; the former says more than one is too crowded. The rank world of the farmer says, "Don't back down." "Go around," orders the tame cosmos of the urban employee. Is it to be "Let's settle it right now" or "Let's be sure at least to do lunch sometime"? The farmer on his own lives concretely for liberty that can end him, the professional in town for equality that protects him. Country folk fear nature in their midst; city dwellers idolize it at a distance.

No other occupation in America is akin to what we once were than the family farmer, this iconoclast of the countryside, who flees from consensus and equanimity, this oddball who is the stuff of history, with its wars, prejudices, depressions, strifes, and tragedies. The student who visits the farm for a day goes away troubled and unsure whether the world is a logical and kind place, the auditor on campus more than likely hears how steps are being taken to solve the woes of the universe. How odd that Enlightenment thinkers once thought that when America had more teachers than farmers, all the old ignorance would fade, not grow. But grow it did— all the more so.

Clearly, the 99 percent in town do not know in any direct way the 1 percent on the farm; the latter experience the world of the former every day on television, in the mail, and on the phone. And the divide will only get worse, until there is no divide and we are all of the nonfarm world. Agriculture in the millennium to come will for the most part be conducted in vast expanses, away from town—the corporate void where no sane man wishes to live. This vast stretch of latifundia will be as Crèvecoeur's Wild West of grasping, brutal, and antisocial thugs come alive: few Americans in the generation to come will ever see where food is made. Try driving through the millions of uninhabited acres between the 99 and Interstate 5 freeways; that dreary corporate plain of central California and its godless culture nevertheless are responsible for dinner tonight in New York, for the Levi's worn by shoppers in Michigan, and the ketchup at McDonald's in Japan.

In contrast, the agrarian patchwork that surrounds the small town will be absorbed by, or sprinkled with, the suburb, edge city, and ranchette; but, again, real food-making that feeds millions will be safe and far away in factorylike settings. How strange that we will destroy what we love, and save what we despise; how odd— but ultimately sinister—that those who will produce what we eat every day will be mostly invisible, and their workplace unknown. How odd that we will never again know we are urban by not being rural, rural by not being urban. The day is upon us where the eater of a plum, drumstick, or ice cream will have never seen once in his life a fruit tree, chicken, or cow, much less an orchard, henhouse, or dairy. Tell him his steak is made of sawdust and his raisins grown on trees, and he will be as likely to believe it as he will be unperturbed.

Dorothea Lange, "Near Meloland, Imperial Valley. Large Scale Agriculture. Gang labor,
Mexican and white, from the Southwest. Pull, clean, tie and crate carrots for the eastern market for eleven
cents per crate of forty-eight bunches. Many can make barely one dollar a day.
Heavy oversupply of labor and competition for jobs is keen," February 1939.
Courtesy Library of Congress, Washington, DC.

When the land matters little, suburbanites and ranchetters will lose the age-old tension between town and countryside, the energy when rustic came to town and the urbanite drove out to see Gramps. The old clear line between rural and urban was far better than what we are becoming, as we will learn when we are all confused among the sprawl: houses here, empty lots there, shopping centers among a few vineyards, a few orchards somehow in the city limits, organic gardens near city parks—even as the corporate fields of the food business are far away, unmentioned, never visited, and of no aesthetic or cultural interest.

What, then, is so beautiful and thus worth saving in a farm, and how does the agrarian's layout differ from the domain of the magnate? Beauty in farming the countryside comes from symmetry in the aesthetic and human sense, when men organize the wild, which has not yet become the town—the farm becomes the mean between complete artificiality and nature absolute. Tree and vine farming is man's best example of this creation of order from chaos. The planter of trees creates order from nature, in a way even the field-crop grower never can. His duty is to be the border, surrounded by, but separate from, the countryside, the intermediary between the town's concrete and the wild woods and wilderness uninhabitable and unlivable. The farm does not mix the two, but rather separates city and wild. The farmer is civilization's buffer against the untame.

Congruence of orchard, house, alleyway, vineyard, and those who are busy within that cosmos is at the heart of this farm. The borders of trees and vines set off nature and have patterns emulated by floral arrangers, linoleum designers, and wallpaper artists. It turns out that the hard-bitten, sour homesteaders of the past were secretly canvas painters and artistes after all! Their nineteenth-century landscapes did not fade with the harvest, but are still here for the developer to nuance, adopt, and destroy long after they are dead. The most tasteful sojourn in Fresno remains Fig Garden, where private homes in shady lanes were plopped down among a once colossal nineteenth-century fig orchard. No one would call their neighborhood Cotton Acres, Alfalfa Hills, or Wheat Estates. Nor would a Fig Garden have grown up over in the far distant barracks hamlet of Strathmore, a town snuggled in the West Side of the Valley between the anonymous 50,000-acre land empires of the corporate class.

Our own orchards and vineyards about a hundred yards south of this house are what Cicero meant when he said man can create with his own hands another world inside nature's own. Out of bare fields rise order, greenery, culture: squares, rectangles, and triangles of measured arrangement amid natural chaos—human-induced order that tends in turn to have a restraining influence on its creators themselves. Out of open ground you craft a dense canopy of peaches; in response that orchard measures you too, tames you, reinvents the very nature of your own time and space. If you doubt me, walk within a thirty-acre cotton field and feel big and then venture through a like-sized acreage of plums and feel like nothing.

Ten years ago we once planted a plum orchard only to dig out and replant the young trees the next day—the planting wire had proved false in the twilight, and in the clear morning the diagonals of the freshly planted stock appeared out of plumb. Who wants to look for the next thirty years at a slightly crooked row, two trees out of a hundred out of line, a random grove, not an ordained orchard? Children, workers, ourselves, all those who would pick that orchard for the next three decades, might assume from its flaws and imperfections that they too could skip a tree, leave the hard-to-get fruit behind, fail to find order and rhythm from the imperfect consonance of the orchard. Chaos is an infectious agent in agriculture—as it should be in all human endeavors. Farming, then, was the ability to impose order upon nature, to use the thought and labor of mankind to organize nature. The physical outlay of the farm must regiment and thus enhance, not let loose, the power of nature. And even the smallest detail, the lay of the land, the line of a row, matters a great deal.

Near our dry pond is a triangular vineyard that follows the land's contour, 5,000 vines in rows of gradually diminishing length. They are green; the alleyways around are not. The vines are six feet apart within each row, the rows themselves twelve feet from one another. In between is the dirt. There it is over 120 degrees and more in July and August. The ground's dusty color, texture, and temperature are set off by its living verdant antithesis. Nearby are three acres of plums, a rectangle that slopes to the pond, 360 trees in a quincunx pattern—a mosaic of proportioned shapes. "To sell this place, you'd have to pull out all those junky plots, level it, and plant something straight and flat where you weren't turning all day on your tractor," more than one uninvited real estate agent has pontificated. If farming were really but the price of diesel and the wage of the driver, he'd be right. But it is not always—not yet at least. Aristotle, after all, did say that the nicely planned town resembled the pattern of a vineyard, not vice versa.

Likewise, there is contrast among the two acres of pears beside, five acres of pomegranates behind, and also four hundred persimmon trees below. From the air these plots are all jigsaw puzzle pieces of differing hues and shapes. From man's ground view, trunk and limb form diagonals, lines, triangles, and squares in every direction. Is it not the purity of mathematical proportion, Plato's pristine reminder that there is an absolute plan and truth on the other side, discernible only through the divine gift of numbers and proportions, which alone now in our world do not lie?

Trees are planted one after another, no randomness, no missing slot, no three bunched here, one over there alone. Try to imagine an orchard or vineyard not in rows and files. Semicircles? A maze? Squiggley lines? A postmodern hodgepodge? Random groves? It would be impossible to cultivate, difficult to traverse, and a compete refutation of any economical use of space and time. Instead, in their fixed and proportional station, the living grids are constantly rearranged by the eye as parallels, perpendiculars, and tangents. They are man's arithmetic—all on a natural landscape that refreshes and relaxes the soul and mind. I think the latent square, rectangle, and grid are always there; the farmer merely puts his new trees and vines into it. In a way he has no choice.

No wonder the garden of the suburbanite is replete with railroad-tie planters and concrete curbs that set off and accentuate flower and fern from lawn and walkway. Landscapers and gardeners do such things for reasons other than worries over creeping weeds and wasted water. As humans, they too know that the mind abhors chaos, and unwinds only amid order that avoids banal repetition.

But not all the pleasure of the grove's symmetry is mere visual experience. The soul's appreciation of these aesthetic patterns can also arise from the tandem of a working mind and moving body integrated into the natural grid. Drive a tractor among the trees. Tie up vines. Prune an orchard. The labor, at first so very monotonous, turns out not to be drudgery after all—it soon begins to blend you into the grid as well. Plunge into row four, tree eleven, and you learn that you too are part of the quincunx, leaving in your wake borders of pruned peaches next to unpruned, half a vineyard clean, the other half weeded by your disk, rows of vines staked in stark contrast to those left without supports. Instantaneously, the job is one-fifteenth, one-third, or five-eighths finished. By row fifteen you learn that you are more than a vine or tree. You are a man, who alone can find and then create such patterns that last but a few hours or an entire lifetime—inside a vineyard or orchard every day, any day of your life on a farm—through your muscle, craft, and tools. We all, do we not, search for some mathematical dimension, some living proportion that is unchanging and so gives us measure as we grow and pass on?

Soon this recognition of balance and dimension within the field does not even leave when you do, and thus enhances the cadence of an already rhythmical agricultural calendar. The vineyard's pruning, shredding, tying, disking, suckering, irrigating, harrowing, dusting, picking are timed pulsations of their own. Like the vineyard itself, the work becomes a pattern the farmer learns as he mundanely moves from vine to vine, from row thirteen on through twenty. "Up and back, up and back, mister," old Mohinder Brahr from his truck in the alleyway once barked at me for an hour and more as I was pruning vines. With five hundred acres, a million in the bank—and half a diseased heart for the tab—Mr. Brahr pined still for the stoop labor of others, lost recompense for the unhappiness that the transition from agriculture to agribusiness had brought him. All that money had taken him away from the shears, out of the vineyard, and into the hospital—and now on toxic medicine, slumped over in his new truck. "Up and back, up and back, boy; I pruned all day with twenty men way behind. Up and back. Don't quit now. Come on, prune, up and back—in the dark, like I used to. Always in the dark. Give me the shears, boy, the shears right over here." By the time I made it out of the vineyard with the shears and over to his truck, he was asleep.

Right now I am looking out at the shed. Eight kids—nephews, nieces, son, and daughter—are arranged in a line packing fruit as it leaves the sizer. The Santillan brothers are running the machinery, and Rosalio is cruising through the orchard with loaded bins of peaches. Rigo and his men are on ladders, and Roberto has the truck, checking the irrigation water. Somebody, my brother or cousin maybe, is overseeing this bustle. But like the hoplite general on the right horn of the phalanx, they are indistinguishable from their troops, and at the cutting edge seek no reprieve from heat, dust, and the monotony of the struggle. Everyone doing something different, yet too everyone fit into and ordered by the symmetry of orchard and vineyard that guides the tractor, makes the ladders progress along lines, assures Rigo's pickers that there are no missing trees in the orchard, ensures to Roberto that one irrigation valve is after the next, that the water will go down row 40 like it did 41 and 39 before, that each line in the vineyard according to its station does in fact have 102 vines.

It is quiet out here and it is peaceful in the countryside—but men are at work at a frantic pace to tame, order, and exploit nature. And something is going on—men and women are being reenergized even as they work themselves into exhaustion.

Even as they age and wrinkle, the denizens of the farm do so with like kind in the natural world around them, who like them are fighting to nourish themselves one more day and leave something of themselves behind when they leave. Time is both of the essence and irrelevant; we are working for the frantic moment, true, but no differently than the now dead did here five decades ago. No one gets out alive, they say, but we who plant and tame think, in a way, that we do. We like looking at town on our horizon, we like that it is there and not in our grid.

A farm of trees and vines turns out to be a fairly large, ordered, and understandable universe that exists without conscious worry about others in the city. Its orchards and vineyards say to those wearied by town and disappointed by the failure and rejection that is so much now a part of city life, "Come into me, where you have always belonged. Prune, shovel, or water. Come into me and be me, and I will make you forget all else, forget time itself, forget that you were ever far from me—forget that you were once failed and as nothing." I once went back into these Acres of Forgetfulness in 1980 after nine years pretty much on the outside. I only realized it was 1985 when an itinerant barn painter offhandedly remarked, as he cleaned out his airless sprayer, "Aren't you the one who went to school over on the coast and was going to be some professor or something?" I snapped to attention from my dream for the first time in sixty months and mused, "You know, I think I was." Watch out for the sirens of the orchard and vineyard. . . .

III. ANCIENT AND MODERN

. . . This divide between countryside and town is more than mere appearance and also involves the binaries of itinerancy and roots, volatility and permanence, tribalism and the *polis*. Could the following fabricated obituary for a man of the city ever be true now? "Mr. Rex Smith, an executive for an advertising agency, died recently at seventy-eight. He lived his entire life in the house where he was born, and worked for fifty years at the local firm, where his son and daughter, who live on either side of the original Smith home, now are also employed, as well as both of his granddaughters."

Or try to envision the reverse: "Royce Smethers, a farmer, passed away recently at seventy-eight. He was transferred to the Valley from Utah and retired twenty years ago after a successful career in vine service in Fresno, plum management in Selma, and nectarine oversight in Reedly, where he was promoted to executive vice president for field operations. His career started in Bakersfield as assistant for market advancement for melons, before joining Tree World's apricot operations team in Merced. For a decade he was vice president of sales for Agrisun in New York. He leaves behind children in Texas, New York, and Utah."

We expect all others in America to be itinerant and to be in search of cash, always to be buried where they were not born, to live apart from their children, to gauge their worth in the manner of a *cursus honorum* of some lupine Roman aedile. In contrast, the farmer we assume to be a food producer, true. But he is also a man whose birth, death, residence, and progeny are fated at the moment of his conception —if he, in fact, is to be a farmer. And his rootedness stands as a rock amid the tide of others, as they splash by him to and fro, wearing him, eroding him, but always, inevitably retreating before him, to leave him there and alone but a second after they had by all appearances washed him away.

Western culture began with the rise of this ideology of agrarianism. It was the notion that communities of small farmers would craft their own laws, fight their own wars, and own land on which to do as they pleased, inventing the concept of a citizen and freedom itself. No other culture had free citizens before the city-state emerged—the very word "citizen" does not exist in the non-Greek vocabulary of the Mediterranean. Few have had them since. But within two centuries of the discovery of the *polis,* the ancient Greeks were facing the contradictions of their own success. They were now free, increasingly prosperous, often educated—and not hungry. Agricultural surpluses and the stability of the countryside under this revolutionary decentralized and private regime thus gave way to trade, commerce, urbanism, and travel—and then the beneficiaries in turn created leisure, affluence, and liberality, as rural egalitarianism became radical urban democracy, freedom smugness and at times cynicism, independence self-centeredness, and prosperity overindulgence—such are the wages when men are free to think, work, and prosper, by law safe from thug, pharaoh, or prophet. Yet this second and subsequent Western tradition of unmatched material accomplishment—the legacy of the Hellenistic, not the Hellenic, world—saw all moral impediments to economic practice, to parochial and quasi-ritualized infantry warfare, to desire itself, let loose. The natural classical genius that was found in materialism and individualism was left unchecked, and so Western man was on his way to becoming Hellenistic and imperial—on his way to seeing that we were, above all, fat, safe, and complacent.

In other words, the city-state—the city and its surrounding cultivated land—was now to be more the city than the state, and what had once been seen as salvation was now deemed passé, if not harsh and contradictory. The Greeks of the later fifth century and the fourth century did not need the aftermath of World War I to discover something of the nihilism of modernism. There were sophists long before Michel Foucault; ask Aristophanes whether there were enough affluent, nasal-voiced, long-haired elite dandies mad at their parents, their gods, and their culture well before comparative literature graduate seminars at Stanford.

So the rise of an urban culture, as dynamic as it was explosive, was often antithetical to, rather than complementary of, early rural values, themselves rather absolute and blinkered—and, to be frank, confining. Within two centuries of the arrival of the Greek city-state, a paradox appeared that plagues us still in the West: the more stable, the more prosperous, the more law-abiding, and the more humane becomes the society that our founders built with their ideas of individual freedom and constitutional government, the further their offspring move away from the sacrifice and hardship that were responsible for that initial and thereafter necessary bounty—a natural paradox of aging, as the romantic Roman poets like Lucretius, Horace, and Virgil knew, that is best seen in the move from rural to urban. We farmers of the countryside create the pleasant circumstances that can alone bring on our own self-loathing. Not a few historians thought from creation to decline took but three generations. Juvenal, Tacitus, and Petronius, the great skeptical triad of Roman imperial genius, knew of the smugness that arose out of the embarrassment of riches.

Philosophy, oratory, tragedy, high art, and comedy are the dividends of a sound social fabric. But they are also inevitably the products of an urban and elite culture, which sometimes despises the rather unsophisticated arms and backs that make it all possible in the first place. Solon's farmers of Attica would never have given us a gaudy Acropolis, but then they were not the urban mob who voted to send the classical Athenian armada to its ruin at Sicily either, or who slaughtered autonomous islanders to pay for the Parthenon or nodded their heads while they voted to execute Socrates. In any case, it was the yokels outside the city wall and their unremembered creation of free speech, constitutional practice, and economic activity in the first place that made the later more democratic city-state capable of both artistic creation and ruthless destruction. There is no solution to this tragic, endless cycle of creation, enjoyment, and decline in the West. Would that we could arrest the evolution in mid-cycle, and hold on to a well-read agrarian or a callused philosopher. Would that we might retain picture-book city-states, nestled in small valleys, the agrarian patchwork of homestead farms spread about, everyone content to have town and countryside separate but in harmony, none eager to have the two mixed but in cacophony.

Dutiful men and women who unceasingly labor with like kind on the farm, marry, raise children, craft and obey laws, apparently do so in order that others to come may choose not to—and that others may often resent those who allow them that very choice. In the West they create the scaffolds that support the painters and writers above. For the ancients—Hesiod, Virgil, and others were onto it—the cycle of toil, leisure, and degeneration was biological, the state an organism with a very definite life cycle. Culture is created in the countryside through hard work and sacrifice, which leads to a better life in town, where the rough and often cruel edges are sanded, where the resulting polish softens our progeny—often to such a degree that original victory over the wild is now meaningless or forgotten entirely, or finally ridiculed as quaint and silly by a cynical, smug, and immobilized third generation—or worse, still romanticized as a day in the garden at a villa on Capri. The bedraggled yeomen at Cannae were slaughtered so that men like Petronius' Giton and Encolpius one day could freely and without worry prance at the baths with all sexes or practice their farting at elegant dinner parties—or laugh at the industri-ous rag-collectors and strong-jawed centurions still in their midst.

Jefferson's America was 85 percent agrarian, and he alone seems to have predicted that such a society could not exist as it once had when 95 percent lived in the city. How could it when the original idea was to demand of American citizens that they bring to government their distrust of complexity and bureaucracy, their reliance on self and family, their faith in their own arms and head, their knowledge of the fragil-ity of nature harnessed and the chaos of a nature let go, their skepticism of taxation and of the idea that someone who does not grow, manufacture, or build could ever know best how money should be spent? How could it be that Crèvecoeur's man, who had muscles and courage and an innate aesthetic sense of and respect for the wild, would prefer to see those traits become vestigial in the city? Are we surprised that 6 million crowded into Hong Kong, Tokyo, or Singapore are more or less well-behaved, a similar number spread out in Los Angeles—still imbued with the tradition begun by Mr. Crèvecoeur's independent planter yeoman—bored, restless, and dangerous? Ignorant Mr. Crèvecoeur back from the dead would need no more than a glance at L.A. to pronounce a failure of the human spirit—and would he be wrong?

By the fifth century B.C. there was a clear antithesis at Athens between the rus-tic (*agroikos*) and the more urbane (*asteios*). It was a stock theme in Aristophanes' comedies, where naive buffoons frequently outsmart the more sophisticated who had lost touch with natural pragmatism, who were smart but not wise. Aristotle and Plato, who sought an elusive middle between abstraction and pragmatism, must have felt the divide keenly. Although both men were products of the *polis* and the intellectual electricity of urban living, they were nevertheless at heart conservatives. Both may have been maladjusted and reactionary. Both dreamed of a pre-Athenian world, where men still fought as hoplites, worked the soil, and saw the world in absolutes, immune from distortions in language and the loss of shame. So in both their utopias, land-holding is central. Plato and Aristotle even offer the unworkable solution (as my parents learned) that the ideal citizen is to have two homes—one in the countryside, one in the city—as if intellectual contemplation could be tempered and gauged against rural savvy, as if urban sophistication could be grounded with an occasional infusion of rural pragmatism.

The polarity between city and countryside, profit and sustenance, leisure and drudgery, the stuff of all Western civilization, has been but a smaller skirmish in the wider question, "What is wisdom?" Is knowledge—not the accumulation of facts—to be the accrued body of erudition from the ages, abstract and printed, the academic sweep from technology to aesthetics? Or is it found alone in the school of hard knocks, the experience drawn from mechanics and fabrication with the hands, the wisdom gained from thousands of personal misfortunes in the social and natural jungle, uncontaminated by pampered abstraction? Does a man understand the universe because he can read Descartes, or does such insight arise only after he has lost his ripe crop a day before harvest? Is the university to be 1,000 peach trees, black soil, and a sharp March wind from the north, or the dreary seminar room in the Lit. Department? Is the scholar with hands polished by the page the true visionary, or is it old Ray Mix, with his missing limb and the remaining hand just one large callous for all the extra duty? Do you send young Jason or Nicole to yet another year of computer camp and SAT preparatory study, or put them to work in the orchard for summer?

"Both," Socrates, killer at the battle of Delium, inquisitor of Protagoras, would say. To the classical Greeks, for whom our modern notion of an institutionalized academia did not exist, abstraction was to be married within the same person with action, contemplation, and theory immediately substantiated or rejected through praxis. Too much reading created useless and unread data and knowledge of no value that would eventually lead to the creation of scholastic, academic, and pedantic expression of no interest or purpose to anyone. After all, I could write a lengthy and unread history of every telephone pole on this farm and have the monograph blurbed by obsequious peers as "cutting-edge," "visionary," and "much-needed." More blurbs and reviews still if I included the words "rhetoric," "gender," or "construct" in the thesis title. More so even yet, should I footnote all the blurbers and reviewers inside.

A $30,000 fellowship can allow a young scholar the chance to provide America with a fuller appreciation of a fragmented epigram of the Hellenistic poet Callimachus. But that gift does not mean that either the nation's treasury or the young man's time was well spent, or that the scholar or his country was more knowledgeable or humane for the investment. The appreciation of ambiguity and irony that education spawns, much less the reservoir of accumulated data that accrues, is not always an aid to the moral life. Just as likely the wages of erudition are a lost sense of morality: when we learn too much, we can talk away, think away, and rationalize away any vice we like. I am afraid that I now agree with Mr. Crèvecoeur about the ideal of a nation of "scholars of husbandry":

> As I intend my children neither for the law nor the church, but for the cultivation of the land, I wish them no literary accomplishments; I pray heaven that they may be one day nothing more than expert scholars in husbandry; this is the science which made our continent to flourish more rapidly than any other.

Oratory, logic, and abstraction are but a few of the tools the intellectual acquires to make the down up, the good bad. *Sophos,* wise man, is but a hairsbreadth from *sophistiko,* wise guy, checked and kept on the right side only by muscles and toil and unthinking drudgery on occasion. Professors, artists, pundits, you really do need plumbers and carpenters—and farmers—around to tell you that you are often full of crap, that you should not explain away what you can explain away. The muscular classes are more valuable to your work than all the money in Mr. Guggenheim's will.

True, with too much labor, the exhausted mind could not make sense or purpose of the daily wear and damage to the flesh. If you prune, water, and cultivate trees and vines all day for a year, sometimes an hour with a book can cause migraines. No wonder Aristotle labored over the proper mixture of the Man of Action with the Man of Contemplation, and felt farmers had too much to do to become truly politic. Only for a while did the Greeks join the pragmatic with the sophisticated, and only with difficulty did the bumpkin come into town to hear Demosthenes or clap for Sophocles. Believe me, it is hard to do that, and so it is rare to have true family farms and the world of agrarianism around for very long, these strange men and women who themselves must think how to employ their own muscles for their own survival, these strange people who seemingly pop out and rise from the soil we ignore.

Yet that active mind tempered by muscles creates a tension at the heart of knowledge. And it is a balance almost impossible to maintain now, since city and countryside are more often at war rather than at peace with each other—or, worse, since there is increasingly little countryside left. We are now Hittites and Egyptians, perhaps Aztecs too, rarely any longer Argives, Eleans, and Thebans of a teeming rural patchwork.

We country dwellers, you see, can go to town, can see what we need and learn what we lack. But is the opposite any longer true? Can the city boy harrow for a few hours each week, shovel each summer, get peach fuzz down his back as tonic to his urban existence? We know how our peaches are sold, the color of our raisin boxes, how much work it takes to drive plums to New York and cart them off to Food Mart, how to turn on the Internet to learn the hourly price of pomegranates.

But do you, reader, understand the feel of shears in your hands for a month? The trick of picking from a three-legged ladder with fifty pounds of apricots strapped to your belly? Of pulling a tandem disk for ten hours, six inches away from the vines? The frustration of trying to figure out why your five long and tiring days of fertilizing destroyed your plum crop instead of saving it? The embarrassment of explaining to Mr. Ulysses Ponce and his crew of fifteen at 6 A.M. in the morning that, in fact, you have erred, the plums are not quite ripe, and can he and his rather impoverished pickers simply go away for the day unpaid and come back tomorrow, same time? The knowledge that you are doing so to a Mr. Ponce, whose own hands are callused and who pledged faith to those who took him at his word? Do you know how to fail on your own? Do you know that for all your intelligence a weak arm can at this late age bring ruin still? Or that for all those pectorals a simple error of mental reckoning loses your pomegranates? That a day inside with the flu means a day that vines were not watered? Or that a vile, mean man cares not a whit that you are right, rational, and sincere, if he senses you are weak? Vines and orchards, then, are not the 1,000 trees or 20,000 vines to the naked eye, but thousands of mundane tasks and divine revelations that inculcate generation after generation of wisdom that has no calibration, real learning that has no doctorate or master of anything.

Should pruners leave four vine canes or six? Or perhaps even eight? Four should result in good-quality grapes in the fall, but less of a crop—and in some years no crop at all. Six mean if it should freeze or hail after pruning, some grapes will still be there at harvest. When there are eight canes to wrap on the wires, you ensure that even sloppy vine tiers cannot damage your crop by breaking off canes. But then a good spring, free of disaster, just may result in seventy bunches in September, not twenty, making you the greedy fool with a large, sour, and worthless vintage on your now embarrassingly bushy eight canes. In March, should you rise at 3 A.M. to pump expensive warm subterranean water down your vine rows, to guarantee against crop-destroying frost? Some sleepless and exhausted farmers do just that and lose their crop anyway, while their snoring and tightfisted neighbor's vineyard, weedy and unready for the cold nights, comes through unscathed every season—the divine, not man, has decided his place and not yours is by nature always a degree or two warmer than freezing. In your sixties and without a pension, do you "rent" your vineyard *gratis* to your sober, hardworking, and broke son-in-law, who needs 100, not 70, percent of the crop from your ground to feed your grandchildren? And should you rent forty, sixty, or a hundred beehives for your almond grove, knowing that forty might be enough to pollinate a crop, that the safe path of a hundred is well beyond your budget, that sixty gives a good chance of success—and in a warm year perhaps too much success and thus too many nuts? Or is the proper solution—and there is always a proper solution—somewhere between forty-two and fifty-one hives after all? Do you still wish to buy exorbitant hail insurance for your plums from the slick broker of dubious character, whom you have enriched each year with cash premiums on the slight chance that once or not even once in this life a March rain will turn hard and scar and knock off your fruit? The bank said that you should protect its money, but then lends you no money to do so; the weatherman says hail is unlikely, but then will not be foreclosed upon when he is wrong; your brother, you suspect, thinks you gullible for doing so, and just may be idiotic for not doing so; your daughter needs that premium for tuition money now, but won't be in college at all if it hails. And so you always or never buy the Guardian Plum Plan? Or buy it for a decade when the skies are clear, skipping it the one spring the black clouds shred your orchards? And when your vines are scorched and the irrigation district's water is short, out of principle and in fear of divine wrath, do you take exactly and only your rightful turn on the communal ditch? Such rectitude means that you will cover 60 percent of your vineyard, so that your pesky neighbor down the line, who is prone to take a little too much of a little too long a turn, will now have enough hydraulic pressure to hog the ditch and thereby save his grapes and not yours?

Can we Americans, then, as we used to, and as the Greeks taught us, any longer mold the complete citizen, who—like Pericles, Socrates, or Sophocles—could wound, sail, build, plant, and chisel between speeches, plays, and debates? In short, we need town and city, which are nothing without each other. We have the latter of sorts, the increasingly specialized and narrow, to surfeit. But as for the former, is it tapped for its knowledge, is an active rural life even there any longer to be had?

Are there working citizens outside the city limits? Are there city limits at all? The Greeks, who unlike us were seldom obese and occasionally even were hungry, knew that man farms not merely to be fed, but also to learn how his society should be organized. We now farm to eat cheaply (as if America's ongoing problem is famine or an absence of disposable income, as if Americans are too thin), and so have lost the best—and is it the only?—blueprint of how we are to organize a society.

Hayden Carruth

Hayden Carruth was born in 1921 and grew up in Woodbury, Connecticut, during the Great Depression. After serving in World War II, Carruth lived in Chicago and was editor of Poetry *magazine, but he suffered from psychiatric disorders and spent more than a year in the hospital. Eventually progressing, as he put it, from the need for seclusion to mere reclusion, and desperately poor, in 1961 he moved to northern Vermont with his wife and infant son, buying a few acres and a small house. There, at first from necessity as much as by choice, Carruth learned the skills of rural survival—carpentry, plumbing, woodcutting, gardening, fixing engines. He also worked as a hired hand on the dairy farm of his neighbor, Marshall Washer. He wrote his poems and criticism mostly late at night, after chores. He was on hand to witness both the building boom of second homes for city dwellers in Vermont, which he deplored, and the arrival of the counterculture back-to-the-landers, with whom he felt great sympathy, sharing their radical rejection of mainstream American culture and politics.*

Carruth said that he and his wife used to laugh when praised by their new friends for their "voluntary poverty," but that he came to see his difficult life, coping with the rigors and hardships of living by physical labor in a rugged environment, as a blessing. It allowed him to reintegrate his life and his work, and quite literally gave him something to write about. As Carruth's poems garnered recognition, he gradually regained the confidence to appear in public. In 1979 he moved to upstate New York and took a position in the graduate program at Syracuse University in which he remained for the rest of his life.

Carruth's poetry is today widely admired for its great range and inventiveness, both in form and content, and many other poets, including Wendell Berry, have gratefully acknowledged his influence. Some of Carruth's best-loved poems are humorous but hard-edged portraits of his rural neighbors. His poem about Marshall Washer is straightforward and sincere, but its seemingly simple observations gradually open into a larger meditation on the decline of traditional American farming and the break in the ancient cycles of fertility among farmers, animals, and land.

"Marshall Washer" (1978)

1

They are cowshit farmers, these New Englanders
who built our red barns so admired as emblems,
in photograph, in paint, of America's imagined
past (backward utopians that we've become).
But let me tell you how it is inside those barns.
Warm. Even in dead of winter, even in the
dark night solid with thirty below, thanks
to huge bodies breathing heat and grain sacks
stuffed under doors and in broken windows, warm,
and heaped with reeking, steaming manure, running
with urine that reeks even more, the wooden channels
and flagged aisles saturated with a century's
excreta. In dim light, with scraper and shovel,
the manure is lifted into a barrow or a trolley
(suspended from a ceiling track), and moved
to the spreader—half a ton at a time. Grain
and hay are distributed in the mangers, bedding
of sawdust strewn on the floor. The young cattle
and horses, separately stabled, are tended. The cows
are milked; the milk is strained and poured
in the bulk tank; the machines and all utensils
are washed with disinfectant. This, which is called
the "evening chores," takes about three hours.
Next morning, do it again. Then after breakfast
hitch the manure spreader to the old Ferguson
and draw it to the meadows, where the manure
is kicked by mechanical beaters onto the snow.
When the snow becomes too deep for the tractor,
often about mid-January, then load the manure
on a horse-drawn sled and pitch it out by hand.
When the snow becomes too deep for the horses
make your dung heap behind the barn. Yes, a good
winter means no dung heap; but a bad one
may mean a heap as big as a house. And so,
so, night and morning and day, 365 days
a year until you are dead; this is part
of what you must do. Notice how many times
I have said "manure"? It is serious business.
It breaks the farmers' backs. It makes their land.
It is the link eternal, binding man and beast

and earth. Yet our farmers still sometimes say
of themselves, derogatively, that they are "cowshit
farmers."

2
         I see a man with a low-bent back
driving a tractor in stinging rain, or just as he
enters a doorway in his sheepskin and enormous
mittens, stomping snow from his boots, raising
his fogged glasses. I see a man in bib overalls
and rubber boots kneeling in cowshit to smear
ointment on a sore teat, a man with a hayfork,
a dungfork, an axe, a 20-pound maul
for driving posts, a canthook, a grease gun.
I see a man notching a cedar post
with a double-blade axe, rolling the post
under his foot in the grass: quick strokes and there
is a ringed groove one inch across, as clean
as if cut with the router blade down at the mill.
I see a man who drags a dead calf or watches
a barn roaring with fire and thirteen heifers
inside, I see his helpless eyes. He has stood
helpless often, of course: when his wife died
from congenital heart disease a few months before
open-heart surgery came to Vermont, when his sons
departed, caring little for the farm because
he had educated them—he who left school
in 1931 to work by his father's side
on an impoverished farm in an impoverished time.
I see a man who studied by lamplight, the journals
and bulletins, new methods, struggling to buy
equipment, forty years to make his farm
a good one; alone now, his farm the last
on Clay Hill, where I myself remember ten.
He says "I didn't mind it" for "I didn't notice it,"
"dreened" for "drained," "climb" (pronounced *climm*)
for "climbed," "stanchel" for "stanchion,"
and many other unfamiliar locutions; but I
have looked them up, they are in the dictionary,
standard speech of lost times. He is rooted
in history as in the land, the only man I know
who lives in the house where he was born. I see

a man alone walking his fields and woods,
knowing every useful thing about them, moving
in a texture of memory that sustains his lifetime
and his father's lifetime. I see a man
falling asleep at night with thoughts and dreams
I could not infer—and would not if I could—
in his chair in front of the television.

3
                                        I have written
of Marshall often, for his presence is in my poems
as in my life, so familiar that it is not named;
yet I have named him sometimes too, in writing
as in life, gratefully. We are friends. Our friendship
began when I came here years ago, seeking
what I had once known in southern New England,
now destroyed. I found it in Marshall, among others.
He is a friend and neighbor both, an important
distinction. His farm is one-hundred-eighty acres
(plus a separate woodlot of forty more), and one
of the best-looking farms I know, sloping smooth
pastures, elm-shaded knolls, a brook, a pond,
his woods of spruce and pine, with maples and oaks
along the road—not a showplace, not by any means,
but a working farm with fences of old barbed wire;
no pickets, no post-and-rail. His cows are Jerseys.
My place, no farm at all, is a country laborer's
holding, fourteen acres "more or less" (as the deed
says), but we adjoin. We have no fence. Marshall's
cows graze in my pasture; I cut my fuel
in his woods. That's neighborliness. And when
I came here Marshall taught me . . . I don't know,
it seems like everything: how to run a barn,
make hay, build a wall, make maple syrup
without a trace of bitterness, a thousand things.
(Though I thought I wasn't ignorant when I came,
and I wasn't—just three-quarters informed.
You know how good a calf is, born three-legged.)
In fact half my life now, I mean literally half,
is spent in actions I could not perform without
his teaching. Yet it wasn't teaching; he *showed* me.
Which is what makes all the difference. In return

I gave a hand, helped in the fields, started
frozen engines, mended fence, searched for lost calves,
picked apples for the cider mill, and so on.
And Marshall, now alone, often shared my table.
This too is neighborliness.

4
                         As for friendship,
what can I say where words historically fail?
It is something else, something more difficult. Not
western affability, at any rate, that tells
in ten minutes the accommodation of its wife's—well,
you know. Yankees are independent, meaning
individual and strong-minded but also private;
in fact private first of all. Marshall and I
worked ten years together, and more than once
in hardship. I remember the late January
when his main gave out and we carried water,
hundreds and thousands of gallons, to the heifers
in the upper barn (the one that burned next summer),
then worked inside the well to clear the line
in temperatures that rose to ten below
at noonday. We knew such times. Yet never
did Marshall say the thought that was closest to him.
Privacy is what this is; not reticence, not
minding one's own business, but a positive sense
of the secret inner man, the sacred identity.
A man is his totem, the animal of his mind.
Yet I was angered sometimes. How could friendship
share a base so small of mutual substance?
Unconsciously I had taken friendship's measure
from artists elsewhere who had been close to me,
people living for the minutest public dissection
of emotion and belief. But more warmth was,
and is, in Marshall's quiet "hello" than in all
those others and their wordiest protestations,
more warmth and far less vanity.

5
                         He sows
his millet broadcast, swinging left to right,
a half-acre for the cows' "fall tonic" before

they go in the barn for good; an easy motion,
slow swinging, a slow dance in the field, and just
the opposite, right to left, for the scythe
or the brush-hook. Yes, I have seen such dancing
by a man alone in the slant of the afternoon.
At his anvil with his big smith's hammer
he can pound shape back in a wagon iron, or tap
a butternut so it just lies open. When he skids
a pine log out of the woods he stands in front
of his horse and hollers, "Gee-up, goddamn it,"
"Back, you ornery son-of-a-bitch," and then
when the chain rattles loose and the log settles
on the stage, he slicks down the horse's sweaty
neck and pulls his ears. In October he eases
the potatoes out of the ground in their rows,
gentle with the potato-hook, then leans and takes
a big one in his hand, and rubs it clean
with his thumbs, and smells it, and looks
along the new-turned frosty earth to fields,
to hills, to the mountain, forests in their color
each fall no less awesome. And when in June
the mowing time comes around and he fits the wicked
cutter-bar to the Ferguson, he shuts the cats
indoors, the dogs in the barn, and warns
the neighbors too, because once years ago,
many years, he cut off a cat's legs in the tall
timothy. To this day you can see him
squirm inside when he tells it, as he must tell it,
obsessively, June after June. He is tall,
a little gray, a little stooped, his eyes
crinkled with smile-lines, both dog-teeth gone.
He has worn his gold-rimmed spectacles so long
he looks disfigured when they're broken.

6

                       No doubt
Marshall's sorrow is the same as human
sorrow generally, but there is this
difference. To live in a doomed city, a doomed
nation, a doomed world is desolating, and we all,
all are desolated. But to live on a doomed farm
is worse. It must be worse. There the exact

point of connection, gate of conversion, is—
mind and life. The hilltop farms are going.
Bottomland farms, mechanized, are all that survive.
As more and more developers take over
northern Vermont, values of land increase,
taxes increase, farming is an obsolete vocation—
while half the world goes hungry. Marshall walks
his fields and woods, knowing every useful thing
about them, and knowing his knowledge useless.
Bulldozers, at least of the imagination,
are poised to level every knoll, to strip bare
every pasture. Or maybe a rich man will buy it
for a summer place. Either way the link
of the manure, that had seemed eternal, is broken.
Marshall is not young now. And though I am only
six or seven years his junior, I wish somehow
I could buy the place, merely to assure him
that for these few added years it might continue—
drought, flood, or depression. But I am too
ignorant, in spite of his teaching. This is more
than a technocratic question. I cannot smile
his quick sly Yankee smile in sorrow,
nor harden my eyes with the true granitic resistance
that shaped this land. How can I learn the things
that are not transmissible? Marshall knows them.
He possesses them, the remnant of human worth
to admire in this world, and I think to envy.

WES JACKSON

*Wes Jackson is president of the Land Institute in Salina, Kansas, best known for
its pioneering work on perennial polyculture and natural systems agriculture. A native
Kansan, he was born in 1936 near Topeka. He earned a BA from Kansas Wesleyan, an
MA from the University of Kansas, and his PhD in genetics from North Carolina State
University. After a stint teaching back at Kansas Wesleyan, Jackson went to California
State University–Sacramento, where he established one of the nation's first programs
in environmental studies. He returned to Salina in 1976 to raise his family and pursue
research on sustainable living. There, with his former wife, Dana, he cofounded the Land
Institute, a private research effort devoted to ecologically appropriate technology and de-*

sign similar to others of that era such as the New Alchemy Institute on Cape Cod and the Farallones Institute in California.

The Land Institute soon focused on sustainable agriculture. In 1980 Jackson published New Roots for Agriculture, *a book with strong intellectual ties to Hugh Hammond Bennett, Paul Sears, Aldo Leopold, J. Russell Smith, and the Friends of the Land, who had campaigned for a permanent ecological agriculture a generation before. Jackson argues that soil erosion is not only a problem in agriculture but the problem of agriculture, and that in the long run a viable replacement for tillage production of annual grains will be required for civilization to endure. Researchers at the Land Institute work with native prairie species and domesticated crops in an attempt to create such a new type of agriculture, modeled on the native prairie: perennial, grain-producing crops grown intermixed with legumes and composites so that they hold the soil, control pests, and sponsor their own nutrient fertility with minimal fossil fuel–based inputs.*

As the foremost spokesman for agriculture that uses nature as a model, Jackson has traveled the world promoting the research of the Land Institute. Once aptly described as a cross between a Methodist preacher and a bison (showing characteristic hybrid vigor), Jackson has also become an important spokesman for sustainable agriculture in general and for a new agrarian vision for rural America. He received a Macarthur Fellowship in 1992, a Right Livelihood Award in 2000, and was named by Life *magazine one of the one hundred "most important Americans of the twentieth century," among many other awards and prizes. In this selection, Jackson describes the work of "homecoming" that he believes will be necessary to rebuild a sustainable future.*

From "Becoming Native to Our Places" (1994)

It seems to be a characteristic of life that no matter what the level of organization, the juvenile stage is characterized by an excess of potential energy and an inefficiency in use of that energy. This seems to be as true of the early stages of an ecosystem as of a teenager. But we have seldom considered a corollary—that an excess of potential energy can *generate* a juvenile condition. The industrial revolution really hit its stride after World War II. It was only then that we became a truly affluent society. The Depression and the war contributed to making us a disciplined people, but after the war economic growth and invention really took off. We came to believe that comfort and security were the solutions to the human condition.

But what this excess of potential energy has yielded us, beyond the throughput of goods, is a decrease in our maturity. Our culture is now like a time machine running backward. National polls frequently show that when the issues are framed as value questions, the public will give what in my view is the responsible answer. Then they'll vote otherwise. We saw this during the Reagan and Bush years. Environment gets a high rating because it is the right answer, but people want government to do it without raising taxes or having a national discussion about getting rid of the automobile. It is reminiscent of a child who can give the answer the parents want and then goes on and does what he or she wants to do.

This is not what Madison and the founding fathers expected. They believed the maturity of people's judgment would expand. Worried about corruption, they assumed that eventually our judgment would be larger, more diverse, and therefore more stable. Instead, we have gone the other way. We have become a more juvenile culture. We have become a childish "me, me, me" culture with fifteen-second attention spans. The global village that television was supposed to bring is less a village than a playground. We'd rather gossip about President Clinton's sex life than talk about the issues. And so few of the issues are really being dealt with. We seem satisfied to keep tossing around that vague term "environment" without talking about our relationship with nature. The destruction of wilderness is not even a secondary consideration. Community destruction is scarcely mentioned. Destruction of our agricultural communities may as well go unnoticed; little is done about it. Widespread if not universal child neglect is less discussed than "the economy."

Nearly all of the suggestions for change are off the mark. We educate kids to take tests. We make the assumption that better organized education will be better education. But what of the content? Teachers don't even know how to talk about community responsibility. Little attempt is made to pass on our cultural inheritance, and our moral and religious traditions are neglected except in the shallow "family values" arguments. In our universities there is good reason to believe that the Declaration of Independence would not be passed by university professors if it were brought to a vote today. Unlike those who signed that document, most modern scholars are less servants of the people.

A necessary part of our intelligence is on the line as the oral tradition becomes less and less important. There was a time throughout our land when it was common for stories to be told and retold, a most valuable exercise, for the story retold is the story reexamined over and over again at different levels of intellectual and emotional growth. *Huck Finn* read at the fifth-grade level is different from *Huck Finn* read in high school or college or as a young parent or grandparent. That is true with almost any story. But "news" as displayed on television appears once only, unlike the story in the oral tradition with its many levels of meaning.

Entire neighborhoods are more accessible to the world than their members are to one another. Is this part of our nature? It is always easier to think of a better way to produce food or a consumer item than to think of how to avoid using that food or that gadget wastefully. We waste, I believe, largely because of our fallen condition. We employ human cleverness to make the earth yield an unbounded technological array, which in turn produces countless more technologies, more things. In agriculture, we hot-wire the landscape, bypassing nature's control devices. We do this in the face of abundant evidence that we are destroying our habitat because of our "unwitting accessibility" to the world. I should explain my interpretation of this phrase.

A few years ago on the last page of *Life* magazine I saw a memorable photograph of a near-naked, well muscled tribesman of Indonesian New Guinea who was staring at a parked airplane in a jungle clearing. The caption noted the Indonesian government's attempt to bring such "savages" into the money economy. A stand had been set up at the edge of the jungle and was reportedly doing a brisk business in beer, soda pop, and tennis shoes. We can imagine what must have followed for the members of the tribe, what the wages of their "sin," their "fall," must have been— decaying teeth, anxiety in a money system, destruction of their social structure. If they were like what we know of most so-called primitive peoples, then in spite of its hierarchical structure, their society was much more egalitarian than today's industrialized societies.

Unlike Adam and Eve, the New Guinea tribesmen received no explicit commandment to avoid the goodies of civilization. They simply accepted, unwittingly, the proffered accessibility to beer, soda pop, and tennis shoes. In Genesis, the primal sin involves disobedience, an exercise of free will. In our modern non-Paradise version, "original sin" is our unwitting acceptance of the material things of the world. I perceive this to be the largest threat to our planet and to our ability to accept nature-as-measure.

In *Beyond the Hundredth Meridian* (1953), Wallace Stegner describes the breakdown of American Indian culture:

> However sympathetically or even sentimentally a white American viewed the Indian, the industrial culture was certain to eat away at the tribal cultures like lye. One's attitude might vary, but the fact went on regardless. What destroyed the Indian was not primarily political greed, land hunger, or military power, not the white man's germs or the white man's rum. What destroyed him was the manufactured products of a culture, iron and steel, guns, needles, woolen cloth, *things that once possessed could not be done without.* [Italics added.]
>
> It was not the continuity of the Indian race that failed; what failed was the continuity of the diverse tribal cultures. These exist now only in scattered, degenerated reservation fragments among such notably resistant peoples as the Pueblo and Navajo of the final, persistent Indian Country. And here what has protected them is aridity, the difficulties in the way of dense white settlement, the accident of relative isolation, as much as the stability of their own institutions. Even here a Hopi dancer with tortoise shells on his calves and turquoise on his neck and wrists and a kirtle of fine traditional weave around his loins may wear down his back as an amulet a nickel-plated Ingersoll watch, or a Purple Heart medal won in a white man's war. Even here, in Monument Valley where not one Navajo in ten speaks any English, squaws may herd their sheep through the shadscale and rabbitbrush in brown and white saddle shoes and Hollywood sun-

glasses, or gather under a juniper for gossip and bubblegum. The lye still corrodes even the resistant cultures.

This reality—things once possessed that cannot be done without—is so powerful that it occupies our unconscious. And yet we know that nature, in Milton's words, "means her provision only to the good / That live according to her sober laws / And holy dictate of spare Temperance."

At work on my houses at Matfield Green [in the Flint Hills of eastern Kansas], I've had great fun tearing off the porches and cleaning up the yards. But it has been sad, as well, going through the abandoned belongings of families who lived out their lives in this beautiful, well-watered, fertile setting. In an upstairs bedroom, I came across a dusty but beautiful blue padded box labeled "Old Programs—New Century Club." Most of the programs from 1923 to 1964 were there. Each listed the officers, the Club Flower (sweet pea), the Club Colors (pink and white), and the Club Motto ("Just Be Glad"). The programs for each year were gathered under one cover and nearly always dedicated to some local woman who was special in some way.

Each month the women were to comment on such subjects as canning, jokes, memory gems, a magazine article, guest poems, flower culture, misused words, birds, and so on. The May 1936 program was a debate: "Resolved that movies are detrimental to the young generation." The August 1936 program was dedicated to coping with the heat. Roll call was "Hot Weather Drinks"; next came "Suggestions for Hot Weather Lunches"; a Mrs. Rogler offered "Ways of Keeping Cool."

The June roll call in 1929 was "The Disease I Fear Most." That was eleven years after the great flu epidemic. Children were still dying in those days of diphtheria, whooping cough, scarlet fever, pneumonia. On August 20, the roll call question was "What do you consider the most essential to good citizenship?" In September that year it was "Birds of our country." The program was on the mourning dove.

What became of it all?

From 1923 to 1930 the program covers were beautiful, done at a print shop. From 1930 until 1937, the effects of the Depression are apparent; programs were either typed or mimeographed and had no cover. The programs for two years are now missing. In 1940, the covers reappeared, this time typed on construction paper. The print shop printing never came back.

The last program in the box dates from 1964. I don't know the last year Mrs. Florence Johnson attended the club. I do know that Mrs. Johnson and her husband Turk celebrated their fiftieth wedding anniversary, for in the same box are some beautiful white fiftieth anniversary napkins with golden bells and the names Florence and Turk between the years "1920" and "1970." A neighbor told me that Mrs. Johnson died in 1981. The high school had closed in 1967. The lumber yard and hardware store closed about the same time but no one knows when for sure. The last gas station went after that.

Back to the programs. The Motto never changed. The sweet pea kept its standing. So did the pink and white club colors. The club collect which follows persisted month after month, year after year:

A Collect for Club Women

> Keep us, O God, from pettiness;
> Let us be large in thought, in word,
> in deed.
> Let us be done with fault-finding
> and leave off self-seeking.
> May we put away all pretense and
> meet each other face to face, without
> self-pity and without prejudice.
> May we never be hasty in judgment
> and always generous.
> Let us take time for all things;
> make us grow calm, serene, gentle.
> Teach us to put into action our
> better impulses; straightforward
> and unafraid.
> Grant that we may realize it is
> the little things that create differences;
> that in the big things of life
> we are as one.
> And may we strive to touch and
> to know the great common woman's
> heart of us all, and oh, Lord God,
> let us not forget to be kind.
> Mary Stewart

By modern standards, these people were poor. There was a kind of naïveté among these relatively unschooled women. Some of their poetry was not good. Some of their ideas about the way the world works seem silly. Some of their club programs don't sound very interesting. Some sound tedious. But their monthly agendas were filled with decency, with efforts to learn about everything from the birds to our government, and with coping with their problems, the weather, diseases. Here is the irony: they were living up to a far broader spectrum of their potential than most of us do today!

I am not suggesting that we go back to 1923 or even to 1964. But I will say that those people in that particular generation, in places like Matfield Green, were farther along in the necessary journey to become native to their places, even as they were losing ground, than we are.

Why was their way of life so vulnerable to the industrial economy? What can we do to protect such attempts to be good and decent, to live out modest lives responsibly? I don't know. This is the discussion we need to have, for it is particularly problematic. Even most intellectuals who have come out of such places as Matfield Green have not felt that their early lives prepared them adequately for the "official" culture.

I want to quote from two writers. The first is Paul Gruchow, who grew up on a farm in southern Minnesota:

> I was born at mid-century. My parents, who were poor and rural, had never amounted to anything, and never would, and never expected to. They were rather glad for the inconsequence of their lives. They got up with the sun and retired with it. Their routines were dictated by the seasons. In summer they tended; in fall they harvested; in winter they repaired; in spring they planted. It had always been so; so it would always be.
>
> The farmstead we occupied was on a hilltop overlooking a marshy river bottom that stretched from horizon to horizon. It was half a mile from any road and an eternity from any connection with the rest of the culture. There were no books there; there was no music; there was no television; for a long time, no telephone. Only on the rarest of occasions—a time or two a year—was there a social visitor other than the pastor. There was no conversation in that house.

Similarly, Wallace Stegner, the great historian and novelist, confesses to his feeling of inadequacy in coming from a small prairie town in the Cypress Hills of Saskatchewan. In *Wolf Willow* he writes:

> Once, in a self-pitying frame of mind, I was comparing my background with that of an English novelist friend. Where he had been brought up in London, taken from the age of four onward to the Tate and the National Gallery, sent traveling on the Continent in every school holiday, taught French and German and Italian, given access to bookstores, libraries, and British Museums, made familiar from infancy on with the conversation of the eloquent and the great, I had grown up in this dung-heeled sagebrush town on the disappearing edge of nowhere, utterly without painting, without sculpture, without architecture, almost without music or theater, without conversation or languages or travel or stimulating instruction, without libraries or museums or bookstores, almost without books. I was charged with getting in a single lifetime, from scratch, what some people inherit as naturally as they breathe air.
>
> How, I asked this Englishman, could anyone from so deprived a background ever catch up? How was one expected to compete, as a cultivated man, with people like himself? He looked at me and said dryly, "Perhaps you got something else in place of all that."
>
> He meant, I suppose that there are certain advantages to growing up

a sensuous little savage, and to tell the truth I am not sure I would trade my childhood of freedom and the outdoors and the senses for a childhood of being led by the hand past all the Turners in the National Gallery. And also, he may have meant that anyone starting from deprivation is spared getting bored. You may not get a good start, but you may get up a considerable head of steam.

Countless writers and artists have been vulnerable to the "official" culture, as vulnerable as the people of Matfield Green. Stegner comments:

> I am reminded of Willa Cather, that bright girl from Nebraska, memorizing long passages from the *Aeneid* and spurning the dust of Red Cloud and Lincoln with her culture-bound feet. She tried, and her education encouraged her, to be a good European. Nevertheless she was a first-rate novelist only when she dealt with what she knew from Red Cloud and the things she had "in place of all that." Nebraska was what she was born to write; the rest of it was got up. Eventually, when education had won and nurture had conquered nature and she had recognized Red Cloud as a vulgar little hold, she embraced the foreign tradition totally and ended by being neither quite a good American nor quite a true European nor quite a whole artist.

It seems that we still blunt our sensitivities about our local places by the likes of learning long passages from the *Aeneid* while wanting to shake from us the dust of Red Cloud or Matfield Green. The extractive economy cares for neither Virgil nor Mary Stewart. It lures just about all of us to its shopping centers on the edges of major cities. And yet, for us, the *Aeneid* is as essential to becoming native to the Matfield Greens as the bow and arrow were to the paleolithic Asians who walked here across the Bering land bridge of the Pleistocene.

Our task is to build cultural fortresses to protect our emerging nativeness. They must be strong enough to hold at bay the powers of consumerism, the powers of greed and envy and pride. One of the most effective ways for this to come about would be for our universities to assume the awesome responsibility to both validate and educate those who want to be homecomers—not necessarily to go home but to go someplace and dig in and begin the long search and experiment to become native.

It will be a struggle, but a worthy one. The homecomer will not learn the likes of Virgil to adorn his talk, to show off, but will study Virgil for insight, for utility, as well as for pleasure.

We can then hope for a resurrection of the likes of Mrs. Florence Johnson and her women friends who took their collect seriously. Unless we can validate and promote the sort of "cultural information in the making" that the New Century Club featured, we are doomed. An entire club program devoted to coping with the heat of August is being native to a place. That club was more than a support group; it was cultural information in the making, keyed to place. The alternative, one might suggest, is mere air conditioning—not only yielding greenhouse gases but contributing to global warming and the ozone hole as well, and, if powered with nuclear power, to future Chernobyls.

Becoming native to this place means that the creatures we bring with us—our domesticated creatures—must become native, too. Long ago they were removed from the original relationships they had with their ecosystems and pressed into our service. Our interdependency has now become so complete that, if proprietorship is the subject, we must acknowledge that in some respects they own us. We humans honor knowledge of where we came from, counting that as baseline information, essential to our journey toward nativeness. Similarly we must acknowledge that our domesticated creatures are descendants of wild things that were shaped in an ecological context *not* of our making when we found them. The fence we built to keep their relatively tame and curious wild ancestors out of our Early Neolithic gardens eventually became the barbed wire that would contain them. At the moment of first containment, those fences must have enlarged our idea of property and property lines. When we brought that notion of property lines with us to this distant and magnificent continent, it was a short step to the invisible grid that in turn created the tens of thousands of hard and alien lines that dominate our thoughts today. Did the natives at Lone Tree Massacre foresee this? Those lines will be with us forever, probably. But we can soften them. We'll have to, for the hardness of those lines is proportional to our sense of the extent to which we own what we use. Our becoming native will depend on our emerging consciousness of how we are to use the gifts of the creation. We must think in terms of different relationships. Perhaps we *will* come to think of the chicken as fundamentally a jungle fowl. The hog will once again be regarded as a descendant of a roaming and rooting forest animal. Bovines will be seen as savanna grazers.

An extractive economic system to a large degree is a derivative of our perceptions and values. But it also controls our behavior. We have to loosen its hard grip on us, finger by finger. I am hopeful that a new economic system can emerge from the homecomer's effort—as a derivative of right livelihood rather than of purposeful design. It will result from our becoming better ecological accountants at the community level. If we must as a future necessity recycle essentially all materials and run on sunlight, then our future will depend on accounting as the most important and interesting discipline. Because accountants are students of boundaries, we are talking about educating a generation of students who will know how to set up the books for their ecological community accounting, to use three-dimensional spreadsheets. But classroom work alone won't do. They will need a lifetime of field experience besides, and the sacrifices they must make, by our modern standards, will be huge. They won't be regarded as heroic, at least not in the short run. Nevertheless, that will be their real work. Despite the daily decency of the women in the Matfield Greens, decency could not stand up against the economic imperialism that swiftly and ruthlessly plowed them and their communities under.

The agenda of our homecoming majors is already beyond comprehensive vision. They will have to be prepared to think about such problems as balances between efficiency and sufficiency. This will require informed judgment across our entire great ecological mosaic. These graduates in homecoming will be unable to hide in bureaucratic niches within the major program initiatives of public policy that big government likes to sponsor. Those grand solutions are inherently anti-native because they are unable to vary across the varied mosaic of our ecosystems, from the cold deserts of eastern Washington to the deciduous forests of the East, with the Nebraska prairie in between. The need is for each community to be coherent. Knowing this, we must offer our homecomers the most rigorous curriculum and the best possible faculty, the most demanding faculty of all time.

Professor J. Stan Rowe describes what we might see by imagining that we humans could make ourselves small enough to enter some average-sized cell and, once there, continue to miniaturize to the point that we need binoculars to examine various parts off in the distance. The parts with dynamic processes we would undoubtedly designate as living. There would be streaming cytoplasm, replicating DNA, and amino acids being hitched to one another in precise order, forming polypeptide chains of varying lengths. These along with the mitochondria, the cells' powerhouses we, with our human minds, would likely designate as living, too. But if we have lost our perspective during miniaturization we would inevitably designate some things—crystals, some membranes—as dead. If we used this living versus dead taxonomy, it would only be because we had, during our shrinkage, lost our ability to see the larger perspective. Lacking our former, more comprehensive mind, we would think that some things count more than others.

Now imagine that a proportionally large creature were to arrive in our solar system and, after some shopping, pick the Earth to visit. Imagine such a creature able to shrink to our size but, unlike us, keeping the larger perspective in mind. He or she or it would soon discover that a most amazing species had changed the face of the earth in some dramatic ways to grow food and fiber. Our visitor, at first glance, might think that humans are basically artists bent on dressing the earth with their own designs.

Now imagine that to our visitor the Earth is a sort of field site, an object of study for something like a doctoral dissertation. Maybe it is a long-term study in which the investigator visits our planet every two hundred years or so beginning ten thousand years ago. On this, the fiftieth visit, this student of the earth would realize that the population of one particular species, self-named *Homo sapiens*, is rapidly increasing in number and that as it does so its members pollute and destroy more and more of the parts of the planet necessary for the maintenance of what they call life.

Many humans in thousands of places over the globe *have* been truly artful. But, as our visitor would realize, there is cause for alarm. During the two hundred years between the last and the present visit, many deposits of energy-rich car-

Intensive market gardening at Land's Sake, a community and educational farm in Weston, Massachusetts, 2006. Courtesy Sara Gregg.

bon have been discovered in pools and seams and employed to power the human enterprise.

From the outside perspective, when the traditional cultures relied exclusively on contemporary sun power, the agricultural/cultural activity around and below could best be characterized as a form of intimacy. This activity was at once artistic and cultural. The body and the mind with the eye worked as one. There was no separation. Food was not just fuel then, and the tools necessary to capture sunlight and to provide food—air, water, soil—if they were dead, were dead in the same way as a crystal or a membrane is dead in the cell. This is no mere detail. The designation non-living invites a prejudice. Since air and water and soil are just dead stuff lying around, we act as though we can pollute or destroy them at will.

For the artistic farmer, the tool is not the brush but rather the pitchfork, the hoe, the rake, the shovel, the pruning shears, a team of horses, even a diesel tractor if it is run on vegetable oil. Whether in China, Peru, Africa, Sicily, or among the Hopi of the Southwest, the agricultural artist prefers wisdom over cleverness. He or she experiments, but experimentation is subordinate to tradition. The true artist honors balance of emotion and technique, people and land, individual and community, plant and animal.

Bruce Wooster picks chard at Land's Sake, Weston, Massachusetts, 2005.
Courtesy Nina Danforth.

Since the 1930s industrialized agriculture has been increasingly promoted by the industrial mind. But now a small but growing minority realizes that high energy destroys information of both the cultural and the biological varieties. This approach not only pollutes landscapes, it rooster-tails the finite supply of nutrients from our agricultural lands into the supermarkets, into the kitchen sinks, onto the chopping boards, onto the tables, and into the human gut, and, once there, more or less heads only one way, downstream into the sewers and graveyards.

Maybe our problem is that we are unable to keep the perspective of that outside observer, for we fail to absorb what we know: that nutrient cycles must be closed; that if we introduce into the environment chemicals with which we have not evolved, we must regard them as guilty until proven innocent; that fossil fuels are finite; that agriculture—not agri-business—is the source of a healthy culture; that all parts are important to the whole and to other parts.

The realities of industrialization are all around us. No "ain't-it-awful" checklist is necessary. What we must think about, therefore, is an agriculture with a human face. We must give standing to the new pioneers, the homecomers bent on the most important work for the next century—a massive salvage operation to save the vulnerable but necessary pieces of nature and culture and to keep the good and artful examples before us. It is time for a new breed of artists to enter front and center, for the point of art, after all, is to connect. This is the homecomer I have in mind: the scientist, the accountant who converses with nature, a true artist devoted to the building of agriculture and culture to match the scenery presented to those first European eyes.

WENDELL BERRY

Wendell Berry was born in Henry County, Kentucky, in 1934. His father, John Berry, was a prominent lawyer involved in the organization of the Burley Tobacco Growers Cooperative and the drafting of New Deal price supports. Both his father and his mother, Virginia, were of farm families deeply rooted in the area, and Berry grew up steeped in the local culture of the small, mixed farming of his grandfathers' generation, having been born just in time to learn to work with horses and mules. He attended the University of Kentucky, studying English, and received his BA and MA in 1956 and 1957. After studying and teaching writing at Stanford, New York University, and in Europe, Berry returned to Kentucky in 1965 to assume a professorship of English at his alma mater, and he and his wife, Tanya, bought and restored what is now a 125-acre farm near Port Royal. By this return, Berry committed himself not only to a life of farming, but also to expressing and defending the life of his native place in his writing.

Berry's farm is a touchstone in his work toward sustainable agriculture. He wrote in the foreword to The Gift of Good Land *(1982) that "agriculture is an integral part of the structure, both biological and cultural, that sustains human life, and . . . you cannot disturb one part of that structure without disturbing all of it; . . . in short, though there may be specialized causes, there are no specialized effects." Berry's writing has explored the effects of that profound disturbance to American farming during his lifetime. He has published dozens of books of poetry, essays, and fiction. Berry's novels and short stories are set in the fictional Port William, Kentucky, where a dwindling "membership" of farmers and townspeople struggle to maintain bonds of family, community, and memory stretching back generations to the white settlement of the surrounding countryside. His strongest, most fully realized characters—Jack Beechum, Mat Feltner, Burley Coulter— are farmers of the late nineteenth and early twentieth centuries, when the community, for all its flaws, was intact and thriving. Berry's essays cover a much wider range than farming alone, but much of his work is an attack on the industrial ideal and a defense of the best of traditional agrarian culture.* The Unsettling of America *(1977), perhaps his most famous book, is an indictment of the modern system of industrial agriculture. Berry is without question the leading intellectual architect of modern agrarianism as a worldview consciously opposed to corporate industrialism, drawing upon the thinking of Thomas Jefferson, Liberty Hyde Bailey, and the Southern Agrarians.*

The following essay, originally written for New Farm *magazine and included in* The Gift of Good Land, *celebrates one of many examples of careful farming that Berry has written about, the farms belonging to the Yoder family in Indiana. Berry continues to urge American farmers, citizens, and policymakers to embrace a more sustainable approach to life on the land.*

"Seven Amish Farms" (1981)

In typical Midwestern farming country the distances between inhabited houses are stretching out as bigger farmers buy out their smaller neighbors in order to "stay in." The signs of this "movement" and its consequent specialization are everywhere: good houses standing empty, going to ruin; good stock barns going to ruin; pasture fences fallen down or gone; machines too large for available doorways left in the weather; windbreaks and woodlots gone down before the bulldozers; small schoolhouses and churches deserted or filled with grain.

In the latter part of March this country shows little life. Field after field lies under the dead stalks of last year's corn and soybeans, or lies broken for the next crop; one may drive many miles between fields that are either sodded or planted in winter grain. If the weather is wet, the country will seem virtually deserted. If the ground is dry enough to support their wheels, there will be tractors at work, huge machines with glassed cabs, rolling into the distances of fields larger than whole farms used to be, as solitary as seaborne ships.

The difference between such country and the Amish farmlands in northeast Indiana seems almost as great as that between a desert and an oasis. And it is the *same* difference. In the Amish country there is a great deal more life: more natural life, more agricultural life, more human life. Because the farms are small—most of them containing well under a hundred acres—the Amish neighborhoods are more thickly populated than most rural areas, and you see more people at work. And because the Amish are diversified farmers, their plowed croplands are interspersed with pastures and hayfields and often with woodlots. It is a varied, interesting, healthy looking farm country, pleasant to drive through. When we were there, on the twentieth and twenty-first of last March, the spring plowing had just started, and so you could still see everywhere the annual covering of stable manure on the fields, and the teams of Belgians or Percherons still coming out from the barns with loaded spreaders.

Our host, those days, was William J. Yoder, a widely respected breeder of Belgian horses, an able farmer and carpenter, and a most generous and enjoyable companion. He is a vigorous man, strenuously involved in the work of his farm and in the life of his family and community. From the look of him and the look of his place, you know that he has not just done a lot of work in his time, but has done it well, learned from it, mastered the necessary disciplines. He speaks with heavy stress on certain words—the emphasis of conviction, but also of pleasure, for he enjoys the talk that goes on among people interested in horses and in farming. But unlike many people who enjoy talking, he speaks with care. Bill was born in this community, has lived there all his life, and he has grandchildren who will probably live there all their lives. He belongs there, then, root and branch, and he knows the history and the quality of many of the farms. On the two days, we visited farms belonging to Bill himself, four of his sons, and two of his sons-in-law.

The Amish farms tend to divide up between established ones, which are prosperous looking and well maintained, and run-down, abused, or neglected ones, on which young farmers are getting started. Young Amish farmers *are* still getting started, in spite of inflation, speculators' prices, and usurious interest rates. My impression is that the proportion of young farmers buying farms is significantly greater among the Amish than among conventional farmers.

Bill Yoder's own eighty-acre farm is among the established ones. I had been there in the fall of 1975 and had not forgotten its aspect of cleanness and good order, its well-kept white buildings, neat lawns, and garden plots. Bill has owned the place for twenty-six years. Before he bought it, it had been rented and row cropped, with the usual result: it was nearly played out. "The buildings," he says, "were nothing," and there were no fences. The first year, the place produced five loads (maybe five tons) of hay, "and that was mostly sorrel." The only healthy plants on it were the spurts of grass and clover that grew out of the previous year's manure piles. The corn crop that first year "might have been thirty bushels an acre," all nubbins. The sandy soil blew in every strong wind, and when he plowed the fields his horses' feet sank into "quicksand potholes" that the share uncovered.

The remedy has been a set of farming practices traditional among the Amish since the seventeenth century: diversification, rotation of crops, use of manure, seeding of legumes. These practices began when the Anabaptist sects were disfranchised in their European homelands and forced to the use of poor soil. We saw them still working to restore farmed-out soils in Indiana. One thing these practices do is build humus in the soil, and humus does several things: increases fertility, improves soil structure, improves both water-holding capacity and drainage. "No humus, you're in trouble," Bill says.

After his rotations were established and the land had begun to be properly manured, the potholes disappeared, and the soil quit blowing. "There's something in it now—there's some substance there." Now the farm produces abundant crops of corn, oats, wheat, and alfalfa. Oats now yield 90–100 bushels per acre. The corn averages 100–125 bushels per acre, and the ears are long, thick, and well filled.

Bill's rotation begins and ends with alfalfa. Every fall he puts in a new seeding of alfalfa with his wheat; every spring he plows down an old stand of alfalfa, "no matter how good it is." From alfalfa he goes to corn for two years, planting thirty acres, twenty-five for ear corn and five for silage. After the second year of corn, he sows oats in the spring, wheat and alfalfa in the fall. In the fourth year the wheat is harvested; the alfalfa then comes on and remains through the fifth and sixth years. Two cuttings of alfalfa are taken each year. After curing in the field, the hay is hauled to the barn, chopped, and blown into the loft. The third cutting is pastured.

Unlike cow manure, which is heavy and chunky, horse manure is light and breaks up well coming out of the spreader; it interferes less with the growth of small seedlings and is less likely to be picked up by a hay rake. On Bill's place, horse manure is used on the fall seedings of wheat and alfalfa, on the young alfalfa after the wheat harvest, and both years on the established alfalfa stands. The cow manure goes on the corn ground both years. He usually has about 350 eighty-bushel spreader loads of manure, and each year he covers the whole farm—cropland, hayland, and pasture.

With an abundance of manure there obviously is no *dependence* on chemical fertilizers, but Bill uses some as a "starter" on his corn and oats. On corn he applies 125 pounds of nitrogen in the row. On oats he uses 200–250 pounds of 16-16-16, 20-20-20, or 24-24-24. He routinely spreads two tons of lime to the acre on the ground being prepared for wheat.

His out-of-pocket costs per acre of corn last year were as follows:

Seed (planted at a rate of seven acres per bushel)	$7.00
Fertilizer	7.75
Herbicide (custom applied, first year only)	16.40

That comes to a total of $31.15 per acre—or, if the corn makes only a hundred bushels per acre, a little over $0.31 per bushel. In the second year his per acre cost is $14.75, less than $0.15 per bushel, bringing the two-year average to $22.95 per acre or about $0.23 per bushel.

The herbicide is used because, extra horses being on the farm during the winter, Bill has to buy eighty to a hundred tons of hay, and in that way brings in weed seed. He had no weed problem until he started buying hay. Even though he uses the herbicide, he still cultivates his corn three times.

His cost per acre of oats came to $33.00 ($12.00 for seed and $21.00 for fertilizer)—or, at ninety bushels per acre, about $0.37 per bushel.

Of Bill's eighty acres, sixty-two are tillable. He has ten acres of permanent pasture, and seven or eight of woodland, which produced the lumber for all the building he has done on the place. In addition, for $500 a year he rents an adjoining eight acres of "hill and woods pasture" which provides summer grazing for twenty heifers; and on another neighboring farm he rents varying amounts of cropland.

All the field work is done with horses, and this, of course comes virtually free—a by-product of the horse-breeding enterprise. Bill has an ancient Model D John Deere tractor that he uses for belt power.

At the time of our visit, there were twenty-two head of horses on the place. But that number was unusually low, for Bill aims to keep "around thirty head." He has a band of excellent brood mares and three stallions, plus young stock of assorted ages. Since October 1 of last year, he had sold eighteen head of registered Belgian horses. In the winter he operates a "urine line," collecting "pregnant mare urine," which is sold to a pharmaceutical company for the extraction of various hormones. For this purpose he boards a good many mares belonging to neighbors; that is why he must buy the extra hay that causes his weed problem. (Horses are so numerous on this farm because they are one of its money-making enterprises. If horses were used only for work on this farm, four good geldings would be enough.)

One bad result of the dramatic rise in draft horse prices over the last eight or ten years is that it has tended to focus attention on such characteristics as size and color to the neglect of less obvious qualities such as good feet. To me, foot quality seems a critical issue. A good horse with bad feet is good for nothing but decoration, and at sales and shows there are far too many flawed feet disguised by plastic wood and black shoe polish. And so I was pleased to see that every horse on Bill Yoder's place had sound, strong-walled, correctly shaped feet. They were good horses all around, but their other qualities were well founded; they stood on good feet, and this speaks of the thoroughness of his judgment and also of his honesty.

Though he is a master horseman, and the draft horse business is more lucrative now than ever in its history, Bill does not specialize in horses, and that is perhaps the clearest indication of his integrity as a farmer. Whatever may be the dependability of the horse economy, on this farm it rests upon a diversified agricultural economy that is sound.

He was milking five Holstein cows; he had fifteen Holstein heifers that he had raised to sell; and he had just marketed thirty finished hogs, which is the number that he usually has on fence. All the animals had been well wintered—Bill quotes his father approvingly: "Well wintered is half summered"—and were in excellent condition. Another saying of his father's that Bill likes to quote—"Keep the horses on the side of the fence the feed is on"—has obviously been obeyed here. The feeding is careful, the feed is good, and it is abundant. Though it was almost spring, there were ample surpluses in the hayloft and in the corn cribs.

Other signs of the farm's good health were three sizable garden plots, and newly pruned grapevines and raspberry canes. The gardener of the family is Mrs. Yoder. Though most of the children are now gone from home, Bill says that she still grows as much garden stuff as she ever did.

All seven of the Yoders' sons live in the community. Floyd, the youngest, is still at home. Harley has a house on nearly three acres, works in town, and returns in the afternoons to his own shop where he works as a farrier. Henry, who also works in town, lives with Harley and his wife. The other four sons are now settled on farms that they are in the process of paying for. Richard has eighty acres, Orla eighty, Mel fifty-seven, and Wilbur eighty. Two sons-in-law also living in the community are Perry Bontrager, who owns ninety-five acres, and Ervin Mast, who owns sixty-five. Counting Bill's eighty acres, the seven families are living on 537 acres. Of the seven farms, only Mel's is entirely tillable, the acreages in woods or permanent pasture varying from five to twenty-six.

These young men have all taken over run-down farms, on which they are establishing rotations and soil husbandry practices that, being traditional, more or less resemble Bill's. It seemed generally agreed that after three years of this treatment the land would grow corn, as Perry Bontrager said, "like anywhere else."

These are good farmers, capable of the intelligent planning, sound judgment, and hard work that good farming requires. Abused land heals and flourishes in their care. None of them expressed a wish to own more land; all, I believe, feel that what they have will be enough—when it is paid for. The big problems are high land prices and high interest rates, the latter apparently being the worst.

The answer, for Bill's sons so far, has been town work. All of them, after leaving home, have worked for Redman Industries, a manufacturer of mobile homes in Topeka. They do piecework, starting at seven in the morning and quitting at two in the afternoon, using the rest of the day for farming or other work. This, Bill thinks, is now "the only way" to get started farming. Even so, there is "a lot of debt" in the community—"more than ever."

With a start in factory work, with family help, with government and bank loans, with extraordinary industry and perseverance, with highly developed farming skills, it is still possible for young Amish families to own a small farm that will eventually support them. But there is more strain in that effort now than there used to be, and more than there should be. When the burden of usurious interest becomes too great, these young men are finding it necessary to make temporary returns to their town jobs.

The only one who spoke of his income was Mel, who owns fifty-seven acres, which, he says, *will be* enough. He and his family milk six Holsteins. He had nine mares on the urine line last winter, seven of which belonged to him. And he had twelve brood sows. Last year his gross income was $43,000. Of this, $12,000 came from hogs, $7,000 from his milk cows, the rest from his horses and the sale of his wheat. After his production costs, but *before* payment of interest, he netted $22,000. In order to cope with the interest payments, Mel was preparing to return to work in town.

These little Amish farms thus become the measure both of "conventional" American agriculture and of the cultural meaning of the national industrial economy.

To begin with, these farms give the lie direct to that false god of "agribusiness": the so-called economy of scale. The small farm is not an anachronism, is not unproductive, is not unprofitable. Among the Amish, it is still thriving, and is still the economic foundation of what John A. Hostetler (in *Amish Society,* third edition) rightly calls "a healthy culture." Though they do not produce the "record-breaking yields" so touted by the "agribusiness" establishment, these farms are nevertheless highly productive. And if they are not likely to make their owners rich (never an Amish goal), they can certainly be said to be sufficiently profitable. The economy of scale has helped corporations and banks, not farmers and farm communities. It has been an economy of dispossession and waste—plutocratic, if not in aim, then certainly in result.

What these Amish farms suggest, on the contrary, is that in farming there is inevitably a scale that is suitable both to the productive capacity of the land and to the abilities of the farmer; and that agricultural problems are to be properly solved, not in expansion, but in management, diversity, balance, order, responsible maintenance, good character, and in the sensible limitation of investment and overhead. (Bill makes a careful distinction between "healthy" and "unhealthy" debt, a "healthy debt" being "one you can hope to pay off in a reasonable way.")

Most significant, perhaps, is that while conventional agriculture, blindly following the tendency of any industry to exhaust its sources, has made soil erosion a national catastrophe, these Amish farms conserve the land and improve it in use.

And what is one to think of a national economy that drives such obviously able and valuable farmers to factory work? What value does such an economy impose upon thrift, effort, skill, good husbandry, family and community health?

In spite of the unrelenting destructiveness of the larger economy, the Amish—as Hostetler points out with acknowledged surprise and respect—have almost doubled in population in the last twenty years. The doubling of a population is, of course, no significant achievement. What is significant is that these agricultural communities have doubled their population *and yet remained agricultural communities* during a time when conventional farmers have failed by the millions. This alone would seem to call for a careful look at Amish ways of farming. That those ways have, during the same time, been ignored by the colleges and the agencies of agriculture must rank as a prime intellectual wonder.

Amish farming has been so ignored, I think, because it involves a complicated structure that is at once biological and cultural, rather than industrial or economic. I suspect that anyone who might attempt an accounting of the economy of an Amish farm would soon find himself dealing with virtually unaccountable values, expenses, and benefits. He would be dealing with biological forces and processes not always measurable, with spiritual and community values not quantifiable; at certain points he would be dealing with mysteries—and he would be finding that these unaccountables and inscrutables have results, among others, that are economic. Hardly an appropriate study for the "science" of agricultural economics.

The economy of conventional agriculture or "agribusiness" is remarkable for the simplicity of its arithmetic. It involves a manipulation of quantities that are all entirely accountable. List your costs (land, equipment, fuel, fertilizer, pesticides, herbicides, wages), add them up, subtract them from your earnings, or subtract your earnings from them, and you have the result.

Suppose, on the other hand, that you have an eighty-acre farm that is not a "food factory" but your home, your given portion of Creation which you are morally and spiritually obliged "to dress and to keep." Suppose you farm, not for wealth, but to maintain the integrity and the practical supports of your family and community. Suppose that, the farm being small enough, you farm it with family work and work exchanged with neighbors. Suppose you have six Belgian brood mares that you use for field work. Suppose that you also have milk cows and hogs, and that you raise a variety of grain and hay crops in rotation. What happens to your accounting then?

To start with, several of the costs of conventional farming are greatly diminished or done away with. Equipment, fertilizer, chemicals all cost much less. Fuel becomes feed, but you have the mares and are feeding them anyway; the work ration for a brood mare is not a lot more costly than a maintenance ration. And the horses, like the rest of the livestock, are making manure. Figure that in, and figure, if you can, the value of the difference between manure and chemical fertilizer. You can probably get an estimate of the value of the nitrogen fixed by your alfalfa, but how will you quantify the value to the soil of its residues and deep roots? Try to compute the value of humus in the soil—in improved drainage, improved drought resistance, improved tilth, improved health. Wages, if you pay your children, will still be among your costs. But compute the difference between paying your children and paying "labor." Work exchanged with neighbors can be reduced to "man-hours" and assigned a dollar value. But compute the difference between a neighbor and "labor." Compute the value of a family or a community to any one of its members. We may, as we must, grant that among the values of family and community there is economic value—but what is it?

In the Louisville *Courier-Journal* of April 5, 1981, the Mobil Oil Corporation ran an advertisement which was yet another celebration of "scientific agriculture." American farming, the Mobil people are of course happy to say, "requires *more petroleum products than almost any other industry.* A gallon of gasoline to produce a single bushel of corn, for example. . . ." This, they say, enables "each American farmer to feed sixty-seven people." And they say that this is "a-maizing."

Well, it certainly is! And the chances are good that an agriculture totally dependent on the petroleum industry is not yet as amazing as it is going to be. But one thing that is already sufficiently amazing is that a bushel of corn produced by the burning of one gallon of gasoline has already cost more than *six times* as much as a bushel of corn grown by Bill Yoder. How does Bill Yoder escape what may justly be called the petroleum tax on agriculture? He does so by a series of substitutions: of horses for tractors, of feed for fuel, of manure for fertilizer, of sound agricultural methods and patterns for the exploitive methods and patterns of industry. But he has done more than that—or, rather, he and his people and their tradition have done more. They have substituted themselves, their families, and their communities for petroleum. The Amish use little petroleum—and need little—because they have those other things.

I do not think that we can make sense of Amish farming until we see it, until we become willing to see it, as belonging essentially to the Amish practice of Christianity, which instructs that one's neighbors are to be loved as oneself. To farmers who give priority to the maintenance of their community, the economy of scale (that is, the economy of *large* scale, of "growth") can make no sense, for it requires the ruination and displacement of neighbors. A farm cannot be increased except by the decrease of a neighborhood. What the interest of the community proposes is invariably an economy of *proper* scale. A whole set of agricultural proprieties must be observed: of farm size, of methods, of tools, of energy sources, of plant and animal species. Community interest also requires charity, neighborliness, the care and instruction of the young, respect for the old; thus it assures its integrity and survival. Above all, it requires good stewardship of the land, for the community, as the Amish have always understood, is no better than its land. "If treated violently or exploited selfishly," John Hostetler writes, the land "will yield poorly." There could be no better statement of the meaning of the *practice* and the practicality of charity. Except to the insane narrow-mindedness of industrial economics, selfishness does not pay.

The Amish have steadfastly subordinated economic value to the values of religion and community. What is too readily overlooked by a secular, exploitive society is that their ways of doing this are not "empty gestures" and are not "backward." In the first place, these ways have kept the communities intact through many varieties of hard times. In the second place, they conserve the land. In the third place, they yield economic benefits. The community, the religious fellowship, has many kinds of value, and among them is economic value. It is the result of the practice of neighborliness, and of the practice of stewardship. What moved me most, what I liked best, in those days we spent with Bill Yoder was the sense of the continuity of the community in his dealings with his children and in their dealings with their children.

Bill has helped his sons financially so far as he has been able. He has helped them with his work. He has helped them by sharing what he has—lending a stallion, say, at breeding time, or lending a team. And he helps them by buying good pieces of equipment that come up for sale. "If he ever gets any money," he says of one of the boys, for whom he has bought an implement, "he'll pay me for it. If he don't, he'll just use it." He has been their teacher, and he remains their advisor. But he does not stand before them as a domineering patriarch or "authority figure." He seems to speak, rather, as a representative of family and community experience. In their respect for him, his sons respect their tradition. They are glad for his help, advice, and example, but there is nothing servile in this. It seems to be given and taken in a kind of familial friendship, respect going both ways.

Everywhere we went, when school was not in session, the children were at the barns, helping with the work, watching, listening, learning to farm in the way it is best learned. Wilbur told us that his eleven-year-old son had cultivated twenty-three acres of corn last year with a team and a riding cultivator. That reminded Bill of the way he taught Wilbur to do the same job.

Wilbur was little then, and he loved to sit in his father's lap and drive the team while Bill worked the cultivator. If Wilbur could drive, Bill thought, he could do the rest of it. So he got off and shortened the stirrups so the boy could reach them with his feet. Wilbur started the team, and within a few steps began plowing up the corn.

"Whoa!" he said.

And Bill, who was walking behind him, said, "Come up!"

And it went that way for a little bit:

"Whoa!"

"Come up!"

And then Wilbur started to cry, and Bill said:

"Don't cry! Go ahead!"

Slow Food
Nation'08
Come to the Table

Logo to promote the Slow Food Nation celebration over Labor Day weekend in 2008,
evoking the centrality of agriculture to American ideas about freedom. Part of the international
Slow Food movement, this event was designed to spotlight the importance of healthy,
whole food to the modern nation. Courtesy Slow Food USA.

Conclusion: American Agrarianism in the Twenty-first Century

The dawn of the twenty-first century has revealed what might possibly be mistaken for a new wave of agrarianism sweeping the nation. Within the past decade, a renewed appreciation for connection with soil and community has captured the interest of a growing and influential segment of American food culture. Recent years have brought upsurges in community-supported agriculture (CSA) farms that distribute local produce to urban and suburban subscribers, as well as in urban community gardens from Chicago's City Farm to corner lots in New York City. Every week the *New York Times* seems to feature another article about a Wall Street broker or corporate lawyer who has chucked it all to move to a farm upstate and raise pigs on acorns; an urban farmer (whose father was a sharecropper) growing worms, fish, and salad with kids on vacant lots in Milwaukee; upscale exurban housing developments being designed around working ranches and CSA farms instead of golf courses; young urban farmers raising chickens behind their apartments; recent graduates from prestigious colleges heading off to work for next to nothing as interns on organic farms—to the mixed pride and dismay of their debt-ridden parents.

Best-selling books about food like Eric Schlosser's *Fast Food Nation*, Michael Pollan's *The Omnivore's Dilemma* and *In Defense of Food*, Mark Kurlansky's *The Food of a Younger Land*, and Barbara Kingsolver's *Animal, Vegetable, Miracle* have joined rows of cookbooks that indict the industrial food system. These books and the astonishing (but gratifying, to old agrarians) acclaim they have received have created a new generation of adults who are thinking about the origins of their food. Ironically, the movement toward thoughtful consumption has arisen mainly in affluent urban and suburban neighborhoods inhabited by the descendants of those who were the first to turn away from the farm. Many of them have embraced a new brand of agrarian philosophy with the zealousness of converts, making the connection between consumption and the thoughtful *production* of food, raised as locally as possible. But the growing distrust of industrial food production has moved beyond the middle class. Gourmet restaurants and public school cafeterias alike have begun to rebel against the standardization of food and the adulteration of its nutritional value, and movements toward healthy consumption have sprung up even in the most unlikely places. Farmers' markets now accept WIC and food stamps, celebrity chefs such as Alice Waters have thrown their weight behind integrating gardens and food programs into urban schools, and First Lady Michelle Obama has planted an organic garden at the White House with local schoolchildren.

The principles that drive this movement are well summarized in the 2008 "Declaration for Healthy Food and Agriculture," organized by Roots for Change in California and Slow Food USA:

> We, the undersigned, believe that a healthy food system is necessary to meet the urgent challenges of our time. Behind us stands a half-century of industrial food production, underwritten by cheap fossil fuels, abundant land and water resources, and a drive to maximize the global harvest of cheap calories. Ahead lie rising energy and food costs, a changing climate, declining water supplies, a growing population, and the paradox of widespread hunger and obesity.
>
> These realities call for a radically different approach to food and agriculture. We believe that the food system must be reorganized on a foundation of health: for our communities, for people, for animals, and for the natural world. The quality of food, and not just its quantity, ought to guide our agriculture. The ways we grow, distribute, and prepare food should celebrate our various cultures and our shared humanity, providing not only sustenance, but justice, beauty and pleasure. . . .
>
> Governments have a duty to protect people from malnutrition, unsafe food, and exploitation, and to protect the land and water on which we depend from degradation. Individuals, producers, and organizations have a duty to create regional systems that can provide healthy food for their communities. We all have a duty to respect and honor the laborers of the land without whom we could not survive. The changes we call for here have begun, but the time has come to accelerate the transformation of our food and agriculture and make its benefits available to all.
>
> We believe that the following twelve principles should frame food and agriculture policy, to ensure that it will contribute to the health and wealth of the nation and the world. A healthy food and agriculture policy:
>
> *Forms the foundation of secure and prosperous societies, healthy communities, and healthy people.*
> *Provides access to affordable, nutritious food to everyone.*
> *Prevents the exploitation of farmers, workers, and natural resources; the domi-*

*nation of genomes and markets; and the cruel treatment of animals, by any
nation, corporation, or individual.*

Upholds the dignity, safety, and quality of life for all who work to feed us.

*Commits resources to teach children the skills and knowledge essential to food
production, preparation, nutrition, and enjoyment.*

*Protects the finite resources of productive soils, fresh water, and biological
diversity.*

*Strives to remove fossil fuel from every link in the food chain and replace it with
renewable resources and energy.*

Originates from a biological rather than an industrial framework.

*Fosters diversity in all its relevant forms: diversity of domestic and wild species;
diversity of foods, flavors, and traditions; diversity of ownership.*

*Requires a national dialog concerning technologies used in production, and
allows regions to adopt their own respective guidelines on such matters.*

*Enforces transparency so that citizens know how their food is produced, where
it comes from, and what it contains.*

*Promotes economic structures and supports programs to nurture the development
of just and sustainable regional farm and food networks.*

Our pursuit of healthy food and agriculture unites us as people and
as communities, across geographic boundaries, and social and economic
lines. We pledge our votes, our purchases, our creativity, and our energies
to this urgent cause.

This declaration did not come out of nowhere. It represents a kind of coming-of-
age celebration for the sustainable agriculture and organic food movements of the
1970s. Many of the key framers of the declaration are leaders who emerged with
that earlier movement, some of whom were featured in the last section of this book:
Joan Gussow, Fred Kirschenmann, Alice Waters, Winona LaDuke, Wes Jackson,
and Wendell Berry, to name a few. But alongside these well-known figures are
legions of others who, in one way or another, went "back to the land" in the 1970s.
They are now running successful organic farms and restaurants; working for land
trusts, nonprofit organizations, and government environmental and social services
agencies; and serving on local town boards. Many are well entrenched in positions
of influence within their communities, having persevered through several difficult
decades. And beyond them are millions more who shared similar sensibilities in the
1970s and then moved on to other lives and careers, but with whom the message of
healthy food and sustainable farming still strongly resonates. These are the readers
of Pollan and Kingsolver, the members of CSA farms, and the buyers of local organic
produce, grass-fed meat, and artisanal cheese. And after them comes a fresh wave
of enthusiasm from younger people discovering these ideas from Pollan and Kingsolver, their professors, or their parents. So to many of those who were part of it, the
1970s return to the land doesn't look like a naïve drug trip that quickly collapsed in

the face of the realities of rural life. It looks more like an enduring movement that, while it has not yet changed the overall course of mainstream American industrial agriculture, has succeeded in generating a powerful new agrarian constituency.

The recent local food movement is being driven by a recurrence of the same sense of ecological and economic crisis that drove the 1970s movement. A new element of this movement is a powerful backlash against the relentless expansion of industrial agriculture that has taken place since then, which has fed us a steady diet of food safety scandals and environmental crises. An instinctive (and increasingly scientifically documented) distaste for food produced industrially has spread beyond the health food fringe to the larger public. But more broadly, after a lapse of several decades, we have seen a return of the high energy prices and serious economic downturns that marked the 1970s. One wonders whether every economic rebound for the foreseeable future will be beaten back by a corresponding rise in fuel costs. Compounding that possibility is the largest, potentially, of all ecological crises— greenhouse gas emissions and global warming—which, one way or another, is likely to undermine the global economic growth of which industrial agriculture is a key component. Sustainable local farming and forestry fit very well within the urge to reduce our carbon footprint.

But the current sustainable farming movement invites comparison not only to the 1970s era of crisis and counterculture but also to the 1930s, which gave us the drive for a permanent ecological agriculture, and perhaps to even earlier crises and responses in America's long transition from an agrarian to an industrial nation. It is obviously a long way to the "Declaration for Healthy Food and Agriculture" from the People's Party's Omaha Platform of 1892, let alone from the Revolutionary-era views of Jefferson and Crèvecoeur. Yet as we look at that evolution, it is instructive to trace the continuities and changes along the lines with which we began this collection: the economic, political, social, and ecological dimensions of agrarianism.

The fundamental economic assumption of American agrarianism was initially that small farmers would be self-sufficient by owning their own land, producing a large measure of their own sustenance through diversified household economies and local exchange, and staying clear of ruinous debts or undue dependence on commercial markets over which they had no control. These standards were steadily compromised in the course of the nineteenth century, and all but banished from mainstream agriculture in the twentieth. The result has been a bonanza of cheap foodstuffs for American processors and consumers but ongoing economic hardship and periodic catastrophe for generations of farmers. Given this reality, neoagrarians face the daunting task of developing a workable alternative economic base for food production that rewards agrarian values. There is always the enduring Amish example to remind us that it can be done, given sufficient discipline and communal support to dictate the terms of engagement between the farm household and American consumer society. But what else comes to hand?

Today, agrarians are promoting an array of economic mechanisms. One is always to encourage some degree of household production, such as a revival of home gardening and backyard chicken flocks. Beyond these practices are not only community gardens but also a variety of community and educational farms—typically nonprofit organizations with more funding sources than sales alone, but with a real capacity to provide substantial amounts of locally grown food. There are CSAs, raw milk buying clubs, farmers' markets, and other forms of direct retail links between consumers looking for healthy food and the farmers they would like to support. All well and good, but seasoned agrarians understand that the more these efforts succeed, the more they will be imitated and undermined by industrial food suppliers. Many small farmers simply accept that another source of income is almost always a necessity for a farm to survive, and that it makes sense to structure the farming part of the household economy around what they really want for their families and enjoy doing, and not to bet the farm on the success of the enterprise. Gene Logsdon calls these "cottage farmers," and to a historian they bear an uncanny resemblance to many colonial farmer/artisan households, with their diverse economic organization. Not all the farmers moving to restructure their operations along more sustainable lines are newcomers, either—a small but steady stream of traditional farm families are making similar choices, as exemplified by groups such as the Practical Farmers of Iowa. Finally, at the level of policy, neoagrarians strive to reform government research, regulatory, and subsidy systems so that they support sustainable farming rather than agriculture geared solely toward cheap commodity production. All of these efforts raise the basic question that has always dogged American agrarianism: how to remain engaged with and make use of the market for a decent return, without becoming its slave.

The original political ideal of American agrarianism was that widespread dispersal of farm ownership among most of the populace would safeguard political independence. This possibility waned, and the survival of democracy itself will have to be otherwise ensured. But let us imagine that the agrarian dream of moving American farming toward the Amish model of farms under two hundred acres, and in many cases under one hundred acres, were somehow realized. That approach might democratize food production and repopulate the countryside in a healthy way, but it would still involve only a relatively small percentage of the population directly in farming. This conclusion returns us to the fundamental political questions in American agrarianism: Who will own and have access to farmland? Can excessive consolidation or fragmentation of ownership be guarded against? Will the large majority of citizens who are not farmers themselves have any meaningful connection to the ownership and care of farmland? The private market in land, as shaped by existing laws and policies, has worked against agrarian aims for many generations, leading toward consolidated control of productive land on the one side and sprawling residential development on the other. Many of the points in the "Declaration for Healthy Food and Agriculture," by contrast, imply the use of government intervention at many levels to ensure agrarian aims in landownership and use—similar to the New Deal drive for a permanent agriculture. In terms of landownership, for decades now there has been a growing movement for both state and local governments and nonprofit land trusts to hold conservation easements on farmland, thus ensuring that it remains in agriculture, usually at an agrarian scale. This can represent a way of dispersing landownership rights and responsibilities over a much larger part of the community than the private owners alone. In any case, the nature of widespread agrarian landownership in a postindustrial society will be bound to look very different than agrarian ownership in a preindustrial state, when almost everyone was a farmer.

The social ideal of agrarianism was that working the land was a particularly moral and healthy way to live, free from the corruption of commerce and the luxurious dissipation of city life. Many would agree that American rural culture has generally done a fine job of raising healthy, upstanding, hardworking, capable citizens. It has, however, had a good deal more trouble keeping those citizens on the farm. The claim of general moral superiority for farm life was fatally undermined in the early days of the Republic by the expansion of slavery, and in the twentieth century remained compromised by sharecropping, lack of protection for migrant farm workers, the economic and physical stress imposed by the pace of industrial farming, and the isolation caused by declining rural communities. These factors have rendered our rural areas decidedly less socially healthy today, and so one of the aims of modern agrarianism is to restore social health by repopulating the countryside. The fundamental assumption that farm life is good for people has remained at the heart of virtually all strains of American agrarianism to this day, and has certainly helped to drive each successive back-to-the-land movement.

But within this belief in the moral uplift of farm life there has been a shift from the older agrarian faith that farming bred tough, resilient, fiercely independent characters (as celebrated still by Victor Davis Hanson), to a softer, more aspirational faith that farming is good for the soul. This faith in the spiritual benefits of contact with the earth arrived with the Romantic movement of the early nineteenth century, and has only grown stronger over time—especially among nonfarming individuals who want to restore that connection. People feel it particularly strongly regarding children: Who has not heard that kids should "learn where their food comes from" or "get a little dirt under their fingernails"? This softer strain of agrarianism runs some danger of setting up naïve expectations of the reality of farm work, and may be tough for some to harmonize with the older, hard-edged strain. But it also opens many paths for engagement with agrarian values and experience on the part of a largely nonagrarian population.

One of the most striking changes in the social ideal of agrarianism today is the elevated role of women. The yeoman ideal was undergirded by a staggering contribution of female labor that, while it may have been respected and rewarded in most families, was unmistakably subservient. Many historians consider farm-women's desire to escape from their heavy burden of work and improve their status to have been a major force driving American country life toward a more commercial, consumer-based model. Even in the early twentieth-century back-to-the-land movement, a friend of E. B. White (contemplating the household work requirements of Ralph Borsodi's version of robust self-sufficiency) was heard to remark, "Well, my wife isn't Mrs. Borsodi." Helen Nearing set another example of equality: the 1970s back-to-the-land movement coincided with the rise of modern feminism, and no one who is involved in today's local food and sustainable farming movement can mistake the leading role of women.

Agrarianism has also gone through important changes in its ecological dimension. The European mixed-husbandry tradition, as it reached American shores, brought along practical knowledge of how to take good care of the land, and the yeoman ideal included an injunction to "improve" the land and leave it to the next generation in as good condition as it was received, or better. This has never been an easy ideal to fulfill, but in America it was soon eclipsed by the relentless demands of the market and by the abundance of natural resources that swung the economic advantage in agriculture toward soil mining. As a result, the conservation movement of the late nineteenth and early twentieth centuries tended to fix blame for land degradation on uneducated farmers, and was in many ways antiagrarian. But the new agrarian thinking of the twentieth century that emerged as a critique of industrialism began to incorporate ecological and conservation principles, led by the likes of Liberty Hyde Bailey and Aldo Leopold. By the late twentieth century, not so much agriculture per se but the modern industrial form of agriculture was excoriated by conservationists as environmentally destructive. Modern agrarianism, in the form of the sustainable farming movement, has thoroughly embraced environmental values. This is not to say that farming in a way that is "ecologically sound" is an easy thing to define or to accomplish, but it does represent another important evolution in agrarian philosophy.

There is one more feature of modern agrarianism that, while not entirely absent in the past, has leapt to striking prominence: an intense concern with the quality of food. Certainly, vegetarianism and praise for a healthy and simple diet have been part of some earlier strains of agrarianism, from the Transcendentalists onward. And from the mid-twentieth-century work of Albert Howard in England and J. I. Rodale in America, the connection between the health of the land and the health of the human body has been central to the organic food movement. But in the decades since the 1970s, the passion for good, healthy food has gone far beyond broccoli pie, brown rice, and granola. Modern agrarians are just as interested in exquisitely fresh local salad greens, grass-fed lamb, artisanal cheeses aged in caves, and biodynamic wines that express *goût de terroir*—though they hasten to agree that consumers of all personal tastes and income levels ought to have access to quality foods that make eating a healthy joy. This preoccupation with food is driven partly by the growing perception that industrial agriculture has unleashed an epidemic of obesity, diabetes, and eating disorders in the past few decades. It is also driven by the rediscovery of the pleasure of regional cuisines, brought largely from France and Italy through the Slow Food movement and chefs like Alice Waters and Dan Barber, and applied with great success using American ingredients.

This new call for healthy food *and* healthy agriculture cuts across all the dimensions of agrarian philosophy, especially as it addresses the challenge of making connections between farmers and urban consumers. It provides emotional links between good eating and good farming, healthy bodies and healthy land—links that a large part of the population can accept and support politically. In this way the good food mantra resembles the Populists' attempt to reach beyond the farm and build a broader alliance around the unifying idea of fair reward for honest labor. Whether the healthy food movement will lead to fundamental changes in the way in which all but a small subsection of American farming is conducted, or to a broad segment of the American population making meaningful connections to how its food is grown, is unpredictable. We are surely not about to see a mass return of the majority of Americans to small farms, barring some unspeakable collapse of industrial society, which would sweep away most agrarian dreams along with it. Only a handful of agrarians today insist on the Jeffersonian ideal of a nation of small, independent farmers being necessary to democracy, and they are writing elegies. But American agrarianism has proven to be a remarkably resilient creed, and it continues to promote a reform of industrialism that is almost as radical and fundamental as that pure ideal. First, agrarians believe that farming should be done by those who do hold agrarian values, and that they should be in a reasonably secure position to enact those values, rather than being servants to the market and large government programs over which they have no control. Second, agrarians insist that our use of the land and nature in general be guided by what has emerged as a central value of modern agrarianism, which is a healthy respect for the natural creation, from the soil to our own bodies. For that kind of ecological sensibility to be politically sustained in a market economy, agrarians believe that the American people have to be exposed to and come to espouse agrarian values, even if they are not farmers themselves. Readers may judge for themselves whether, if these conditions are ever met, Thomas Jefferson would be satisfied that the healthy parts of the American Republic are once again in greater proportion than the rotten ones.

Bibliography

GENERAL

Atack, Jeremy, and Fred Bateman. *To Their Own Soil: Agriculture in the Antebellum North.* Ames: Iowa University Press, 1987.

Barron, Hal S. *Mixed Harvest: The Second Great Transformation in the Rural North, 1870–1930.* Chapel Hill: University of North Carolina Press, 1997.

———. *Those Who Stayed Behind.* Cambridge: Cambridge University Press, 1984.

Beard, Charles A., and Mary R. Beard. *The Rise of American Civilization.* New York: Macmillan, 1936.

Blum, Jerome, ed. *Our Forgotten Past: Seven Centuries of Life on the Land.* London: Thames and Hudson, 1982.

Carlson, Allan. *The New Agrarian Mind: The Movement toward Decentralist Thought in Twentieth-Century America.* New Brunswick, NJ: Transaction, 2000.

Carman, H. C., ed. *American Husbandry.* New York: Columbia University Press, 1939.

Conlogue, William. *Working the Garden: American Writers and the Industrialization of Agriculture.* Chapel Hill: University of North Carolina Press, 2001.

Conrat, Maisie, and Richard Conrat. *The American Farm: A Photographic History.* San Francisco: California Historical Society / Boston: Houghton Mifflin, 1977.

Danbom, David B. *Born in the Country: A History of Rural America.* Baltimore: Johns Hopkins University Press, 1995.

Douglas, Louis, ed. *Agrarianism in American History.* Lexington, MA: Heath, 1969.

Govan, T. P. "Agrarian and Agrarianism: A Study in the Use and Misuse of Words." *Journal of Southern History* 30 (1964): 35–47.

Griswold, A. W. *Farming and Democracy.* New York: Harcourt, Brace, 1948.

Hanson, Victor Davis. *The Other Greeks.* New York: Free Press, 1995.

Hart, John Fraser. *The Changing Scale of American Agriculture.* Charlottesville: University of Virginia Press, 2003.

————. *The Land That Feeds Us.* New York: Norton, 1991.

Haystead, Ladd, and Gilbert C. Fite. *The Agricultural Regions of the United States.* Norman: University of Oklahoma Press, 1955.

Hesiod. *Works and Days.* Ann Arbor: University of Michigan Press, 1959.

Kamphoefner, Walter D., Wolfgang Helbich, and Ulrike Sommer, eds. *News from the Land of Freedom: German Immigrants Write Home.* Translated by Susan Carter Vogel. Ithaca, NY: Cornell University Press, 1991.

Logsdon, Gene. *At Nature's Pace: Farming and the American Dream.* New York: Pantheon, 1994.

Marx, Leo. *The Machine in the Garden: Technology and the Pastoral Ideal in America.* New York: Oxford University Press, 1964.

Montmarquet, J. A. *The Idea of Agrarianism.* Moscow: University of Idaho Press, 1989.

Mumford, Lewis. *Technics and Civilization.* New York: Harcourt, Brace, 1934.

Nelson, Lynn A. *Pharsalia: An Environmental Biography of a Southern Plantation, 1780–1880.* Athens: University of Georgia Press, 2007.

Paarlberg, Robert, and Don Paarlberg. "Agricultural Policy in the Twentieth Century." *Agricultural History* 74 (Spring 2000): 136–61.

Ratner, Sidney, James Soltow, and Richard Sylla. *The Evolution of the American Economy: Growth, Welfare, and Decision Making.* New York: Basic Books, 1979.

Schlebecker, John T. *Whereby We Thrive: A History of American Farming, 1607–1972.* Ames: Iowa University Press, 1975.

Shi, David E., ed. *In Search of the Simple Life: American Voices, Past and Present.* Layton, UT: Gibbs M. Smith, 1986.

Simpson, Lewis P. *The Dispossessed Garden: Pastoral and History in Southern Literature.* Athens: University of Georgia Press, 1975.

Slicher van Bath, B. H. *The Agrarian History of Western Europe, A.D. 500–1850.* New York: St. Martin's, 1963.

Smith, Henry Nash. *Virgin Land: The American West as Symbol and Myth.* Cambridge, MA: Harvard University Press, 1950.

Stoll, Steven. *Larding the Lean Earth: Soil and Society in Nineteenth-Century America.* New York: Farrar, Straus and Giroux, 2002.

PART 1

Adams, William Howard. *The Paris Years of Thomas Jefferson.* New Haven, CT: Yale University Press, 1997.

Allen, Gay Wilson, and Roger Asselineau. *St. John de Crèvecoeur: The Life of an American Farmer.* New York: Viking, 1987.

Appleby, Joyce. *Liberalism and Republicanism in the Historical Imagination.* Cambridge, MA: Harvard University Press, 1992.

Bailyn, Bernard. *To Begin the World Anew: The Genius and Ambiguities of the American Founders.* New York: Knopf, 2003.

Banning, Lance. *The Jeffersonian Persuasion: Evolution of a Party Ideology.* Ithaca, NY: Cornell University Press, 1978.

Bernstein, R. B. *Thomas Jefferson.* Oxford: Oxford University Press, 2003.

Clark, Christopher. "Economics and Culture: Opening Up the Rural History of the Early American Northeast," *American Quarterly* 43 (June 1991): 279–301.

———. "Rural America and the Transition to Capitalism," *Journal of the Early Republic* 16 (Summer 1996): 223–36.

Cobbett, William. *Selections from Cobbett's Political Works*. Edited by John M. Cobbett and James P. Cobbett. London, n.d.

———. *A Year's Residence in America*. Boston: Small, Maynard, 1817–19.

Craven, Avery O. "John Taylor and Southern Agriculture." *Journal of Southern History* 4 (1938): 137–47.

Danbom, David B. "Romantic Agrarianism in Twentieth Century America." *Agricultural History* 65 (Fall 1991): 1–12.

Edling, Max M., and Mark D. Kaplanoff. "Alexander Hamilton's Fiscal Reform: Transforming the Structure of Taxation in the Early Republic." *William and Mary Quarterly* 61 (October 2004): 713–44.

Ellis, Joseph J. *American Sphinx: The Character of Thomas Jefferson*. New York: Knopf, 1997.

Ferling, John. *A Leap into the Dark: The Struggle to Create the American Republic*. Oxford: Oxford University Press, 2003.

Fox-Genovese, Elizabeth. *The Origins of Physiocracy: Economic Revolution and Social Order in Eighteenth-Century France*. Ithaca, NY: Cornell University Press, 1976.

Hahn, Steven. *The Countryside in the Age of Capitalist Transformation: Essays in the Social History of Rural America*. Chapel Hill: University of North Carolina Press, 1985.

Hamilton, Alexander. *Selections*. Edited by Morton J. Frisch. Washington, DC: AEI Institute for Public Policy Research, 1985.

Henretta, James A. "Families and Farms: *Mentalité* in Pre-industrial America." *William and Mary Quarterly* 35 (January 1978): 3–32.

Howe, Daniel Walker. "European Sources of Political Ideas in Jeffersonian America." *Reviews in American History* 10 (December 1982): 28–44.

———. "Virtue and Commerce in Jeffersonian America." *Reviews in American History* 9 (1981): 347–53.

Irwin, Douglas A. "The Aftermath of Hamilton's *Report on Manufactures*." *Journal of Economic History* 64 (September 2004): 800–821.

Keith, W. J. *The Rural Tradition: William Cobbett, Gilbert White, and Other Non-fiction Writers of the English Countryside*. Hassocks, UK: Harvester, 1975.

Kennedy, Roger G. *Mr. Jefferson's Lost Cause: Land, Farmers, Slavery, and the Louisiana Purchase*. New York: Oxford University Press, 2003.

Kramnick, Isaac. *Bolingbroke and His Circle*. Cambridge, MA: Harvard University Press, 1968.

Logan, Deborah Norris. *Memoir of Dr. George Logan of Stenton*. Philadelphia: Historical Society of Pennsylvania, 1899.

Macleod, Duncan. "The Political Economy of John Taylor of Caroline." *Journal of American Studies* 14 (1980): 387–405.

Madison, James. *The Complete Madison: His Basic Writings*. Edited by Saul K. Padover. New York: Harper and Brothers, 1953.

McCoy, Drew R. *The Elusive Republic: Political Economy in Jeffersonian America*. Chapel Hill: University of North Carolina Press, 1980.

McDonald, Forrest. *Alexander Hamilton: A Biography*. New York: Norton, 1979.

———. *The Presidency of George Washington*. Lawrence: University Press of Kansas, 1974.

———. *The Presidency of Thomas Jefferson*. Lawrence: University Press of Kansas, 1976.

———. *Recovering the Past: A Historian's Memoir*. Lawrence: University Press of Kansas, 2004.

Onuf, Peter, S., ed. *Jeffersonian Legacies*. Charlottesville: University of Virginia Press, 1993.

St. John de Crèvecoeur, J. Hector. *Letters from an American Farmer*. New York: Oxford University Press, 1997.

Shalhope, Robert E. *John Taylor of Caroline: Pastoral Republican*. Columbia: University of South Carolina Press, 1980.

———. "Thomas Jefferson's Republicanism and Antebellum Southern Thought." *Journal of Southern History* 42 (Fall 1976): 529–56.

Sylla, Richard. "Hamilton and the Federalist Financial Revolution, 1789–1795." *New York Journal of American History* 3 (Spring 2004): 32–39.

Taylor, John. *Arator*. Indianapolis: Liberty Fund, 1977.

———. *Tyranny Unmasked*. Indianapolis: Liberty Fund, 1992.

Wermuth, Thomas S. *Rip Van Winkle's Neighbors: The Transformation of Rural Society in the Hudson River Valley, 1720–1850*. Albany: State University of New York Press, 2001.

Wilson, Douglas. "The American Agricola: Jefferson's Agrarianism and the Classical Tradition." *South Atlantic Quarterly* 80 (1981): 339–54.

———. "The Fate of Jefferson's Farmers." *North Dakota Quarterly* 56 (Fall 1988): 22–34.

Wright, Benjamin F. "The Philosopher of Jeffersonian Democracy." *American Political Science Review* 22 (November 1928): 870–92.

Yarbrough, Jean M. *American Virtues: Thomas Jefferson on the Character of a Free People*. Lawrence: University Press of Kansas, 1998.

PART 2

"Agrarianism." *Atlantic Monthly*, April 1859, 393–403.

Allmendinger, David F., Jr. *Ruffin: Family and Reform in the Old South*. New York: Oxford University Press, 1990.

Benton, Thomas Hart. *Speech of Mr. Benton, of Missouri, in Reply to Mr. Webster: The Resolution Offered by Mr. Foote, relative to the Public Lands, Being under Consideration. Delivered in the Senate, Session 1829–30*. Washington, DC: Gales and Seaton, 1830.

Buel, Jesse. *Jesse Buel, Agricultural Reformer: Selections from His Writings*. Edited by Harry J. Carman. New York: Columbia University Press, 1947.

Burlend, Rebecca, and Edward Burlend. *A True Picture of Emigration*. Lincoln: University of Nebraska Press, 1987.

Craven, Avery O. *Edmund Ruffin, Southerner: A Study in Secession*. D. Appleton, 1932.

Demaree, A. L. *The American Agricultural Press, 1819–1860*. New York: Columbia University Press, 1941.

Faragher, John Mack. *Sugar Creek: Life on the Illinois Prairie*. New Haven, CT: Yale University Press, 1986.

Fitzhugh, George. *Cannibals All! Or, Slaves without Masters*. Cambridge, MA: Harvard University Press, 1960.

Foner, Eric. *Free Soil, Free Labor, Free Men: The Ideology of the Republican Party Before the Civil War.* Oxford: Oxford University Press, 1970.

Howe, Daniel Walker. *The American Whigs: An Anthology.* New York: Wiley, 1973.

———. *What Hath God Wrought: The Transformation of America, 1815–1848.* New York: Oxford University Press, 2007.

Julian, George Washington. *Speeches on Political Questions.* New York: Hurd and Houghton, 1872.

MacDonald, Allan. "Lowell: A Commercial Utopia." *New England Quarterly* 10 (March 1937): 37–62.

McClelland, Peter D. *Sowing Modernity: America's First Agricultural Revolution.* Ithaca, NY: Cornell University Press, 1997.

Nelson, Lynn A. *Pharsalia: An Environmental Biography of a Southern Plantation, 1780–1880.* Athens: University of Georgia Press, 2007.

Peskin, Lawrence A. "How the Republicans Learned to Love Manufacturing: The First Parties and the 'New Economy.'" *Journal of the Early Republic* 22 (Summer 2002): 235–62.

Robinson, Solon. *Solon Robinson, Pioneer and Agriculturalist: Selected Writings.* Edited by Herbert Kellar. Indianapolis: Indiana Historical Bureau, 1936.

Ruffin, Edmund. *Incidents of My Life: Edmund Ruffin's Autobiographical Essays.* Charlottesville: University of Virginia Press, 1990.

Sellers, Charles Grier. *The Market Revolution: Jacksonian America, 1815–1846.* New York: Oxford University Press, 1991.

Stokes, Melvyn, and Stephen Conway, eds. *The Market Revolution in America: Social, Political, and Religious Expressions, 1800–1880.* Charlottesville: University of Virginia Press, 1996.

Stoll, Steven. *Larding the Lean Earth: Soil and Society in Nineteenth-Century America.* New York: Hill and Wang, 2002.

Walker, Timothy. "A Defense of Mechanical Philosophy." *North American Review* 33 (1831): 124.

Wilentz, Sean. *Chants Democratic: New York City and the Rise of the American Working Class, 1788–1850.* Oxford: Oxford University Press, 1984.

PART 3

Brisbane, Albert. "False Association." *Harbinger,* November 14, 1846, 365–68.

Cooper, Susan Fenimore. *Rural Hours.* Syracuse, NY: Syracuse University Press, 1968.

Crowe, Charles. *George Ripley, Transcendentalist and Utopian Socialist.* Athens: University of Georgia Press, 1967.

———. "This Unnatural Union of Phalansteries and Transcendentalists." *Journal of the History of Ideas* 20 (October–December 1959): 495–502.

Delano, Sterling. *Brook Farm: The Dark Side of Utopia.* Cambridge, MA: Harvard University Press, Belknap Press, 2004.

———. *The Harbinger and New England Transcendentalism: A Portrait of Associationism in America.* Rutherford, NJ: Fairleigh Dickinson University Press, 1983.

Donahue, Brian. *The Great Meadow: Farmers and the Land in Colonial Concord.* New Haven, CT: Yale University Press, 2004.

Downing, Andrew Jackson. *Rural Essays.* New York: Leavitt and Allen, 1853.

Emerson, Ralph Waldo. *Essays and Lectures.* New York: Library of America, 1983.

Foster, David R. *Thoreau's Country: Journey through a Transformed Landscape.* Cambridge, MA: Harvard University Press, 1999.

Francis, Richard. *Ann the Word: The Story of Ann Lee, Female Messiah, Mother of the Shakers, the Woman Clothed with the Sun.* New York: Arcade, 2000.

———. "The Ideology of the Brook Farm." In *Studies in the American Renaissance,* edited by Joel Myerson. Boston: Twayne, 1977.

———. *Transcendental Utopias: Individual and Community at Brook Farm, Fruitlands, and Walden.* Ithaca, NY: Cornell University Press, 1997.

Golemba, Henry L. *George Ripley.* Boston: Twayne, 1977.

Guarneri, Carl J. "Importing Fourierism to America." *Journal of the History of Ideas* 43 (October–December 1982): 581–94.

Hawthorne, Nathaniel. *The Blythedale Romance.* New York: Modern Library, 2001.

Marshall, Megan. *The Peabody Sisters: Three Women Who Ignited American Romanticism.* Boston: Houghton Mifflin, 2005.

Miller, Perry. *The Transcendentalists: An Anthology.* Cambridge, MA: Harvard University Press, 1950.

Montrie, Chad. "'I Think Less of the Factory Than of My Native Dell': Labor, Nature, and the Lowell 'Mill Girls.'" *Environmental History* 9 (April 2004): 275–95.

Myerson, Joel. *The Brook Farm Book.* New York: Garland, 1987.

Richardson, Robert D., Jr. *Emerson: The Mind on Fire.* Berkeley: University of California Press, 1995.

———. *Henry Thoreau: A Life of the Mind.* Berkeley: University of California Press, 1986.

Rose, Anne C. *Transcendentalism as a Social Movement, 1830–1860.* New Haven, CT: Yale University Press, 1981.

Sams, Henry W., ed. *Autobiography of Brook Farm.* Englewood Cliffs, NJ: Prentice-Hall, 1958.

Thoreau, Henry David. *Writings.* Boston: Houghton Mifflin, 1949.

PART 4

Barnes, Donna A. *Farmers in Rebellion: The Rise and Fall of the Southern Farmers Alliance and People's Party in Texas.* Austin: University of Texas Press, 1984.

Bogue, Allan G. *From Prairie to Cornbelt: Farming on the Illinois and Iowa Prairies in the Nineteenth Century.* Chicago: University of Chicago Press, 1963.

Buck, Solon J. *The Agrarian Crusade.* New Haven, CT: Yale University Press, 1920.

Cather, Willa. *My Ántonia.* Boston: Houghton Mifflin, 1918.

Clanton, Gene. *Populism: The Humane Preference in America, 1890–1900.* Boston: Twayne, 1991.

Fite, Gilbert C. *The Farmer's Frontier, 1865–1900.* Albuquerque: University of New Mexico Press, 1974.

Garland, Hamlin. *A Son of the Middle Border.* New York: Macmillan, 1962.

George, Henry. *Progress and Poverty.* New York: D. Appleton, 1880.

Gjerde, Jon. *From Peasants to Farmers: The Migration from Balestrand, Norway, to the Upper Middle West.* Cambridge: Cambridge University Press, 1985.

Goodwin, Lawrence. *The Populist Moment: A Short History of the Agrarian Revolt in America.* Oxford: Oxford University Press, 1978.

Hahn, Steven. *The Roots of Southern Populism: Yeoman Farmers and the Transformation of the Georgia Upcountry.* Oxford: Oxford University Press, 2006.

Hofstadter, Richard. *The Age of Reform, from Bryan to F.D.R.* New York: Knopf, 1974.

Holmes, William F. "Populism: In Search of Context." *Agricultural History* 64 (Fall 1990): 26–58.

Howard, David. *People, Pride, and Progress: 125 Years of the Grange in America.* Washington, DC: National Grange, 1992.

Ise, John. *Sod and Stubble.* Lawrence: University Press of Kansas, 1996.

Jordan, Terry. *German Seed in Texas Soil: Immigrant Farmers in Nineteenth-Century Texas.* Austin: University of Texas Press, 1966.

Kelley, Oliver Hudson. *Origin and Progress of the Patrons of Husbandry in the United States: A History from 1866 to 1873.* Westport, CT: Hyperion, 1975.

Kellie, Luna. *A Prairie Populist: The Memoirs of Luna Kellie.* Edited by Jane Taylor Nelson. Iowa City: University of Iowa Press, 1992.

McMath, Robert C., Jr. *American Populism: A Social History, 1877–1898.* New York: Hill and Wang, 1993.

———. *The Populist Vanguard: A History of the Southern Farmers' Alliance.* Chapel Hill: University of North Carolina Press, 1976.

Noblin, Stuart. *Leonidas LaFayette Polk, Agrarian Crusader.* Chapel Hill: University of North Carolina Press, 1949.

Nugent, Walter T. K. *The Tolerant Populists; Kansas, Populism, and Nativism.* Chicago: University of Chicago Press, 1963.

Ostler, Jeffrey. *Prairie Populism: The Fate of Agrarian Radicalism in Kansas, Nebraska, and Iowa, 1880–1892.* Lawrence: University Press of Kansas, 1993.

Palmer, Bruce. *"Man over Money": The Southern Populist Critique of American Capitalism.* Chapel Hill: University of North Carolina Press, 1980.

Pisani, Donald J. *From the Family Farm to Agribusiness: The Irrigation Crusade in California and the West, 1850–1931.* Berkeley: University of California Press, 1984.

Pollack, Norman. *The Just Polity: Populism, Law, and Human Welfare.* Urbana: University of Illinois Press, 1987.

———. *The Populist Response to Industrial America.* Cambridge, MA: President and Fellows of Harvard College, 1962.

Rolvaag, O. E. *Giants in the Earth.* Translated by Lincoln Colcord and O. E. Rolvaag. New York: Harper and Row, 1965.

Ruede, Howard. *Sod-House Days: Letters from a Kansas Homesteader.* Edited by John Ise. Lawrence: University Press of Kansas, 1983.

Shannon, Fred A. *The Farmer's Last Frontier: Agriculture, 1860–1897.* New York: Farrar and Rinehart, 1945.

Stoll, Steven. *The Fruits of Natural Advantage: Making the Industrial Countryside in California.* Berkeley: University of California Press, 1998.

Taylor, Henry C. *Tarpleywick: A Century of Iowa Farming.* Ames: Iowa State University Press, 1970.

Tindall, George B., ed. *A Populist Reader: Selections from the Works of American Populist Leaders.* New York: Harper Torchbooks, 1966.

Walton, Gary, and Hugh Rockoff. *History of the American Economy.* Boston: Houghton Mifflin, 1990.

Watson, Thomas E. *The Life and Speeches of Thos. E. Watson.* Nashville, TN, 1908.

Whitaker, James W., ed. *Farming in the Midwest, 1840–1900.* Washington, DC: Agricultural History Society, 1974.

Woodward, C. Vann. *Tom Watson, Agrarian Rebel.* New York: Macmillan, 1938.

———. "Tom Watson and the Negro in Agrarian Politics." *Journal of Southern History* 4 (February 1938): 14–33.

PART 5

Bailey, Liberty Hyde. *The Harvest of the Year to the Tiller of the Soil.* New York: Macmillan, 1927.

———. *The Holy Earth.* New York: Charles Scribner's Sons, 1915.

———. *The Nature-Study Idea.* New York: Doubleday, Page, 1903.

———. *The Outlook to Nature.* New York: Macmillan, 1911.

Baker, Oliver Edwin, Ralph Borsodi, and M. L. Wilson, eds. *Agriculture in Modern Life.* New York: Harper and Brothers, 1939.

Baker, Ray Stannard. *American Chronicle.* New York: Charles Scribner's Sons, 1945.

Beeman, Randal S., and James A. Pritchard. *A Green and Permanent Land: Ecology and Agriculture in the Twentieth Century.* Lawrence: University Press of Kansas, 2001.

Bennett, Hugh Hammond. *Soil Conservation.* New York: McGraw-Hill, 1939.

———. "Soil Erosion Takes $200,000,000 Yearly from U.S. Farmers." In *Yearbook of Agriculture, 1927,* 591–93. Washington, DC: Government Printing Office, 1928.

———. *Thomas Jefferson: Soil Conservationist.* USDA Misc. Publ. 548. Washington, DC: Government Printing Office, 1944.

———. "The Wasting Heritage of the Nation." *Scientific Monthly,* August 1928, 97–124.

Bogue, Margaret Beattie. "Liberty Hyde Bailey, Jr. and the Bailey Family Farm." *Agricultural History* 63 (Winter 1989): 26–48.

Borsodi, Ralph. *Flight from the City: An Experiment in Creative Living on the Land.* New York: Harper and Row, 1933.

———. *This Ugly Civilization.* New York: Simon and Schuster, 1929.

Bowers, William L. *The Country Life Movement in America, 1900–1920.* Port Washington, NY: Kennikat, 1974.

Bromfield, Louis. *Malabar Farm.* New York: Harper and Brothers, 1947.

———. *Pleasant Valley.* New York: Harper and Brothers, 1943.

Butterfield, Kenyon L. *Chapters in Rural Progress.* Chicago: University of Chicago Press, 1908.

Caldwell, John C., James L. Bailey, and Richard W. Watkins. *Our Land and Our Living.* Syracuse, NY: L. W. Singer, 1941.

Chase, Stuart. *Rich Land, Poor Land: A Study of Waste in the Natural Resources of America.* New York: Whittlesey House, 1936.

Danbom, David B. *The Resisted Revolution: Urban America and the Industrialization of Agriculture, 1900–1930.* Ames: Iowa State University Press, 1979.

———. *"The World of Hope": Progressives and the Struggle for an Ethical Public Life.* Philadelphia: Temple University Press, 1987.

Ellsworth, Clayton S. "Ceres and the American Men of Letters since 1929." *Agricultural History* 24 (October 1950): 177–81.

Fink, Deborah. *Agrarian Women: Wives and Mothers in Rural Nebraska, 1880–1940.* Chapel Hill: University of North Carolina Press, 1992.

Fitzgerald, Deborah. *Every Farm a Factory: The Industrial Ideal in American Agriculture.* New Haven, CT: Yale University Press, 2003.

Goldberg, Robert Alan. *Back to the Soil: The Jewish Farmers of Clarion, Utah, and Their World.* Salt Lake City: University of Utah Press, 1986.

Grayson, David [Ray Stannard Baker]. *Adventures in Contentment.* New York: Grosset and Dunlap, 1907.

Hall, Bolton. *Three Acres and Liberty.* New York: Macmillan, 1907.

Hamlin, Christopher, and John T. McGreevy. "The Greening of America, Catholic Style, 1930–1950." *Environmental History* 11 (July 2006): 464–99.

Hampsten, Elizabeth. *Read This Only to Yourself: The Private Writings of Midwestern Women, 1880–1910.* Bloomington: Indiana University Press, 1982.

Henderson, Henry L., and David B. Wolner, eds. *FDR and the Environment.* New York: Palgrave Macmillan, 2005.

Hudson, Lois Phillips. *The Bones of Plenty.* Boston: Little, Brown, 1962.

Issel, William H. "Ralph Borsodi and the Agrarian Response to Modern America." *Agricultural History* 41 (April 1967): 155–66.

McConnell, Grant. *The Decline of Agrarian Democracy.* Berkeley: University of California Press, 1959.

Nourse, Edwin. *Agricultural Economics.* Chicago: University of Chicago Press, 1916.

Rawe, John C. "Agrarianism: The Basis for a Better Life." *American Review,* December 1935, 176–92.

Rawe, John C., and Luigi G. Ligutti. *Rural Roads to Security: America's Third Struggle for Freedom.* Milwaukee: Bruce, 1940.

Report of the Commission on Country Life. New York: Sturgis and Walton, 1911.

Schapsmeier, Edward L., and Frederick H. Schapsmeier. *Henry A. Wallace of Iowa: The Agrarian Years, 1910–1940.* Ames: Iowa State University Press, 1968.

Smith, J. Russell. *North America: Its People and the Resources, Development, and Prospects of the Continent as an Agricultural, Industrial, and Commercial Area.* New York: Harcourt, Brace, 1925.

Turner, Frederick Jackson. "The Significance of the Frontier in American History." *Report of the American Historical Association for 1893* (1893): 199–227.

Wallace, Henry A. *New Frontiers.* New York: Reynal and Hitchcock, 1934.

Wallace, Henry C. *Our Debt and Duty to the Farmer.* New York: Century, 1925.

Worster, Donald. *Dust Bowl: The Southern Plains in the 1930s.* New York: Oxford University Press, 1979.

PART 6

Burgess, R. L. "Farming: A Variety of Religious Experience." *American Review,* October 1934, 591–607.

Cauley, Troy J. *Agrarianism: A Program for Farmers.* Chapel Hill: University of North Carolina Press, 1935.

Cobb, Ned. *All God's Dangers: The Life of Nate Shaw.* Edited by Theodore Rosengarten. New York: Knopf, 1974.

Couch, W. T. "An Agrarian Programme for the South." *American Review,* Summer 1934, 313–26.

Cowan, Louise. *The Fugitive Group: A Literary History.* Baton Rouge: Louisiana State University Press, 1959.

Daniel, Pete. *Breaking the Land: The Transformation of Cotton, Tobacco, and Rice Cultures since 1880.* Chicago: University of Illinois Press, 1985.

Davidson, Donald. "*I'll Take My Stand:* A History." *American Review,* Summer 1935, 301–21.

———. *Southern Writers in the Modern World.* Athens: University of Georgia Press, 1958.

——— "Why the Modern South Has a Great Literature." In *Still Yankees, Still Rebels and Other Essays,* 159–79. Baton Rouge: Louisiana State University Press, 1957.

Fain, John Tyree, and Thomas Daniel Young, eds. *The Literary Correspondence of Donald Davidson and Allen Tate.* Athens: University of Georgia Press, 1974.

Fifteen Southerners. *Why the South Will Survive: Fifteen Southerners Look at Their Region a Half Century after "I'll Take My Stand."* Athens: University of Georgia Press, 1981.

Havard, William C., and Walter Sullivan, eds. *A Band of Prophets: The Vanderbilt Agrarians after Fifty Years.* Baton Rouge: Louisiana State University Press, 1982.

Hobson, Fred C. *Serpent in Eden: H. L. Mencken and the South.* Chapel Hill: University of North Carolina Press, 1974.

Inge, M. Thomas. "The Fugitives and the Agrarians: A Clarification." *American Literature* 62 (September 1990): 486–93.

Karanikas, Alexander. *Tillers of a Myth: Southern Agrarians as Social and Literary Critics.* Madison: University of Wisconsin Press, 1966.

Kirby, Jack Temple. *Rural Worlds Lost: The American South, 1920–1960.* Baton Rouge: Louisiana State University Press, 1987.

Lowell, Robert. "Visiting the Tates." *Sewanee Review* (Fall 1959): 557–59.

Lucas, Mark. *The Southern Vision of Andrew Lytle.* Baton Rouge: Louisiana State University Press, 1986.

Lytle, Andrew. "John Taylor and the Political Economy of Agriculture." *American Review,* September 1934, 432–47; October 1934, 630–43; November 1934, 84–99.

———. "The State of Letters in a Time of Disorder." *Sewanee Review* 79 (Autumn 1971): 477–97.

———. *A Wake for the Living.* Nashville, TN: J. S. Sanders, 1992.

Lytle, Andrew, and Allen Tate. *The Lytle-Tate Letters.* Jackson: University Press of Mississippi, 1987.

Millgate, Michael. "An Interview with Allen Tate." *Shenandoah* 12 (Spring 1961): 32.

Montgomery, Marion. *John Crowe Ransom and Allen Tate: At Odds about the Ends of History and the Mystery of Nature.* Jefferson, NC: McFarland, 2003.

Newby, Idus A. "The Southern Agrarians: A View after Thirty Years." *Agricultural History* 37 (July 1963): 143–55.

Owsley, Frank L. "The Pillars of Agrarianism." *American Review*, March 1935, 529–47.

Purdy, Rob Roy, ed. *Fugitives' Reunion: Conversations at Vanderbilt, May 3–5, 1956*. Nashville, TN: Vanderbilt University Press, 1959.

Ransom, John Crowe. *God without Thunder*. New York: Harcourt, Brace, 1930.

———. *Selected Essays of John Crowe Ransom*. Edited by Thomas Daniel Young. Baton Rouge: Louisiana State University Press, 1984.

Rubin, Louis D., Jr. "The Gathering of the Fugitives: A Recollection." *Southern Review*, Fall 1994, 658–73.

———. *The Wary Fugitives: Four Poets of the South*. Baton Rouge: Louisiana State University Press, 1978.

Sullivan, Walter. *Allen Tate: A Recollection*. Baton Rouge: Louisiana State University Press, 1988.

Tate, Allen. "Notes on Liberty and Property." *American Review*, March 1936, 596–611.

———. "What Is a Traditional Society?" *American Review*, September 1936, 376–87.

Twelve Southerners. *I'll Take My Stand: The South and the Agrarian Tradition*. New York: Harper, 1962.

Underwood, Thomas A. *Allen Tate: Orphan of the South*. Princeton, NJ: Princeton University Press, 2000.

Vinh, Alphonse, ed. *Cleanth Brooks and Allen Tate: Collected Letters, 1933–1976*. Columbia: University of Missouri Press, 1998.

Winchell, Mark Royden. *Where No Flag Flies: Donald Davidson and the Southern Resistance*. Columbia: University of Missouri Press, 2000.

Young, Thomas Daniel. *Gentleman in a Dustcoat: A Biography of John Crowe Ransom*. Baton Rouge: Louisiana State University Press, 1976.

———. *Waking Their Neighbors Up: The Nashville Agrarians Rediscovered*. Athens: University of Georgia Press, 1982.

PART 7

Benedict, Murray R. *Can We Solve the Farm Problem? An Analysis of Federal Aid to Agriculture*. New York: Twentieth Century Fund, 1955.

Berry, Wendell. *The Gift of Good Land: Further Essays Cultural and Agricultural*. New York: North Point, 1982.

———. *Life Is a Miracle*. Washington, DC: Counterpoint, 2000.

———. *A Place on Earth*. New York: Harcourt, Brace, 1967.

———. *Standing by Words*. San Francisco: North Point, 1983.

———. *The Unsettling of America: Culture and Agriculture*. San Francisco: Sierra Club Books, 1977.

———. *What Are People For?* San Francisco: North Point, 1990.

Beus, Curtis E., and Riley E. Dunlap. "Conventional versus Alternative Agriculture: The Paradigmatic Roots of the Debate." *Rural Sociology* 55 (Winter 1990): 590–616.

Breimyer, Harold. *Farm Policy: 13 Essays*. Ames: Iowa State University Press, 1977.

Butz, Earl L. "Food, Farm Programs, and the Future." Address delivered to the National Press Club, Washington, DC, April 3, 1973.

Carlson, Allan. "Agrarianism Reborn: On the Curious Return of the Small Family Farm." *Intercollegiate Review* 43 (Spring 2008): 13–23.

Danbom, David B. "Romantic Agrarianism in Twentieth Century America." *Agricultural History* 65 (Fall 1991): 1–12.

Davidson, Osha Gray. *Broken Heartland: The Rise of America's Rural Ghetto.* New York: Free Press, 1990.

Fish, Charles. *In Good Hands: The Keeping of a Family Farm.* New York: Farrar, Straus and Giroux, 1995.

Graves, John. *Hard Scrabble: Observations on a Patch of Land.* New York: Knopf, 1974.

Hall, Donald. *Here at Eagle Pond.* Boston: Houghton Mifflin, 1990.

———. *String Too Short to Be Saved.* New York: Viking, 1961.

Hanson, Victor Davis. *Fields without Dreams: Defending the Agrarian Idea.* New York: Free Press, 1996.

———. *The Land Was Everything: Letters from an American Farmer.* New York: Free Press, 2000.

Hassebrock, Kenneth. *Rural Reminiscences: The Agony of Survival.* Ames: Iowa University Press, 1990.

Jackson, Wes. *Becoming Native to this Place.* Lexington: University Press of Kentucky, 1994.

———. *New Roots for Agriculture.* San Francisco: Friends of the Earth, 1980.

Kingsolver, Barbara. *Animal, Vegetable, Miracle: A Year of Food Life.* New York: Harper-Collins, 2007.

Kramer, Mark. *Three Farms: Making Milk, Meat and Money from the American Soil.* Boston: Little, Brown, 1977.

Leopold, Aldo. *The River of the Mother of God and Other Essays by Aldo Leopold.* Edited by Susan L. Flader and J. Baird Callicott. Madison: University of Wisconsin Press, 1991.

———. *A Sand County Almanac, and Sketches Here and There.* New York: Oxford University Press, 1949.

Mueller, William. "How We're Gonna Keep 'Em off of the Farm." *American Scholar* 56 (Winter 1987): 57–67.

Muir, Edwin. *An Autobiography.* London: Hogarth, 1954.

Nearing, Helen, and Scott Nearing. *Living the Good Life: How to Live Sanely and Simply in a Troubled World.* New York: Schocken Books, 1970.

———. *The Maple Sugar Book; Together with Remarks on Pioneering as a Way of Living in the Twentieth Century.* New York: Schocken Books, 1970.

Nearing, Scott. *The Making of a Radical: A Political Autobiography.* New York: Harper and Row, 1972.

Owens, Virginia Stem. *Assault on Eden.* Grand Rapids, MI: Baker Books, 1995.

Pollan, Michael. *The Omnivore's Dilemma: A Natural History of Four Meals.* New York: Penguin, 2006.

Saltmarsh, John A. *Scott Nearing: An Intellectual Biography.* Philadelphia: Temple University Press, 1991.

Smiley, Jane. "Losing the Farm." *New Yorker,* June 3, 1996, 88–92.

———. *A Thousand Acres.* New York: Fawcett Columbine, 1991.

Staten, Jay. *The Embattled Farmer*. Golden, CO: Fulcrum, 1987.

Strange, Marty. *Family Farming: A New Economic Vision*. Lincoln: University of Nebraska Press, 1988.

Tweeten, Luther. *Terrorism, Radicalism, and Populism in Agriculture*. Ames: Iowa State University Press, 2003.

CONCLUSION

Bell, Michael Mayerfeld. *Farming for Us All: Practical Agriculture and the Culture of Sustainability*. University Park: Pennsylvania State University Press, 2004.

Logsdon, Gene. *The Contrary Farmer*. White River Junction, VT: Chelsea Green, 1995.

Selection Credits

Index

Numbers in *italics* indicate figures.